MOLECULAR MODELING APPLICATIONS IN CRYSTALLIZATION

Crystallization is an important purification process used in a broad range of industries, including pharmaceuticals, foods, and bulk chemicals. In recent years, molecular modeling has emerged as a useful tool in the analysis and solution of problems associated with crystallization. Modeling allows more focused experimentation based on structural and energetic calculations, instead of intuition and trial and error.

This book is the first to offer a general introduction to molecular modeling techniques and their application in crystallization. After explaining the basic concepts of molecular modeling and crystallization, the book goes on to discuss how modeling techniques are used to solve a variety of practical problems related to crystal size, shape, internal structure, and properties.

With chapters written by leading experts and an emphasis on problem-solving, this book will appeal to scientists, engineers, and graduate students involved in research and the production of crystalline materials.

Allan S. Myerson is the Joseph J. and Violet J. Jacobs Professor of Chemical Engineering at Polytechnic University, Brooklyn, New York.

T0215845

CAMBRIDGE UNIVERSITY PRESS
Cambridge, New York, Melbourne, Madrid, Cape Town, Singapore, São Paulo

Cambridge University Press
The Edinburgh Building, Cambridge CB2 2RU, UK

Published in the United States of America by Cambridge University Press, New York

www.cambridge.org
Information on this title: www.cambridge.org/9780521552974

© Cambridge University Press 1999

First published 1999
This digitally printed first paperback version 2005

A catalogue record for this publication is available from the British Library

ISBN-13 978-0-521-55297-4 hardback
ISBN-10 0-521-55297-4 hardback

ISBN-13 978-0-521-01951-4 paperback
ISBN-10 0-521-01951-6 paperback

Contents

List of Contributors   vii

Preface   ix

1   Introduction to Molecular Modeling   1
ALEXANDER F. IZMAILOV AND ALLAN S. MYERSON

2   Crystallization Basics   55
ALLAN S. MYERSON

3   The Study of Molecular Materials Using Computational Chemistry   106
ROBERT DOCHERTY AND PAUL MEENAN

4   Towards an Understanding and Control of Nucleation, Growth,
Habit, Dissolution, and Structure of Crystals Using "Tailor-Made"
Auxiliaries   166
ISABELLE WEISSBUCH, MEIR LAHAV, AND LESLIE
LEISEROWITZ

5   Ionic Crystals in the Hartman–Perdok Theory with Case Studies:
ADP ($NH_4H_2PO_4$)-type Structures and Gel-Grown Fractal
Ammonium Chloride (NH_4Cl)   228
C. S. STROM, R. F. P. GRIMBERGEN, P. BENNEMA, H. MEEKES,
M. A. VERHEIJEN, L. J. P. VOGELS, AND MU WANG

6   The Solid-State Structure of Chiral Organic Pharmaceuticals   313
G. PATRICK STAHLY AND STEPHEN R. BYRN

Index   347

Contributors

P. Bennema
Research Institute of Materials, Laboratory of Solid State Chemistry, University of Nijmegen, 6525 ED Nijmegen, The Netherlands

Stephen R. Byrn
School of Pharmacy, Purdue University, West Lafayette, IN 47907, USA

Robert Docherty
Zeneca Specialties Research Centre, Blackley, Manchester M9 8ZS, UK

R. F. P. Grimbergen
DSM Research, Geleen, The Netherlands

Alexander F. Izmailov
Department of Chemical Engineering, Polytechnic University, Brooklyn, NY 11201, USA

Meir Lahav
Department of Materials and Interfaces, The Weizmann Institute of Science, Rehovot 76100, Israel

Leslie Leiserowitz
Department of Materials and Interfaces, The Weizmann Institute of Science, Rehovot 76100, Israel

H. Meekes
Research Institute of Materials, Laboratory of Solid State Chemistry, University of Nijmegen, 6525 ED Nijmegen, The Netherlands

Paul Meenan
E. I. du Pont de Nemours & Co., Inc., Wilmington, DE 19880-0304, USA

Allan S. Myerson
Department of Chemical Engineering, Polytechnic University, Brooklyn, NY 11201, USA

G. Patrick Stahly
SSCI, Inc., Purdue Research Park, 1291 Cumberland Avenue, Suite E, West Lafayette, IN 47906, USA

C. S. Strom
SeriousToo Crystal Morphology Software, P.A. 30002, 1000 PC Amsterdam, The Netherlands and Research Institute of Materials, Laboratory of Solid State Chemistry, University of Nijmegen, 6525 ED Nijmegen, The Netherlands

M. A. Verheijen
Philips Centre for Manufacturing Technology, Prof. Holstlaan 4, 5656 AA Eindhoven, The Netherlands

L. J. P. Vogels
Philips Semiconductors, Gerstweg 2, 6534 AE Nijmegen, The Netherlands

Mu Wang
National Laboratory of Solid State Microstructure, Nanjing University, and Center for Advanced Studies of Science and Technology of Microstructures, Nanjing 210093, China

Isabelle Weissbuch
Department of Materials and Interfaces, The Weizmann Institute of Science, Rehovot 76100, Israel

Preface

Crystallization from solution is an important separation and purification process used in a wide variety of industries. In addition to product purity, it is often necessary to control the external shape and size of the crystals and to produce the desired polymorph and/or optical isomer. In some applications it is necessary to inhibit the formation of crystals or control their size and shape by the use of an additive.

Molecular modeling or calculational chemistry are terms that are used to describe techniques that employ quantum mechanics and statistical mechanics in conjunction with computer simulation to study the chemical and physical properties of materials.

In recent years, molecular modeling has emerged as a useful tool in the solution of a number of the crystallization problems mentioned above. In particular, modeling allows more focused experimentation based on structural and energetic calculations instead of intuition and trial and error. The availability of commercial modeling packages has led to its widespread use in many industrial and academic laboratories. These packages often lead to investigators treating the modeling process as a "black box" without understanding the basic principles underlying the methods. The purpose of this book is to explain the basic concepts of molecular modeling and their application to problems in crystallization.

The first two chapters introduce basic molecular modeling concepts (Chapter 1) and the basics of crystals and crystallization (Chapter 2). These chapters are aimed at the nonspecialist who needs an introduction to the area. The remaining chapters deal with molecular modeling techniques for molecular crystals (Chapter 3), the role and selection of additives (Chapter 4), modeling techniques in ionic systems (Chapter 5), and chirality (Chapter 6).

Allan S. Myerson
Brooklyn, NY

1

Introduction to Molecular Modeling

ALEXANDER F. IZMAILOV AND ALLAN S. MYERSON

1.1. Introduction

In recent years modeling methods based on computer simulation have become a useful tool in solving many scientific and engineering problems. Moreover, with the introduction of powerful workstations the impact of applications of computer simulation is expected to increase enormously in the next few years. To some extent, computer-based modeling methods have filled the long existing gap between experimental and theoretical divisions of natural sciences such as physics, chemistry, and biology. Such a dramatic role for computer simulation methods is due to the *statistically exact character* of information that they provide about the *exactly defined model* systems under study. The term "statistically exact information" means "information known within the range defined by standard deviation of some statistical distribution law." This deviation can usually be reduced to an extent required by the problem under study. The term "exactly defined model" means that all parameters required to specify the model *Hamiltonian* are known exactly.

Let us specify the role of computer-based simulation with respect to information obtained by analytical derivation and experiment:

1. The analytically exact information is available only for a few theoretical models that allow exact analytical solutions. The most celebrated example of such a model in statistical physics is the two-dimensional Ising model for the nearest-neighbor interacting spins in the absence of external fields. Its analytically exact solution was obtained by Onsager (1944). However, in the majority of other cases, where exact analytical solutions are not known, it is customary to use different approximations. And it is often the case that these approximations are uncontrollable. The same Ising model in the three-dimensional case does not have an exact solution. Needless to say, even less is known about the models with more realistic

intermolecular potentials. Therefore, computer simulation is often used to verify different approximations involved in analytical solutions.

2. The experimentally exact information is also never available because it is impossible to completely define the Hamiltonian of an experimental system. In some important experimental problems, such as nucleation, the Hamiltonian is very often unknown because one or more mechanisms can be triggering the observed result. Nucleation can be triggered either by the intrinsic system properties (metastability) or by unknown impurities present in experimental systems. Therefore, modeling models based on computer simulation are required in order to understand the relative role of metastability and impurities on the nucleation phenomenon.

This brief discussion illustrates that very often there is not enough information to link theory and experiment. Thus, analytical approximations involved in the description of an experimentally observed phenomenon are not necessarily related to the real driving forces of this phenomenon. Computer-based modeling methods provide a unique opportunity to build a "coherence bridge" between the analytical approximations involved in the solution of major problems and the experimental information available.

This chapter serves a much simpler goal than review of the modern modeling methods based on computer simulation. We will try to present some basic concepts of statistical mechanics (Section 1.2), thermodynamics (Section 1.3), intermolecular interactions (Section 1.4) and the Monte Carlo method (Section 1.5) which underlie all major modeling methods based on computer simulation (Section 1.6).

1.2. Basics of Statistical Mechanics

1.2.1. *Statistical Distributions*

Let a given macroscopic system have $2N$ degrees of freedom, that is each of N particles (molecules) constituting the system of volume V, will be described by its coordinate q_n and moment $p_n (n = 1, \ldots, N)$. Therefore, all various states of the system under consideration can be represented by points in the *phase space* P with $2N$ coordinates $\{q_n, p_n; n = 1, \ldots, N\}$. A state of the system changes with time and, consequently, the point in space P representing this state, also known as the *phase point*, moves along a curve known as a *phase trajectory*.

Let us also introduce the probability $\omega(q; p) = \omega(q_1, \ldots, q_N; p_1, \ldots, p_N)$ that the system state is represented by a point belonging to the infinitesimal $2N$-dimensional interval $\{[q_n; q_n + dq_n] \oplus [p_n; p_n + dp_n]; n = 1, \ldots, N\}$:

$$d\omega(q; p) = \rho(q; p) dq dp; \; \rho(q; p) = \rho(q_1, \ldots, q_N; p_1, \ldots p_N), \; dq dp = \prod_{n=1}^{N} dq_n dp_n,$$

$$(1.1)$$

where $\rho(q; p)$ is the probability density in space P also known as the *statistical distribution function*. This function must obviously satisfy the *normalization condition*:

$$\int_P d\omega(q, p) = \int_P \rho(q, p)dqdp = 1, \tag{1.2}$$

which simply expresses the fact that the sum of probabilities of all states must be unity.

It is clear that at any time instant, t, phase points, representing states of a closed system, are distributed in phase space according to the same statistical distribution function $\rho(q; p)$. Therefore, the movement of these points in phase space can be formally described by applying the *continuity equation* expressing the fact of constancy of the total number of phase points (system states):

$$\frac{d\rho(q; p)}{dt} = \sum_{n=1}^{N}\left(\frac{\partial\rho(q; p)}{\partial q_n}\frac{dq_n}{dt} + \frac{\partial\rho(q; p)}{\partial p_n}\frac{dp_n}{dt}\right) = 0 \tag{1.3}$$

This equation, known as the *Lioville equation*, states that the distribution function is constant along the phase trajectories.

If the statistical distribution is known, one can calculate the probabilities of various values of any physical quantities depending on the system states. One also can calculate the mean of any such quantity $f(q; p)$:

$$\langle f \rangle = \int_P f(q; p)d\omega(q; p) = \int_P f(q, p)\rho(q; p)dqdp, \tag{1.4}$$

where the sign $\langle\ldots\rangle$ designates the average over an ensemble of phase points.

It is understandable that in a sufficiently long time, τ, the phase trajectory passes many times through each infinitesimal volume $\Delta q\Delta q$ of the phase space. Let $\Delta t(q; p)$ be the part of the total time τ during which the system states belong to the phase space volume $\Delta q\Delta q$. Then, when the total time τ goes to infinity, the ratio $\Delta t(q; p)/\tau$ tends to the limit:

$$\Delta w(q; p) = \lim_{\tau\to\infty}\frac{\Delta t(q; p)}{\tau}. \tag{1.5}$$

It is important to note that, by virtue of relations (1.4) and (1.5), the averaging with respect to the distribution function, also known as *statistical averaging* allows one to drop the necessity of following the variation of physical quantity $f(q; p)$ with time in order to determine its mean value:

$$\langle f \rangle = \lim_{\tau\to\infty}\frac{1}{\tau}\int_0^\tau f(t)dt. \tag{1.6}$$

Relationship (1.6) is known as the *ergodic theorem* and was first formulated by Boltzmann (see textbook by Landau and Lifshitz (1988) for details).

1.2.2. *Ensembles*

The statistical distribution function $\rho(q; p)$ can be expressed entirely in terms of some combinations of the variables q and p, which remain constant when a system moves as a closed system. These combinations are known as *mechanical invariants* or *integrals of motion*, which remain constant during the motion of a system state in phase space. The distribution function, being constructed from mechanical invariants, is, therefore, itself a mechanical invariant.

It is possible to restrict significantly the number of mechanical invariants on which the function $\rho(q; p)$ can depend. In order to demonstrate this, we have to take into account the fact that the distribution function $\rho_{1+2}(q; p)$ for a non-overlapping combination of two subsystems is equal to the product of distribution functions $\rho_1(q; p)$ and $\rho_2(q; p)$ describing each subsystem:

$$\rho_{1+2}(q; p) = \rho_1(q; p)\rho_2(q; p)$$

Hence, we obtain:

$$\log[\rho_{1-2}(q; p)] = \log[\rho_1(q; p)] + \log[\rho_2(q; p)], \tag{1.7}$$

that is, the logarithm of the distribution function is an additive quantity. We reach the conclusion, therefore, that the logarithm of the distribution function must not be merely a mechanical invariant, but an additive mechanical invariant. As we know from mechanics, there exist only seven independent additive mechanical invariants: the energy $E(q; p)$, three components of the momentum vector $\boldsymbol{P}(q; p)$, and three components of the angular momentum vector $\boldsymbol{M}(q; p)$. The only additive combination of these quantities is a linear combination of the form:

$$\log[\rho_j(q; p)] = \alpha_j + \alpha_{j,E}E_j(q; p) + \boldsymbol{\alpha}_{j,P} \cdot \boldsymbol{P}_j(q; p) + \boldsymbol{\alpha}_{j,M} \cdot \boldsymbol{M}_j(q; p), \quad j = 1, 2, \tag{1.8}$$

where coefficients α_j, $\alpha_{j,E}$, $\boldsymbol{\alpha}_{j,P}$ and $\boldsymbol{\alpha}_{j,M}$ are some constants of the j subsystem. Conditions required for the determination of these constants include normalization for α_j and seven constant values of the additive mechanical invariants for $\alpha_{j,E}$, $\boldsymbol{\alpha}_{j,P}$ and $\boldsymbol{\alpha}_{j,M}$. Therefore, the values of additive mechanical invariants (energy, momentum, and angular momentum) completely define the statistical properties of a closed system, that is, the statistical distribution function $\rho_j(q; p)$ and the mean values of any physical quantity related to them.

The correct way of determining the distribution function for a closed system is:

$$\rho(q; p) = \text{const } \delta[E_0 - E(q; p)] \, \delta[\boldsymbol{P}_0 - \boldsymbol{P}(q; p)] \, \delta[\boldsymbol{M}_0 - \boldsymbol{M}(q; p)], \tag{1.9}$$

where E_0, $\boldsymbol{P}_0$ and $\boldsymbol{M}_0$ are some given values of $E(q; p)$, $\boldsymbol{P}(q; p)$ and $\boldsymbol{M}(q; p)$. The presence of the *Dirac delta function* ensures that the distribution function $\rho(q; p)$ is zero at all points in phase space where one or more of seven quantities $E(q; p)$, $\boldsymbol{P}(q; p)$ or $\boldsymbol{M}(q; p)$ is not equal to the given values E_0, $\boldsymbol{P}_0$ or $\boldsymbol{M}_0$. The statistical distribution $\rho(q; p)$ defined by expression (1.9) is known as *microcanonical distribution* describing the *microcanonical ensemble* of system states. The momentum and angular momentum of the closed system depend on its motion as a whole (uniform translation and uniform rotation). Therefore, one may conclude that a statistical state executing a given motion depends only on its energy. In consequence, energy is of great importance in statistical physics. Thus, the *j*-subsystem distribution function acquires the following simple form:

$$\ln[\rho_j(q; p)] = \alpha_j + \alpha_{j,E}E_j(q; p), \quad j = 1, 2. \tag{1.10}$$

Therefore, the microcanonical distribution function can be simplified as follows:

$$\rho(q; p) = Z_{\mathrm{mc}}^{-1}(V; N)\,\delta[E_0 - E(q; p)],\ Z_{\mathrm{mc}}(V; N) = \rho(N) \int \int \delta[E_0 - E(q; p)]\mathrm{d}q\mathrm{d}p, \tag{1.11}$$

where $Z_{\mathrm{mc}}(V; N)$ is some normalization constant and $\rho(N)$ is the *degeneracy factor* giving the relative number of indistinguishable configurations in phase space (see Section 2.2.3). The closed systems considered above are not of substantial interest in physics. It is much more important to study systems that interact with their environment, by energy or/and matter exchange. Therefore, let us first consider systems that can exchange energy with their environment. In this case, equation (1.11) is no longer satisfied and an expression for the distribution function $\rho(q; p)$ can be derived directly from relationships given by equations (1.7) and (1.10), which yield:

$$\rho(q; p) = Z_{\mathrm{c}}^{-1}(\beta; V; N)e^{-\beta E(q;p)},\ Z_{\mathrm{c}}(\beta; V; N) = \rho(N) \int \int e^{-\beta E(q;p)}\mathrm{d}q\mathrm{d}p, \tag{1.12}$$

where β is a constant, whose physical meaning will be described later. The result given by expression (1.12) is one of the most important in statistical physics. It gives the statistical distribution of any macroscopic subsystem that is a comparatively small part of a large closed system. The distribution (1.12), known as the *Gibbs distribution* or *canonical distribution*, describes the *canonical ensemble* of system states (phase points).

So far, we have always tacitly assumed that the number of particles in a system is some given constant and have deliberately passed over the fact that in reality particles may be exchanged between a system and its environment. In other words, the number of particles N in the system will fluctuate about its mean value $\langle N \rangle$. The distribution function now depends not only on the energy of the state but also on the number of particles N in the system. Equation (1.10), therefore, acquires the form:

$$\ln[\rho_j(q; p; N)] = \alpha_j + \alpha_{j,E}E_j(q; p) + \alpha_{j,N}N, \quad j = 1, 2. \qquad (1.13)$$

An expression for the distribution function $\rho(q; p; N)$ can be derived directly from relationship (1.13) yielding:

$$\rho(q; p; N) = Z_{gc}^{-1}(\beta; V)e^{\beta[\mu N - E(q;p;N)]}, \ Z_{gc}(\beta; V; \mu)$$

$$= \sum_{N=0}^{\infty} \rho(N) \int \int e^{\beta[\mu N - E(q;p;N)]} dq dp, \qquad (1.14)$$

where μ is a constant that will be defined later. The distribution $\rho(q; p; N)$, known as the *grand canonical distribution*, describes the *grand canonical ensemble* of system states.

1.2.3. *Partition Functions*

The normalization constants $Z_{mc}(V; N)$, $Z_c(\beta; V; N)$ and $Z_{gc}(\beta; V; \mu)$ introduced by relationships (1.11), (1.12), and (1.13) respectively are known as *partition functions*. The partition function can be obtained by summing up the Boltzmann factors, $\exp[-\beta E(q; p)]$, over all states. For systems possessing continuous spectra, the summation over states is substituted by integration over phase space. As will be demonstrated later, the partition function serves as the connection between macroscopic quantities and microscopic states.

The states of a classical system are continuously distributed in phase space and, therefore, cannot be counted. In order to determine the *classical analog* of a quantum state, let us take into account that the uncertainty $\Delta q \Delta q$ in the determination of any quantum state is restricted from below (see Landau and Lifshitz (1980) for details):

$$\Delta q \Delta p \sim h^{sN} (\textit{Heisenberg uncertainty principle}), \qquad (1.15)$$

where h is the Planck constant ($h = 6.6256 \cdot 10^{-27}$ erg $\cdot$ sec) and s is a degree of freedom of each molecule constituting the system ($s = 3$ for a one-atom molecule, $s = 5$ for a rigid two-atom molecule, etc.). In other words, any phase point located in a cell of volume h^{sN} cannot be distinguished from any other point located in the same cell. Moreover, there is another source of uncertainty. In quantum mechanics it is impossible to distinguish phase-space configurations such as (A) state S_1 belongs to cell C_1 and state S_2 belongs to cell C_2 and (B) state S_1 belongs to cell C_2 and state S_2 belongs to cell C_1. Therefore, the system containing N particles can possess $N!$ indistinguishable configurations ($N!$ different permutations over N cells) of the same states in phase space. The expression for the degeneracy factor $\rho(N)$, first introduced in expression (1.11), has the form:

$$\rho(N) = \frac{1}{h^{sN} N!}. \tag{1.16}$$

1.2.3.1. *The Microcanonical Ensemble*

In this case the system under consideration is closed (there is neither energy nor matter exchange with the environment). Given expressions (1.10) and (1.11), it is straightforward to derive the following expression for the partition function in the case of a microcanonical ensemble:

$$Z_{\text{mc}}(V; N) = \frac{1}{h^{sN} N!} \int \int \delta[E_0 - E(q; p)]\mathrm{d}q\mathrm{d}p. \tag{1.17}$$

1.2.3.2. *The Canonical Ensemble*

In this case, the system under consideration is "semi-closed" (there is energy but no matter exchange with the environment). Given expressions (1.10) and (1.11), it is straightforward to derive the following expression for the partition function in the case of a canonical ensemble:

$$Z_{\text{c}}(\beta; V; N)\frac{1}{h^{sN} N!} \int \int e^{-\beta E(q;p)}\mathrm{d}q\mathrm{d}p. \tag{1.18}$$

In classical statistics, the energy $E(q, p)$ can be written as the sum of the potential $U(q)$ and kinetic $K(p)$ energies. The potential energy is dependent on the interaction between the system molecules and is a function of their coordinates. The kinetic energy is a quadratic function of the momenta of the molecules. Therefore, the probability $\mathrm{d}\omega(q; p)$ introduced by relations (1.1) and (1.2) can be presented as the product of two factors:

$$\mathrm{d}\omega(q; p) = \text{const} \cdot e^{-U(q)}e^{-K(p)}\mathrm{d}q\mathrm{d}p, \tag{1.19}$$

where the first factor depends only on the molecular coordinates and the second only on their momenta. Equation (1.14) allows the conclusion that the probabilities for molecular coordinates and momenta are independent and the probability of one does not influence probability of the various values of the other. Thus, the probabilities $\mathrm{d}\omega(q)$ and $\mathrm{d}\omega(p)$ of the various values of the molecular coordinates and momenta can be written in the form:

$$\mathrm{d}\omega(q) = \text{const}_U \cdot e^{-U(q)}\mathrm{d}q, \mathrm{d}\omega(p) \, \text{const}_K \cdot e^{-K(p)}\mathrm{d}p, \tag{1.20}$$

where const $= \text{const}_U\text{const}_K$. Since the sum of probabilities of all possible values of the momenta and coordinates must be unity, each of the probabilities $\mathrm{d}\omega(q)$ and $\mathrm{d}\omega(p)$ has to be normalized separately by integrating over the coordinates and momenta, respectively. Therefore, normalization allows the determination of constants const_U and const_K.

Let us consider the probability distribution for the molecular momenta. The kinetic energy of the entire system is equal to the sum of the kinetic energies of all molecules in the system. This means that the probability $d\omega(p)$ can be expressed as the product of factors, with each factor dependent on the momentum of only one molecule. This means that the momentum probabilities of different molecules are independent (the momentum of one molecule does not affect the probabilities of various momenta of any other molecule). The probability distribution for the momentum of each individual molecule can, therefore, be written.

For a molecule of mass m the kinetic energy $K(p)$ is: $K(p) = K(p_x, p_y, p_z) = (p_x^2 + p_y^2 + p_z^2)/(2m)$, where p_x, p_y, and p_z are the Cartesian coordinates of its momentum. Thus, the probability $d\omega(p) = d\omega(p_x, p_y, p_z)$ acquires the form:

$$d\omega(p_x, p_y, p_z) = \text{const}_K \exp\left(-\frac{p_x^2 + p_y^2 + p_z^2}{2mk_BT}\right)dp_x dp_y dp_z. \tag{1.21}$$

The normalization constant const_K can be easily found to be $\text{const}_K = (2\pi m k_B T)^{-3/2}$. Changing from momenta to velocities, $p = mv$, one can write the corresponding velocity probability $d\omega(v) = d\omega(v_x, v_y, v_z)$ as follows:

$$d\omega(v_x, v_y, v_z) = \rho(v_x, v_y, v_z)dv_x, dv_y, dv_z, \rho(v_x, v_y, v_z)$$
$$= \left(\frac{m}{2\pi k_B T}\right)^{3/2} \exp\left(-\frac{m(p_x^2 + p_y^2 + p_z^2)}{2k_BT}\right). \tag{1.22}$$

The distribution function $\rho(v_x, v_y, v_z)$ is known in the literature as the *Maxwell-Boltzmann distribution*.

1.2.3.3. *The Grand Canonical Ensemble*

In this case the system under consideration is open (there are both the energy and matter exchanges with the environment). Given expressions (1.21) and (1.22), the following expression can be derived for the partition function in the case of a grand canonical ensemble:

$$Z_{\text{gc}}(\beta; V; \mu) = \sum_{N=0}^{\infty} \frac{e^{\beta\mu N}}{h^{sN} N!} \int \int e^{-\beta E(q;p;N)}dqdp. \tag{1.23}$$

This relationship for partition function is sometimes rewritten in terms of the *fugacity*, $z = e^{\beta\mu}$:

$$Z_{\text{gc}}(\beta; V; z) = \sum_{N=0}^{\infty} \frac{z^N}{h^{sN} N!} \int \int e^{-\beta e(q;p;N)}dqdp. \tag{1.24}$$

In this section we have introduced three different ensembles that describe a system in thermodynamic equilibrium. The microcanonical ensemble describes a closed system, the internal energy $E(q; p)$ of which is fixed: $E(q; p) = E_0$. The canonical ensemble describes a "semi-closed" system, the energy of which is not fixed. However, the *mean internal energy*, $\langle E \rangle$ of the canonical ensemble is fixed. The grand canonical ensemble describes an open system, the energy and number of particles of which are not fixed. However, the mean energy $\langle E \rangle$ and the number of particles $\langle N \rangle$ of a grand canonical ensemble are fixed.

It is important to note that the forms of partition function (1.17), (1.18), and (1.24) obtained for the equilibrium microcanonical, canonical, and grand canonical ensembles correspond to the maximum entropy achievable in a state of thermodynamic equilibrium (see Sections 1.3.1 and 1.3.2).

1.2.4. *Relationship between Statistical Mechanics and Thermodynamics*

Classical thermodynamics is based on many empirical results, which have been studied, systemized, generalized, and formulated in the form of the *Three Laws of Thermodynamics*. These laws allow derivation of many useful relationships between different quantities characterizing various mechanical and thermal processes. However, there is a flaw inherent in thermodynamics. Thermodynamics provides relationships between various quantities, but does not provide methods to determine their absolute values. For example, thermodynamics establishes a functional relationship between the heat capacity C_V at constant volume V and the heat capacity C_P at constant pressure P. If C_P is known, then C_V can be determined theoretically without need for experiments. However, thermodynamics alone provides no method to allow determination of the value of C_P itself. This is the role of statistical mechanics.

Therefore, statistical mechanics allows solution of the following two major problems:

1. derivation of expressions for macroscopic thermodynamic quantities from microscopic mechanics (for example, from the molecular energy levels that can be determined by spectroscopic methods);
2. derivation of microscopic properties (for example, the nature of intermolecular interactions) from the measurable macroscopic quantities.

All quantities used in thermodynamics can be divided into three major groups. The first group contains *external quantities*, parameters such as volume V, number of particles N, external fields (gravitational, electromagnetic), etc., the absolute values of which are fixed either by the external environment or by an experimentalist. The second group contains *mechanical quantities*, such as internal energy E and pressure P. The third group is specific to thermodynamics and contains *thermal quantities*, parameters such as temperature, entropy, etc. These quantities do not have any microscopic meaning and can be defined exclusively on the macroscopic level. For example, one can define the molecular energy, but it is impossible to define the temperature of a single molecule. Therefore, thermal quantities are defined by the entire ensemble.

For the sake of simplicity, let us consider the most widely used case of canonical ensemble and introduce the function $H_c(\beta; V; N)$:

$$Z_c(\beta; V; N) = e^{-\beta H_c(\beta; V; N)}. \tag{1.25}$$

There exists the following important relationship:

$$\left[\frac{\partial}{\partial \beta} (\beta H_c(\beta; V; N)) \right]_{V,N} = -\frac{\partial}{\partial \beta} \ln[Z_c(\beta; V; N)]$$

$$= Z_c^{-1}(\beta; V; N) \frac{1}{h^{sN} N!} \int \int E(q; p) e^{-\beta E(q;p)} dq dp = \langle E \rangle, \tag{1.26}$$

where the brackets $[...]_{V,N}$ designate an expression taken at constant V and N. Relationship (1.26) is easily recognizable in traditional thermodynamics if one accepts the following treatments:

1. $\beta = 1/k_B T$ is the inverse temperature ($k_B = 1.38054 \times 10^{-16}$ erg/K is the *Boltzmann constant*).
2. $H_c(\beta; V; N)$ is the *Helmholtz free energy* of the canonical ensemble (see Section 1.3.8).

Thus, within the framework of statistical mechanics the Helmholtz free energy of canonical ensemble $H_c(\beta; V; N)$ can be defined as follows:

$$H_c(\beta; V; N) = -k_B T \ln[Z_c(\beta; V; N)]. \tag{1.27}$$

The same considerations carried out for the grand canonical ensemble allow derivation of another important relationship:

$$\mu = \left[\frac{\partial H_{gc}(\beta; V; \mu)}{\partial N} \right]_{T,V}, \tag{1.28}$$

where μ is the *chemical potential* for one molecule. Therefore, basic relationships (1.26) and (1.28) provide the linkage between statistical mechanics and thermodynamics.

1.2.5. Fluctuations

Let us consider the most widely utilized case of the canonical ensemble. The normalization condition for this ensemble can be rewritten as:

$$\frac{1}{h^{sN} N!} \int \int e^{\beta[H_c(\beta; V; N) - E(q;p)]} dq dp = 1. \tag{1.29}$$

Differentiation of this relationship twice with respect to β gives:

$$\frac{1}{h^{sN}N!}\int\int e^{\beta[H_c(\beta;V;N)-E(q;p)]}\left[\left(\frac{\partial}{\partial\beta}(\beta H_c(\beta;V;N))-E(q;p)\right)^2\right.$$

$$\left.+\frac{\partial^2}{\partial\beta^2}[\beta H_c(\beta;V;N)]\right]dqdp=0. \tag{1.30}$$

Now, taking into account relationship (1.26) for $\langle E\rangle$, we obtain:

$$\langle(E-\langle E\rangle)^2\rangle\equiv\langle E^2\rangle-\langle E\rangle^2=-\frac{\partial^2}{\partial\beta^2}[\beta H_c(\beta;V;N)]=-\frac{\partial\langle E\rangle}{\partial\beta}, \tag{1.31}$$

where

$$\langle E^2\rangle=\frac{1}{h^{sN}N!}\int\int E^2(q;p)e^{[H_c(\beta;V;N)-E(q;p)]}dqdp.$$

It is important to note that differentiation with respect to β in expressions (1.30) and (1.31) is carried out at constant volume V. Therefore, expression (1.31) can be rewritten as follows:

$$\langle(E-\langle E\rangle)^2\rangle\equiv\langle E^2\rangle-\langle E\rangle^2=k_BT^2C_V,\text{ where }C_V=\left[\frac{\partial\langle E\rangle}{\partial T}\right]_V. \tag{1.32}$$

Relationship (1.32) is fundamental since it relates fluctuations of the microscopic energy to a macroscopic measurable quantity, such as the heat capacity at constant volume. C_V and E are additive quantities (see Section 1.2.2 for the definition of additivity), that is, they are proportional to N. Therefore, the relative contribution of fluctuations can be quantitatively estimated from the ratio:

$$\frac{1}{\langle E\rangle}\langle(E-\langle E\rangle)^2\rangle^{1/2}\sim\frac{N^{1/2}}{N}=N^{-1/2}. \tag{1.33}$$

This ratio demonstrates that, even though the energy fluctuations are quite sizable in absolute value ($\sim N^{1/2}$), their contribution is negligible compared to the mean internal energy $\langle E\rangle\sim N$ itself.

To conclude this section let us note that the problem of fluctuations is not always trivial because of their small relative value. There is a non-zero probability of the existence of huge local fluctuations confined within small volumes. Such huge fluctuations are extremely important in consideration of many physical phenomena such as light scattering, nucleation, phase transition, plasma oscillations, etc.

1.3. Equilibrium Thermodynamics

Thermodynamic variables can be defined as variables describing macroscopic states of matter. They include those that have both thermodynamic and purely mechanical significance, such as energy and volume. There are also variables of another kind, such as entropy and temperature, that originate from statistical laws and have no meaning when applied to nonmacroscopic systems. Thermodynamic variables are subjected to fluctuations. However, these fluctuations are negligible because thermodynamic variables are varying only with the macroscopic state of matter (see Section 1.2.5).

1.3.1. *Entropy*

Let us consider a closed system of energy E in *statistical equilibrium*, that is in a state obtained in the result of all possible relaxation processes which could occur in the system. Its N_s subsystems of energies $E_j (j = 1, \ldots, N_s)$ can be described in terms of the statistical distributions $\rho(E_j)$ considered as some functions of energy. Therefore, the probability $\Delta w(E_j)$ for the j subsystem to have energy in the interval $[E_j, E_j + \Delta E_j]$ can be given by the following expression:

$$\Delta w(E_j) = \rho(E_j)\Delta\Gamma(E_j), \ \Delta\Gamma(E_j) = \frac{d\Gamma(E_j)}{dE_j}\Delta E_j, \quad j = 1, \ldots, N_s, \qquad (1.34)$$

where $\Gamma(E_j)$ is a number of the j-subsystem states with energies less than or equal to E_j and $\Delta\Gamma(E_j)$ is a number of the j-subsystem states corresponding to the energy interval ΔE_j. The interval ΔE_j is equal in order of magnitude to the mean energy fluctuation of the j subsystem. The quantity $\Delta\Gamma(E_j)$ represents the "degree of broadening" of a macroscopic state of energy E with respect to microscopic states and by virtue of this is usually called the *statistical weight* of the macroscopic state of the j subsystem. The normalization condition for the probability $\Delta w(E_j)$ is:

$$\int dw(E_j) = \int \rho(E_j)d\Gamma(E_j) = 1. \qquad (1.35)$$

The concept of statistical equilibrium implies that the function $\rho(E_j)$ has a very sharp maximum at $E_j = \langle E_j \rangle$, being appreciably different from zero only in the immediate neighborhood of this point. Therefore, condition (1.35) can be rewritten in the following simple form:

$$\rho(\langle E_j \rangle)\Delta\Gamma(\langle E_j \rangle) = 1, \ \text{where} \ \Delta\Gamma(\langle E_j \rangle) = \frac{d\Gamma(\langle E_j \rangle)}{dE_j}\Delta E_j. \qquad (1.36)$$

The definition (equation (1.10)) of the distribution function for the j subsystem, $\rho(\langle E_j \rangle)$, acquires the form:

$$\log[\rho(\langle E_j \rangle)] = \alpha_j + \alpha_{j,E} \langle E_j \rangle. \tag{1.37}$$

It is important to emphasize that $\log[\rho(\langle E_j \rangle)]$ is a linear function of $\langle E_j \rangle$.

The logarithm of statistical weight $\Delta\Gamma(E_j)$:

$$S_j = \log[\Delta\Gamma(E_j)], \tag{1.38}$$

is known as the *j*-subsystem *entropy*. The entropy, like the statistical weight itself, is dimensionless. Because the number of states $\Delta\Gamma(E_j)$ is not less than unity, the entropy cannot be negative. Substituting equations (1.36) and (1.37) into equation (1.38) results in an expression for the entropy:

$$S_j = -\log[\rho(\langle E_j \rangle)] = -\langle \log[\rho(E_j)] \rangle = -\rho(E_j) \log[\rho(E_j)]. \tag{1.39}$$

Let us now return to the closed system as a whole. The statistical weight of the system, $\Delta\Gamma(E)$, defined as the product of its subsystem weights, yields the following expression for the entropy of a closed system:

$$S = \sum_{j=1}^{N_s} S_j = -\sum_{j=1}^{N_s} \rho(E_j) \log[\rho(E_j)], \tag{1.40}$$

showing that the entropy is an additive quantity.

The microcanonical ensemble describing a closed system has the form (see equation (1.11)):

$$\Delta w(E) = Z_{mc}^{-1}\delta(E_0 - E)\prod_{j=1}^{N_s} \Delta\Gamma(E_j), \quad \text{where } E = \sum_{j=1}^{N_s} E_j. \tag{1.41}$$

If equation (1.38) is taken into account, this expression can be rewritten as follows:

$$\Delta w(E) = Z_{mc}^{-1}\delta(E_0 - E)e^S\prod_{j=1}^{N_s} \Delta E_j, \quad \text{where } S = \sum_{j=1}^{N_s} S_j(E_j). \tag{1.42}$$

Expression (1.42) emphasizes that the quantity $\Delta w(E)$ is the probability that the subsystems of a closed system have energies in given intervals $\{[E_j, E_j + \Delta E_j]; j = 1, \ldots, N_s\}$. The factor $\delta(E_0 - E)$ ensures that the total system energy E has the given value E_0.

Attention should be drawn to the significance of time in the definition of entropy. The entropy is a quantity that describes the average properties of the system over some non-zero interval of time (Δt). In order to determine the entropy for the given (Δt) we have to imagine the system divided into subsystems so small that their relaxation times are small in comparison with (Δt). Since these subsystems have to be macroscopic, when the intervals (Δt) are too short, the concept of entropy becomes meaningless so that it is not possible to speak of an "instantaneous" value of entropy.

1.3.2. *The Law of Entropy Increase*

If at some initial states a closed system is not in statistical equilibrium, its macroscopic state will vary in time until eventually the system reaches equilibrium. It follows from equation (1.42), which expresses the probability of energy distribution between the system subsystems, that this probability is an increasing function of time. This increase in probability is significant because it is exponential with respect to entropy. On the basis of this result, it may be concluded that any process occurring in a nonequilibrium system develops in such a way that the system continuously passes from states of lower entropy to those of higher entropy until finally the entropy reaches its maximum possible value, which corresponds to complete statistical equilibrium. Therefore, if a closed system is in a nonequilibrium state at some time, the most probable path for the system to take as it moves towards equilibrium is the path that results in an increase in entropy. The concept of the "most probable" evolution path means that the probability of transition to states of higher entropy is much higher than the probability of any other path. If decreases in entropy due to negligible fluctuations are ignored, one can formulate the *law of entropy increase* as:

if at some time instant the entropy of a closed system does not have its maximum value, then at subsequent instants the entropy will not decrease; it will increase or at least remain constant.

It follows from the law of entropy increase that all processes occurring within macroscopic systems can be divided into *irreversible* and *reversible processes*. The irreversible processes are those that are accompanied by an increase in entropy of the closed system as a whole. The reversible processes are those in which the entropy of the closed system as a whole remains constant.

The law of entropy increase or the *second law of thermodynamics* was discovered by R. Clausius in 1865. Its statistical explanation was first given by L. Boltzmann in the 1870s. Other interesting details and consequences of the second law of thermodynamics can be found in Haase (1969).

1.3.3. *Adiabatic Processes*

There is a special class of interactions between a system and its environment that consists only in a change in the external conditions. By external conditions we mean various external fields. In practice, the external conditions are most often determined by the fact that the system must have a prescribed volume V. The presence of finite volume can be regarded as a particular type of external field, because the walls limiting the volume are equivalent to an infinite potential barrier that prevents the molecules of the system from escaping. If the system is subjected only to such interactions which result in changes in external conditions, it is said to be *thermally isolated*. It should be emphasized that, although a thermally isolated system does not interact directly with other systems, its energy may vary with time, that is, the thermally isolated system is "semi-closed" and its description can be given in terms of the canonical ensemble (see Sections 1.2.2 and 1.2.3).

Let us suppose that the system is thermally isolated and is subjected to a process due to variations in external conditions that vary sufficiently slowly. Such a process is known as an *adiabatic process*. The main characteristic of such a process is that the system entropy remains unchanged, that is, the adiabatic process is reversible.

1.3.4. *Temperature*

Let us consider two subsystems in *thermodynamic equilibrium* (statistical equilibrium), forming a closed system. For a given energy E, the entropy S of this system has its maximum value. The energy $E = E_1 + E_2$, where E_1 and E_2 are the energies of the first and second subsystems, respectively, and the entropy S of the system $S = S_1(E_1) + S_2(E_2)$. The entropy of each subsystem is a function of its energy. The condition for the maximum entropy at thermodynamic equilibrium is given by the equation:

$$\frac{dS}{dE_1} = \frac{dS_1}{dE_1} + \frac{dS_2}{dE_2}\frac{dE_2}{dE_1} = \frac{dS_1}{dE_1} - \frac{dS_2}{dE_2} = 0 (E_2 = E - E_1, \quad E = \text{const}). \quad (1.43)$$

Thus, if the system is in a state of thermodynamic equilibrium, the derivative of entropy with respect to energy is the same for every one of its subsystems, and constant throughout the system. Thermodynamic variable, which is the reciprocal of the derivative of system entropy S with respect to its energy E, is known as the system temperature T:

$$\frac{dS}{dE} = \frac{1}{T}. \quad (1.44)$$

The temperatures of subsystems in thermodynamic equilibrium are equal: $T_1 = T_2$. Like entropy, the temperature is a purely statistical quantity having meaning only for macroscopic systems.

1.3.5. *Pressure*

The macroscopic state of a system at rest in thermodynamic equilibrium is entirely determined by only two variables, for example by the energy E and volume V. All other thermodynamic variables can be expressed as functions of these two. To demonstrate this let us derive an expression for pressure in an adiabatic process (constant entropy process). For this purpose, let us first calculate the force F exerted by the system on the surface bounding its volume. The force acting on the surface element dA is:

$$F = \left[\frac{\partial E(r)}{\partial r}\right]_S = -\left[\frac{\partial E}{\partial V}\right]_S \frac{\partial V}{\partial r} = -\left[\frac{\partial E}{\partial V}\right]_S dA, \tag{1.45}$$

where $\partial V = \partial r \cdot dA$, $E(r)$ is the system energy as a function of the radius vector r of the surface element dA and the symbol $[\ldots]_S$ designates an expression taken at constant entropy. The requirement of constant entropy emphasizes that the process resulting in the force F applied to the surface area dA is adiabatic. Hence, the force applied to the surface element is normal to this element and proportional to its area. Its magnitude per unit area is:

$$P = -\left(\frac{\partial E}{\partial V}\right)_S. \tag{1.46}$$

This quantity is known as pressure.

1.3.6. *Internal Energy*

Internal energy E of a thermodynamic system can be defined as the total system energy with the kinetic and potential energies of the system as a whole subtracted. This energy can be experimentally measured in two different ways:

1.3.6.1. *Measuring the Work W Done on the Given System by External Forces*

Negative work ($W < 0$) means that the system itself does work equal to $|W|$ on some external object (for example in expanding). In this case the system is assumed to be thermally isolated. Therefore, any change of its energy is due to the work done on it. If it is assumed that the system is in a state of mechanical equilibrium, i.e. at each time instant the pressure P is constant throughout the system, the work done on the system per unit time is:

$$\frac{dW}{dt} = -P\frac{dV}{dt}. \tag{1.47}$$

In compression $dV/dt < 0$, so that $dW/dt > 0$. Result (1.47) is applicable to both reversible and irreversible conditions.

1.3.6.2. *Measuring the Energy Q which the System Gains or Loses by Direct Transfer*

This energy change occurs in addition to the work done and it assumes that the system is not thermally isolated. The energy quantity Q is known as *heat* gained or lost by the system.

Therefore, the change in the internal energy of the thermodynamic system per unit time may be presented in the form:

$$\frac{dE}{dt} = \frac{dQ}{dt} - P\frac{dV}{dt}. \qquad (1.48)$$

Let us assume that at every time instant during the process the system may be regarded as being in a state of thermal equilibrium at given internal energy and volume. This does not necessarily mean that the process is reversible because the system may not be in equilibrium with its environment. Taking into account equations (1.44) and (1.46) for the partial differentials of function $E = E(S, V)$, we can see that $dQ = TdS$. Therefore, the relationship (1.48) can be rewritten as follows:

$$\frac{dE}{dt} = T\frac{dS}{dt} - P\frac{dV}{dt}, \quad \text{or } dE = TdS - PdV. \qquad (1.49)$$

It is important to note that the work dW and heat dQ gained by the system during an infinitesimal change of the system state are not total differentials. Only their sum $dE = dW + dQ$, that is, the change of internal energy dE is the total differential. It therefore makes sense to introduce the concept of internal energy E at a given state and it is meaningless to speak about the heat Q which system possesses in a given state. The internal energy of the system at equilibrium cannot be divided into mechanical and thermal parts. However, such a division is possible when the system goes from one state into another.

The relationship (1.49) expresses the *first law of thermodynamics*. As can be seen from equation (1.49), this law is just a statement of internal energy conservation, which specifies two ways for the energy to change: as work done on, or by, the system and as heat flowing into, or out of, the system.

1.3.7. *Enthalpy*

If the volume of a system remains constant during a process, then $dE = dQ$ (see equation (1.49)). The quantity of heat gained by the system is equal to the change of its internal energy. If the process occurs at constant pressure, the quantity of heat gained or lost during the process can be written as the following differential:

$$dQ = dR = d(E + PV) = TdS + VdP, \qquad (1.50)$$

since $d(VP) = VdP + PdV$. The quantity $R = E + VP$ is known as the *enthalpy* or *heat function* of the system. The change of the enthalpy in a process occurring at constant pressure is, therefore, equal to the quantity of heat gained or lost by the system. It follows from expression (1.50) that:

$$T = \left(\frac{\partial R}{\partial S}\right)_P, = \left(\frac{\partial R}{\partial P}\right)_S. \qquad (1.51)$$

If the system is thermally isolated (which does not imply that the system is a closed one), then $dQ = 0$. This means that $R = $ const, that is, the enthalpy is conserved in processes occurring at constant pressure and involving a thermally isolated system.

1.3.8. *The Helmholtz Free Energy*

The work done on a thermodynamic system in an infinitesimal isothermal reversible change of the system state can be given as follows:

$$dW = dH = d(E - TS) = -SdT - PdV, \qquad (1.52)$$

since $d(TS) = TdS + SdT$. The quantity $H = E - TS$ is another function of the system state, known as the *Helmholtz free energy* of the thermodynamic system that possesses the total differential. Therefore, the work done on the system in a reversible isothermal process is equal to the change in its Helmholtz free energy. It follows from expression (1.52) that:

$$S = -\left(\frac{\partial H}{\partial T}\right)_V, \quad P = -\left(\frac{\partial H}{\partial V}\right)_T. \qquad (1.53)$$

Utilizing the relationship $E = F + TS$ one can express the internal energy in terms of the Helmholtz free energy as follows:

$$E = H - T\left(\frac{\partial H}{\partial T}\right)_V = -T^2\left[\frac{\partial}{\partial T}\left(\frac{H}{T}\right)\right]_V. \qquad (1.54)$$

1.3.9. *The Gibbs Free Energy*

One additional total differential dF, with respect to pressure P, and temperature T, still remains for consideration. Its derivation is straightforward:

$$dF = d(E - TS + PV) = d(R - TS) = d(H + PV) = -SdT + VdP. \qquad (1.55)$$

The quantity F introduced by this relationship is known as the *Gibbs free energy* of the thermodynamic system. It follows from expression (1.55) that:

$$S = -\left(\frac{\partial F}{\partial T}\right)_P, \quad V = \left(\frac{\partial F}{\partial P}\right)_T. \qquad (1.56)$$

The enthalpy R is expressed in terms of the Gibbs free energy F in the same way as the internal energy E is expressed in terms of the Helmholtz free energy H (see equation (1.54)):

$$R = F - T\left(\frac{\partial F}{\partial T}\right)_P = -T^2\left[\frac{\partial}{\partial T}\left(\frac{F}{T}\right)\right]_P. \qquad (1.57)$$

1.3.10. *"Open" Thermodynamic Systems*

Equations (1.49), (1.50), (1.52), and (1.55) demonstrate that if one knows any of the thermodynamic quantities, internal energy E, enthalpy R, the Helmholtz free energy H, or the Gibbs free energy F, as a function of the corresponding two thermodynamic variables, then by taking the partial derivatives of that quantity one can determine all the remaining thermodynamic quantities.

Thermodynamic quantities R, H, and F also have the property of additivity as previously shown for energy E and entropy S. This property follows directly from their definitions (see equations (1.50), (1.52), and (1.55)) if one assumes that pressure and temperature are constants throughout a system in an equilibrium state. The additivity of a quantity signifies that when the amount of matter (such as the number N of particles) is changed by a given factor, the quantity is changed by the same factor. In other words, one can state that the additive thermodynamic quantities are some linear functions with respect to the additive variables.

Let us express the internal energy E of the system as a function of entropy S, volume V, and number of particles N. Since S and V are themselves additive, this function has to have the following form:

$$E(S, V, N) = N f_E\left(\frac{S}{N}, \frac{V}{N}\right). \tag{1.58}$$

The quantity E is presented as the most general linear function in S, V, and N. The same considerations allow one to conclude that the enthalpy R, which is a function of entropy S, pressure P, and number of particles N, has the following form:

$$R(S, P, N) = N f_R\left(\frac{S}{N}, P\right). \tag{1.59}$$

The Helmholtz and Gibbs free energies, which are the functions of T, V, N and P, T, N, respectively, allow the following forms:

$$H(T, V, N) = N f_H\left(T, \frac{V}{N}\right), (P, T, N) = N f_G(P, T). \tag{1.60}$$

In the foregoing discussions we have assumed that the number of particles N is a parameter with a given constant value. Let us now consider N as an independent variable of an "open" thermodynamic system. Then the expressions for the total differentials of the thermodynamic quantities have to contain terms proportional to $\mathrm{d}N$:

$$\mathrm{d}E = T\mathrm{d}S - P\mathrm{d}V + \mu\mathrm{d}N, \tag{1.61}$$

$$dR = T dS + V dP + \mu dN, \qquad (1.62)$$

$$dH = -S dT - P dV + \mu dN, \qquad (1.63)$$

$$dF = -S dT + V dP + \mu dN, \qquad (1.64)$$

where the thermodynamic quantity

$$\mu = \left(\frac{\partial E}{\partial N}\right)_{S,V} = \left(\frac{\partial R}{\partial N}\right)_{S,P} = \left(\frac{\partial H}{\partial N}\right)_{T,V} = \left(\frac{\partial F}{\partial N}\right)_{P,T} \qquad (1.65)$$

is known as the system *chemical potential*. Differentiating expression (1.64) for the Gibbs free energy F with respect to N we obtain that $\mu = f_G(P, T)$, that is, $F(P, T, N) = N\mu(P, T)$. Therefore, the chemical potential of the system consisting of identical particles is just the Gibbs free energy per molecule.

1.3.11. *Equations of State*

In practice, the most convenient and widely used pairs of thermodynamic variables are two possible pairs of independent variables: "temperature, volume" T, V and "temperature, pressure" T, P. It is therefore necessary to establish the transformation rules from one set of thermodynamic variables into another, both dependent and independent. If temperature T and volume V are used as the first set of independent variables the results of transformation into the second set of independent variables can be expressed in terms of pressure $P = P(T, V)$ and specific heat $C_V = C_V(T, V)$. The first transformation equation $P = P(T, V)$, which relates pressure, volume, and temperature, is known as the *equation of state* for a given system.

An equation of state describes equilibrium properties of the system. Let us derive an equation of state for the simplest case of an ideal gas that is characterized by the following properties:

- Molecules constituting these systems are mutually independent (non-interacting).
- The system is "semi-open," that is, its molecules have to be described in terms of canonical ensemble.
- The system temperature is high enough to make quantum effects negligible.
- There are no external fields applied to the system.

Such systems are known as the *Maxwell–Boltzmann systems* (see equation (1.22)). The total energy $E(q; p)$ of a one-atom molecule of the Maxwell–Boltzmann system can be given by the kinetic energy of its translational motion:

$$E(q; p) = E(p) = \frac{p^2}{2m}, \qquad (1.66)$$

where $p = |\boldsymbol{p}|$. The volume of phase space $\mathrm{d}q\mathrm{d}p$ corresponding to a one-atom molecule of the Maxwell–Boltzmann system has the following form in spherical coordinates:

$$\mathrm{d}q\mathrm{d}p = 4\pi p^2 \mathrm{d}p\mathrm{d}V. \tag{1.67}$$

Substituting these equations into equation (1.18) for the partition function of a canonical ensemble, one can obtain the result:

$$Z_{\mathrm{c}}(\beta; V; N) = \frac{1}{N!} Z_1^N(\beta; V), \quad Z_1(\beta; V) = V\Lambda, \quad \Lambda = \left(\frac{2\pi m}{\beta h^2}\right)^{3/2}. \tag{1.68}$$

Therefore, expression (1.27) for the Helmholtz free energy of canonical ideal ensemble takes the form:

$$H_{\mathrm{c}}(\beta; V; N) = -\frac{1}{\beta}\ln[Z_{\mathrm{c}}(\beta; V; N)] = -\frac{N}{\beta}\ln\left[\frac{e}{N}Z_1(\beta; V)\right], \tag{1.69}$$

where $e = 2.7183$. In derivation of this result we have taken into account that the number of molecules N, being of order 10^{23}, allows usage of the asymptotic relationship $N! \sim N\ln(N/e)$ due to Stirling.

Let us now introduce the concept of the *thermodynamic limit*, which can be characterized by the following statement: $N/V = n = $ const when $N \to \infty$ and $V \to \infty$. Then, the Helmholz free energy per molecule of a canonical ensemble $h_{\mathrm{c}}(\beta; V)$ can be presented in the following simple form:

$$h_{\mathrm{c}}(\beta; V) = \lim_{N\to\infty} N^{-1} H_{\mathrm{c}}(\beta; V; N) = \frac{1}{\beta}\ln(n) - \frac{1}{\beta} - \frac{3}{2\beta}\ln\left(\frac{2\pi m}{\beta h^2}\right). \tag{1.70}$$

Now, utilizing expression (1.53), we can derive the equation of state for an ideal canonical ensemble (gas):

$$P = -\left(\frac{\partial H_{\mathrm{c}}}{\partial V}\right)_T = n^2\left(\frac{\partial h_{\mathrm{c}}}{\partial n}\right)_T = \frac{n}{\beta} = k_{\mathrm{B}}Tn. \tag{1.71}$$

The equations of state for the more complicated cases, which include quantum effects, interaction between molecules, etc., follow the same derivation scheme as the one presented above for an ideal canonical ensemble. However, some computational particularities in these cases are quite cumbersome and are beyond the scope of this chapter. Derivation of the equations of state for other cases can be found, for example, in Isihara (1971), Mayer and Goeppert-Mayer (1977) and Landau and Lifshitz (1988).

1.4. Intermolecular Forces

1.4.1. *Origins of Intermolecular Forces*

The forces of interaction between particles can be classified into four major categories:

1. *gravitational*;
2. *strong nuclear*;
3. *weak nuclear*;
4. *electromagnetic*.

The gravitational force is extremely long-range and might be thought as the source of intermolecular attraction. However, a simple calculation of the gravitational force between two argon atoms at a separation of 0.4 nm gives only 710^{-52} J, which is some thirty orders of magnitude smaller than any realistic intermolecular forces. Therefore, this force cannot contribute to the intermolecular interaction. The strong nuclear forces are responsible for the binding of neutrons and protons inside the nucleus. The range over which these forces are significant is of the order of 10^{-4} nm. The weak nuclear forces, known to be of electromagnetic origin, are also significant within this short range. Since molecular dimensions are typically of the order of $5 \cdot 10^{-1}$ nm these nuclear forces cannot contribute to the intermolecular forces. Consequently, intermolecular forces must be of electromagnetic origin.

The electromagnetic nature of intermolecular forces suggests that they are due to charged particles, electrons and protons, which constitute an atom or a molecule. Since the intermolecular forces are repulsive at short range and attractive at long range, there must be at least two distinctive contributions to the total intermolecular potential. The repulsive intermolecular forces can be qualitatively explained on the basis of the *Pauli Exclusion Principle*, which prohibits some electrons from occupying the overlap region and, by virtue of this, reduces the electron density in this region. This leads to the situation where the positively charged nuclei of the molecules are incompletely shielded from each other and, therefore, exert a repulsive force on each other. Such short-range forces are known as the *overlap forces*. The long-range attractive component of the intermolecular force is significant when the overlapping of electron clouds is small. In general, depending on the nature of the interacting molecules, one can distinguish three different contributions (A,B,C) to the long-range attractive force, electrostatic, induction and dispersion.

1.4.1.1. *Electrostatic Contributions*

Some molecules, such as HCl, possess permanent electric dipole moments due to the electric charge distribution in the molecule electron cloud. Therefore, it is natural to assume that the long-range electrostatic interaction exists between their dipole moments. Some molecules without dipoles, such as CO_2, possess an electric quadrupole moment that contributes to the attractive intermolecular forces in a similar manner.

1.4.1.2. *Induction Contributions*

Let us consider an interaction between two molecules, where the first molecule is dipolar (possesses a dipole moment), and the second molecule is nondipolar. The electric field of the dipolar molecule distorts the electric charge distribution in the electron cloud of the nondipolar molecule and, by virtue of this, induces a dipole moment on the second, initially non-dipolar, molecule. In its turn, the induced dipole moment interacts with the inducing dipole moment. This interaction results in a long-range attractive force. In the case of interaction of two dipolar molecules, the induction contribution is present as well as a major electrostatic contribution.

1.4.1.3. *Dispersion Contribution*

Let us consider an interaction between two nondipolar molecules. In spite of the lack of permanent dipole moments, such molecules possess instantaneous dipole moments, due to the continuous motion of electrons within their electron clouds. This motion, being of random nature, produces fluctuations of electron density in the clouds, which result in the appearance of fluctuating dipole moments. These moments interact with each other both electrostatically and inductively. The attractive interaction produced by the fluctuating dipolar moments is known as the dispersion force. This force characterizes the correlation of the electron-density fluctuations between two molecules. In the case of interaction between two neutral, nondipolar molecules it produces the only contribution to the long-range attractive interaction.

1.4.2. *Energy of the Long-Range Intermolecular Forces*

1.4.2.1. *Electrostatic Energy*

Let us consider the energy of interaction of two molecules possessing permanent dipole moments. In the simple case each molecule can be represented by a linear distribution of electric charge (see Fig. 1.1). This distribution consists of two charges Q_1 and Q_2 arranged about the center of mass O of the molecule. Such a model might, for example, represent the HCl molecule. Evaluation of the electrostatic potential $\varphi_{el}(r)$ at the point O' in terms of the symbols in Fig. 1.1 gives the following result:

$$\varphi_{el}(\boldsymbol{R}) = \frac{1}{4\pi\varepsilon_0}\left(\frac{Q_1}{r_2} + \frac{Q_2}{r_2}\right), \quad r_1 = |r_1|, r_2 = |r_2|, \tag{1.72}$$

where ε_0 is the vacuum permittivity. In the limiting case of long-range interaction where $R \gg |r - r_1|$ and $R \gg |r - r_2|$ expression (1.72) can be expanded in powers of $|\boldsymbol{R} - r_1|/R$ and $|\boldsymbol{R} - r_2|/R$, yielding the following expression for $\varphi_{el}(\boldsymbol{R})$:

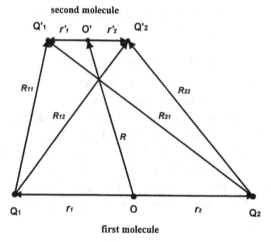

Fig. 1.1. Distribution of electric charge in the system of two two-atom interacting molecules possessing permanent dipole moments.

$$\varphi_{\mathrm{el}}(\boldsymbol{R}) = \frac{1}{4\pi\varepsilon_0}\left[\frac{Q}{R} + \frac{\boldsymbol{dR}}{R^3} + \frac{1}{6}\sum_{\alpha=1}^{3}\sum_{\beta=1}^{3}\Theta_{\alpha\beta}\frac{\partial^2}{\partial R_\alpha \partial R_\beta}\left(\frac{1}{R}\right) + \cdots\right], \qquad (1.73)$$

where $R_1 = X$, $R_2 = Y$, $R_3 = Z$. The quantity $Q = Q_1 + Q_2$ is the total charge of the molecule, also known as the *zeroth moment* of the charge distribution; $\boldsymbol{d} = Q_2 \boldsymbol{r}_2 + Q_1 \boldsymbol{r}_1$ is the dipole moment of the molecule, also known as the *first moment* of the charge distribution; and $\Theta_{\alpha\beta} = Q(3R_\alpha R_\beta - \delta_{\alpha\beta}R^2)$ is the $\alpha\beta$ component of the quadrupole tensor moment of the molecule, also known as the *second moment* of the charge distribution.

Now consider the energy of interaction of two such charge distributions. The center of mass of the second molecule with charges Q_1' and Q_2' is located at the point $\boldsymbol{O}'$, that is, at distance R from that of the first molecule. Therefore, the electrostatic potential energy $U_{\mathrm{el}}(\boldsymbol{R})$ of the two-charge system in terms of the symbols in Fig. 1.1 has the form:

$$\begin{aligned} U_{\mathrm{el}}(\boldsymbol{R}) = {}& Q_1'[\varphi_{\mathrm{el}}(\boldsymbol{R} + \boldsymbol{r}_1' - \boldsymbol{r}_1) + \varphi_{\mathrm{el}}(\boldsymbol{R} + \boldsymbol{r}_2' - \boldsymbol{r}_1)] \\ & + Q_2'[\varphi_{\mathrm{el}}(\boldsymbol{R} + \boldsymbol{r}_1' + \boldsymbol{r}_2) + \varphi_{\mathrm{el}}(\boldsymbol{R} + \boldsymbol{r}_2' - \boldsymbol{r}_2)], \end{aligned} \qquad (1.74)$$

where $\varphi_{\mathrm{el}}(\boldsymbol{R} + \boldsymbol{r}_1' - \boldsymbol{r}_1)$, $\varphi_{\mathrm{el}}(\boldsymbol{R} + \boldsymbol{r}_1' - \boldsymbol{r}_2)$ and $\varphi_{\mathrm{el}}(\boldsymbol{R} + \boldsymbol{r}_2' - \boldsymbol{r}_1)$, $\varphi_{\mathrm{el}}(\boldsymbol{R} + \boldsymbol{r}_2' - \boldsymbol{r}_2)$ are the electrostatic potentials due to the first charge distribution at the locations of the charges Q_1' and Q_2' respectively. As usual, in the limiting case of long-range interaction, where $R \gg |\boldsymbol{r}_1' - \boldsymbol{r}_1|$, $R \gg |\boldsymbol{r}_1' - \boldsymbol{r}_2|$ and $R \gg |\boldsymbol{r}_2' - \boldsymbol{r}_1|$, $R \gg |\boldsymbol{r}_2' - \boldsymbol{r}_2|$, expression (1.74) can be expanded in powers of $|\boldsymbol{R} + \boldsymbol{r}_1' - \boldsymbol{r}_1|/R$, $|\boldsymbol{R} + \boldsymbol{r}_1' - \boldsymbol{r}_2|/R$ and $|\boldsymbol{R} + \boldsymbol{r}_2' - \boldsymbol{r}_1|/R$, $|\boldsymbol{R} + \boldsymbol{r}_2' - \boldsymbol{r}_2|/R$, yielding the following expression for $U_{\mathrm{el}}(\boldsymbol{R})$:

$$U_{\text{el}}(\boldsymbol{R}) = \frac{1}{4\pi\varepsilon_0}\left[\frac{QQ'}{R} + [Qd' - Q'd]\text{grad}\left(\frac{1}{R}\right) + \frac{1}{2}\varphi_{\text{quad}}(\boldsymbol{R})\right], \qquad (1.75)$$

where

$$\varphi_{\text{quad}}(\boldsymbol{R}) = \frac{1}{3}\sum_{i=1}^{2}\sum_{j=1}^{2}Q_i Q_j' \sum_{\alpha=1}^{3}\sum_{\beta=1}^{3}\Theta_{ij,\alpha\beta}\frac{\partial^2}{\partial R_\alpha \partial R_\beta}\left(\frac{1}{R}\right).$$

In this expression $\Theta_{ij,\alpha\beta} = 3(r_i - r_j)_\alpha(r_i - r_j)_\beta - \delta_{\alpha\beta}|r_i - r_j|^2$ is the $\alpha\beta$ component of the quadrupole tensor moment between the charge i of the first distribution and the charge j of the second one. It is important to note that in the derivation of equation (1.75) for the electrostatic potential energy $U_{\text{el}}(\boldsymbol{R})$ of two-charge systems it was assumed that the electrostatic potential $\varphi_{\text{el}}(\boldsymbol{R})$ of each system satisfies the Laplace equation $\Delta\varphi_{\text{el}}(\boldsymbol{R}) = 0$.

If we consider interactions between neutral molecules ($Q = Q' = 0$), then the leading term in the electrostatic energy $U_{\text{el}}(\boldsymbol{R})$ is the dipole–dipole interaction $U_{\text{el,dd}}(\boldsymbol{R})$:

$$U_{\text{el,dd}}(\boldsymbol{R}) = \frac{1}{4\pi\varepsilon_0}\frac{dd' - 3(d\boldsymbol{R})(d'\boldsymbol{R})}{R^5}. \qquad (1.76)$$

This interaction is strongly dependent on the intermolecular orientation varying from attractive to repulsive as one molecule rotates: For example, this interaction is an attractive force between neutral molecules in all their mutual configurations in which $dd' - 3(d\boldsymbol{R})(d'\boldsymbol{R}) < 0$. Therefore, it is necessary to average the result of equation (1.76) for the energy $U_{\text{el,dd}}(\boldsymbol{R})$ over all possible orientations. If all relative orientations were of the same probability, then this average $U_{\text{el,dd}}(R) = \langle U_{\text{el,dd}}(\boldsymbol{R})\rangle_{\text{or}}$ would be zero. However, the probability of observing a configuration of energy $U_{\text{el,dd}}(\boldsymbol{R})$ is proportional to the Boltzmann factor $\exp(-U_{\text{el,dd}}(\boldsymbol{R})/kT)$, which consequently leads to a preferential weighting for configurations of negative energy. The Boltzmann weighted averaging carried out at the approximation of sufficiently high temperatures gives the following expression for the leading term of the energy $U_{\text{el,dd}}(R)$:

$$U_{\text{el,dd}}(R) = -\frac{2(dd')^2}{3(4\pi\varepsilon_0)^2 kTR^6}, \qquad d = |\boldsymbol{d}|, \quad d' = |\boldsymbol{d}'|. \qquad (1.77)$$

Thus, the leading term of the orientation-averaged dipole–dipole contribution to the electrostatic energy of interaction of two neutral molecules is attractive and inversely proportional to the sixth power of their separation. The temperature dependence of $U_{\text{el,dd}}(R)$ is merely due to the orientational averaging with the Boltzmann weighting.

Details of the orientational averaging with Boltzmann weighting can be found in Maitland, Rigby, and Wakeham (1987).

1.4.2.2. *Induction Energy*

Consider a molecule placed into a static electric field E. If the field E is strong enough, it may distort the charge distribution of the molecule and, by virtue of this, induce multipoles. In the simplest case, only a dipole moment d_{ind}, which is proportional and parallel to the field E, is induced:

$$d_{ind} = \alpha(0)E. \qquad (1.78)$$

The scalar quantity $\alpha(0)$, introduced by this equation, is known as the *static polarizability* of the molecule, which is presumed to be isotropic. Therefore, the energy U_{ind} of the dipole d_{ind} in the electric field E is given in terms of the following simple integral:

$$U_{ind} = -\int_0^E d_{ind}dE = -\frac{1}{2}\alpha(0)E^2 \quad \text{where} \quad E = |E|. \qquad (1.79)$$

A molecule possessing non-zero permanent multipole moments gives rise to an electrodynamic potential $\varphi_{el}(R)$ and, consequently, to the electric field $E(R) = -\text{grad}[\varphi_{el}(R)]$. In its turn, this electric field can induce a dipole moment on a second molecule, that is nearby, whether the second molecule is polar or not. Simultaneously, the second molecule induces a dipole moment on the first, which leads to a similar contribution to the induction energy.

An expression for the induced electric field in spherical coordinates has the following form:

$$E(R) = \left[i\frac{\partial}{\partial R} + j\frac{1}{R}\frac{\partial}{\partial\theta} + k\frac{1}{R\sin\theta}\frac{\partial}{\partial\phi}\right]\varphi_{el}(R) = \left[i\frac{\partial}{\partial R} + j\frac{1}{R}\frac{\partial}{\partial\theta}\right]\varphi_{el}(R), \qquad (1.80)$$

where it is assumed that the center of spherical coordinates is chosen to lie on the plane "at the point corresponding to the center of mass of the inducing molecule + the vector corresponding to the induced dipole moment." In expression (1.80) i, j, and k are unit vectors directed along axes X, Y, and Z, respectively and R, θ, and ϕ are spherical polar coordinates. Substituting expression (1.73) for the electrostatic potential $\varphi_{el}(R)$ produced by an electrically neutral molecule ($Q = 0$) into equation (1.80) and keeping only the leading terms, one can derive an expression for the quantity $|E(R)|$ in the form:

$$|E(R)| = \left\{\left[\frac{\partial\varphi_{el}(R)}{\partial R}\right]^2 + \left[\frac{1}{R}\frac{\partial\varphi_{el}(R)}{\partial\theta}\right]^2\right\} = \frac{d}{4\pi\varepsilon_0 R^3}(4\cos^2\theta + \sin^2\theta)^{1/2}. \qquad (1.81)$$

Therefore, the general result (1.79) for the energy $U_{ind} = U_{ind}(R)$ of the induced dipole d_{ind} in the electric field $E(R)$ has the form:

$$U_{\text{ind}}(\boldsymbol{R}) = -\frac{1}{8\pi\varepsilon_0}\frac{\alpha(0)d^2}{R^6}(3\cos^2\theta + 1). \tag{1.82}$$

This interaction is attractive for all configurations and is inversely proportional to the sixth power of the intermolecular separation for a fixed configuration. The orientational averaging of the induction energy $U_{\text{ind}}(\boldsymbol{R})$ with the Boltzmann weighting gives:

$$U_{\text{ind}}(R) = -\frac{2\alpha(0)d^2}{(4\pi\varepsilon_0)^2 R^6}. \tag{1.83}$$

Thus, the leading term of the orientation-averaged dipole–"induced dipole" contribution to the electrostatic energy of interaction of two neutral molecules is attractive and inversely proportional to the sixth power of their separation.

1.4.2.3. *Dispersion Energy*

This contribution to the long-range intermolecular potential is the only one that is present in all intermolecular interactions and is the only contribution to the long-range energy for the interaction of two nonpolar species. Although the dispersion energy is a result of electromagnetic interactions, it cannot be analyzed by means of classical mechanics because its origin is entirely quantum mechanical. Derivation of the expression for dispersion energy U_{disp} is beyond the scope of this chapter since it requires the introduction of concepts of the *Schrödinger equation*. Leaving all details to the references (Drude 1933; Ditchburn 1952; Maitland et al. 1987), we give the final form of this expression for the simplest case of two molecules where the first molecule contains a stationary positive charge $Q_1 = Q$ and the second molecule, with negative charge $Q_2 = |Q_1|$, is oscillating about the first molecule with an angular frequency ω_0:

$$U_{\text{disp}} = U_{\text{disp}}(\boldsymbol{R}) = -\frac{3\alpha^2(0)h\omega_0}{8\pi(4\pi\varepsilon_0)^2 R^6}, \tag{1.84}$$

where R is the distance between centers of mass of the two molecules. Therefore, the dipole–dipole dispersion energy for two molecules in their *ground*, that is, their nonexcited or minimum-energy state is attractive and inversely proportional to the sixth power of the intermolecular separation.

Summarizing the results obtained for the electrostatic, induction, and dispersion energies of two dipolar molecules after orientational averaging with the Boltzmann weighting carried out, one can write down the following expression for the total long-range intermolecular energy $U(R)$:

$$U(R) = U_{\text{el}}(R) + U_{\text{ind}}(R) + U_{\text{disp}}(R)$$
$$= -\frac{1}{(4\pi\varepsilon_0)^2 R^6}\left[\frac{2d^4}{3kT} + 2\alpha(0)d^2 + \frac{3\alpha^2(0)h\omega_0}{8\pi}\right]. \tag{1.85}$$

Table 1.1. *Magnitudes of coefficients belonging to the interaction-energy terms proportional to R^{-6} due to electrostatic, induction, and dispersion interactions, respectively*

			Coefficient of $R^{-6}(10^{-79} \text{ J m}^6)$		
Molecule	10^{30} d(C m)	$10^{30}\alpha(4\pi\varepsilon_0)^{-1}(\text{m}^3)$	Electrostatic	Induction	Dispersion
Ar	0.00	1.63	0.000	0.00	50
Xe	0.00	4.00	0.000	0.00	209
CO	0.40	1.95	0.003	0.06	97
HCl	3.40	2.63	17.000	6.00	150
NH_3	4.70	2.26	64.000	9.00	133
H_2O	6.13	1.48	184.000	10.00	61

Table 1.1 contains a list of magnitudes of three coefficients of the R^{-6} leading term for the different contributions to the interaction of like pairs of simple molecules at temperature $T = 300$ K. The table also provides typical values of dipole moments and polarizabilities of molecules.

1.4.3. *Modeling Intermolecular Potentials*

Based on the previous discussion, it is obvious that intermolecular forces (potentials) can be obtained (estimated) only in a few simple situations. In reality, there is almost nothing known about the form of intermolecular potentials for molecule–molecule separations of the order of one molecular diameter. This explains why empirical thermodynamic and kinetic properties and data serve as a source of information about the character of intermolecular interactions. The opposite situation is very rare, either because of the unknown character of intermolecular forces or because of their very complicated form. In order to overcome this problem in the direct derivation of thermodynamic and kinetic properties from the form of intermolecular forces, use of the *modeling intermolecular potentials* that correctly describe the major features of intermolecular interaction energy has been suggested (Maitland et al. 1987). These potentials usually contain some number of adjustable parameters, the values of which can be determined by comparing the thermodynamic or kinetic properties, calculated on the base of modeling potentials, with experimental data. It is understandable that the introduction of modeling intermolecular potentials is justified only if the number of adjustable parameters is reasonably small, in order to escape the situation of false adjustment.

Several modeling intermolecular potentials that have proved to be the most reliable and widely used will now be discussed.

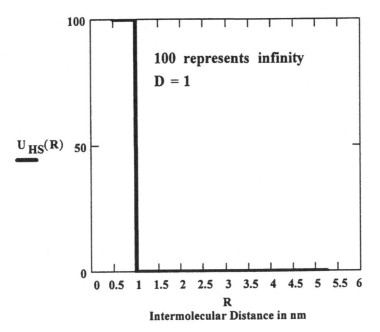

Fig. 1.2. The hard-sphere potential.

1.4.3.1. *The Hard-Sphere (HS) Potential*

In the model of hard spheres it is assumed that molecules behave like billiard balls with diameter D. Therefore, their intermolecular potential $U_{HS}(R)$ has the form:

$$U_{HS}(R) = \begin{cases} \infty \text{ if } R \leq D \\ 0 \text{ if } R > D \end{cases}. \tag{1.86}$$

This repulsive modeling potential (see Fig. 1.2) is the simplest because it contains only one adjustable parameter, D. The major advantage of this potential is in its extreme simplicity, which allows derivation of analytical results for thermodynamic and kinetic properties of matter. For example, all partition functions for a gas of hard spheres can be exactly and easily evaluated. This makes the usage of the HS potential very attractive for the study of complex problems such as the problem of the liquid state. However, the modeling HS potential is far from reality because at small intermolecular separations it grows too fast, whereas at large separations no interaction is taken into account.

1.4.3.2. *The Soft-Sphere (SS) Potential*

This potential is also purely repulsive (see Fig. 1.3):

$$U_{SS}(R) = \left(\frac{D}{R}\right)^{\nu}, \tag{1.87}$$

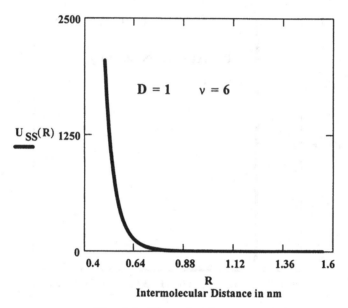

Fig. 1.3. The soft-sphere potential.

where ν is the so-called *exponent of repulsion*, often chosen to be an integer. It is obvious that in the limiting case of $\nu \to \infty$ the SS potential (equation (1.87)) becomes "harder", that is, it converts to the HS potential (equation (1.86)).

The two-parameter SS potential grows at small separations, but not as fast as the HS potential, and allows the molecule compressibility to be taken into account. For most real gases, the exponent of repulsion ν is between $\nu = 9$ (soft molecules) and $\nu = 15$ (hard molecules). It is important to note that there is the particular case of $\nu = 4$ that allows exact analytical evaluations to be carried out for the partition function (see Burnett (1935) for details). This makes this case quite popular in the modern kinetic theory of gases. However, the SS potential still does not have the long-range term corresponding to the forces of attraction.

The major flaw of the HS and SS potentials is the lack of any long-range attractive term. The following two modeling potentials discussed are obtained by combination of either the HS or SS repulsive potential with some attractive potential.

1.4.3.3. *The Sutherland (S) Potential*

Sutherland (1893) suggested a model in which the short-range repulsion was described by the HS potential and the long-range attraction by the power law (see Fig. 1.4):

$$U_S(R) = \begin{cases} \infty \text{ if } R \leq D \\ -\epsilon \left(\dfrac{D}{R}\right)^{\nu} \text{ if } R > D \end{cases} . \qquad (1.88)$$

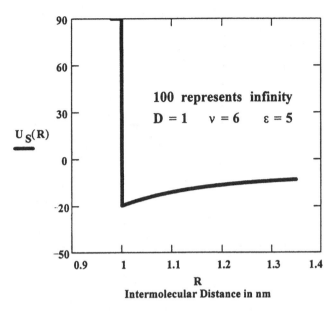

Fig. 1.4. The Sutherland potential.

This modeling potential contains three adjustable parameters ϵ, D, and ν. This makes it more useful in the description of real gases at low temperatures when kinetic coefficients are determined by the long-range attraction rather than by the short-range repulsion. The best agreement with experimental data in this case was obtained with $\nu = 6$. In the repulsion region, however, the S potential still grows too fast, which explains its failure to describe real gases at high temperatures where kinetic coefficients are strongly dependent on the character of the repulsion. This situation was first corrected by Lennard-Jones (1924) by employing the power law for the repulsion interaction.

1.4.3.4. *The Lennard-Jones (LJ) Potential*

Most frequently used at present in intermolecular modeling, the LJ potential has the following form (see Fig. 1.5):

$$U_{\text{LJ}}(R) = \epsilon_1 R^{-\nu_1} - \epsilon_2 R^{-\nu_2}, \tag{1.89}$$

where the first and second terms correspond to repulsion and attraction, respectively. In order to describe repulsion prevailing over attraction at small intermolecular separations and attraction prevailing over repulsion at large intermolecular separations, the powers ν_1 and ν_2 in the LJ potential $U_{\text{LJ}}(R)$ should satisfy the following condition: $\nu_1 > \nu_2$. As in the case of the S potential, the power ν_2 is usually chosen to be equal to 6, whereas for the power ν_1 the most widely used value is 12. Therefore, expression (1.89) is usually given in the form:

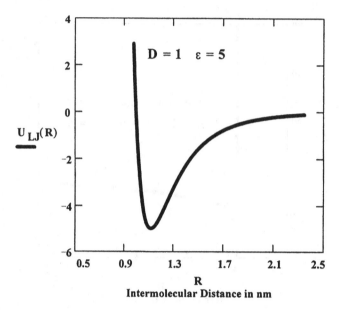

Fig. 1.5. The Lennard-Jones potential.

$$U_{LJ}(R) = 4\epsilon\left[\left(\frac{D}{R}\right)^{12} - \left(\frac{D}{R}\right)^{6}\right]. \tag{1.90}$$

Parameter D in expression (1.90) designates the intermolecular separation at which the LJ potential changes sign and parameter ϵ gives the LJ potential minimum value, $-\epsilon$, achievable at $R = 2^{1/5}D$.

The success of the early applications of the LJ potential is due to its success in agreeing with a variety of experimental data with only two adjustable parameters. This potential has a realistic form almost everywhere, except in the short-range region, where the repulsion energy does not increase rapidly enough. It is known that the real potential function in the short-range region should be close to exponential. In order to partly correct this flaw in the LJ potential it has been suggested that $\nu_1 = 18$ should be used, rather than $\nu_1 = 12$.

1.4.3.5. *The Buckingham (B) Potential*

The B potential $U_B(R)$ eliminates the only substantial flaw of the LJ potential, its failure to describe the fast-growing repulsion at small intermolecular separations. Therefore, it has the form:

$$U_B(R) = \begin{cases} \infty \text{ if } R \leq R_{max} \\ U^*(R) \text{ if } R > R_{max} \end{cases}, \tag{1.91}$$

where

$$U^*(R) = \frac{\epsilon\chi}{\chi - 6}\left[\frac{6}{\chi}e^{\chi(1-R/D)} - \left(\frac{D}{R}\right)^{6}\right].$$

The value of parameter $R_{\max}$ is usually chosen in such a way that $U^*(R_{\max}) =$ minimum. Therefore, the B potential contains three adjustable parameters ϵ, χ and D, which allow the adjustment of the width as well as the depth of the potential energy well.

So far we have discussed the pair intermolecular modeling potentials that depend only on the magnitude of the pair separation R. In general, the potential energy U of a system of N molecules contains contributions coming from individual molecules, their pairs, triplets, etc.:

$$U = \sum_{i=1}^{N} U_1(r_i) + \sum_{i=1}^{N}\sum_{j>i}^{N} U_2(r_i, r_j) + \sum_{i=1}^{N}\sum_{j>i}^{N}\sum_{k>j>i}^{N} U_3(r_i, r_j, r_k) + \dots . \quad (1.92)$$

The first term in equation (1.92) for the system potential energy U represents the effect of an external field (including, for example, container walls) on the system. The remaining terms represent molecular interactions. The second term, consisting of the pair potentials $U_2(r_i, r_j)$, is the most important. The third term involving triplets of molecules should undoubtedly be significant in the liquid state. However, this term is ignored because of the lack of reasonable estimations of its contribution compared to the pair potential energy. The four-molecule and higher terms in equation (1.92) are expected to be small in comparison with pair and triple potential-energy terms and, therefore, are neglected.

1.4.4. *Sources of Information about Intermolecular Forces*

When we discuss the concept of an ideal gas we both regard the intermolecular interaction as causing collisions only and neglect the duration of these collisions and other important details. As a result of these restrictions, intermolecular intermittent collisions lead only to changes of molecular momentum. However, the real effect of intermolecular forces is much more than only a change of momentum. The simplest way to demonstrate the impact of taking into account intermolecular forces is to consider their influence on the equation of state, the concept of which for an ideal gas was introduced in Section 1.3.11. As was emphasized in Section 1.4.3, we will only take into account pair intermolecular forces and neglect triple, quadruple, and higher intermolecular interactions. In addition, in order to simplify formulas, we will consider the case of a monoatomic gas within the framework of the canonical ensemble.

The motion of gas particles can be treated classically, giving the following expression for their energy:

$$E(q; p) = \sum_{i=1}^{N} \frac{p_i^2}{2m} + U(q), \quad (1.93)$$

where the first term is the kinetic energy of N one-atom molecules of the gas and $U(q)$ is the energy of their mutual interaction. In the case of monoatomic gas $U(q)$

is a function of the distances between molecules only. Therefore, the partition function (see equation (1.18)) of the canonical ensemble describing the gas with energy $E(q; p)$ given by expression (1.93) becomes the product of the integral over the momenta of the gas molecules and the integral over their coordinates:

$$Z_c(\beta; V; N) = Z_{c,kin}(\beta; N)Z_{c,conf}(\beta; V; N),$$

$$Z_{c,kin}(\beta; N) = \frac{1}{h^{sN}N!}\int \exp\left[-\beta\sum_{i=1}^{N}\frac{p_i^2}{2m}\right]\prod_{i=1}^{N}dp_i, \qquad (1.94)$$

$$Z_{c,conf}(\beta, V; N) = \int e^{-\beta U(q)}\prod_{i=1}^{N}dV_i,$$

where integration with respect to each $dV_i = dx_i dy_i dz_i$ $(i = 1, \ldots, N)$ is taken over the entire volume V occupied by the gas. The integral $Z_{c,conf}(\beta; V; N)$ is known as the *configuration integral*. Its evaluation, which is crucial in statistical mechanics, will be discussed in the following sections. In the case of an ideal gas, $U(q) = 0$, the configuration integral in equation (1.94) is simply equal to V^N. If the Helmholtz free energy $H_c(\beta; V; N)$ of the gas canonical ensemble is calculated utilizing equation (1.27), one can present it in the following form:

$$H_c(\beta; V; N) = H_{id,c}(\beta; V; N) - \frac{1}{\beta}\ln\left(1 + \frac{1}{V^N}\int [e^{-\beta U(q)} - 1]\prod_{i=1}^{N}dV_i\right), \quad (1.95)$$

where $H_{id,c}(\beta; V; N)$ is the Helmholtz free energy of an ideal gas. To keep the formulas simple, let us assume that the gas is not only sufficiently rarefied but is small in quantity as well, so that not more than one pair of molecules can be colliding at any given instant. This assumption does not affect the generality of the resulting formulas since we know from the additivity of the Helmholtz free energy that it has to have the form: $H_c(\beta; V; N) = Nf_H(\beta, V/N)$ (see equation (1.60)), that is, the formulas deduced for a small quantity of gas are necessarily valid for any quantity.

The interaction between molecules is very small except when any two molecules are very close to each other, that is, almost colliding. Therefore, the integrand in expression (1.95) is appreciably different from zero only when some pair of molecules can satisfy this condition at any given time. Taking into account that such a pair can be chosen from N molecules in $N(N - 1)/2$ different ways, one can rewrite expression (1.95) for $H_c(\beta; V; N)$ as follows:

$$H_c(\beta; V; N) = H_{id,c}(\beta; V; N) - \frac{1}{\beta}\ln\left(1 + \frac{N(N - 1)}{2V^N}\int [e^{-\beta U(r_{12})} - 1]\prod_{i=1}^{N}dV_i\right),$$

$$(1.96)$$

where $U(r_{12})$ is the energy of interaction of two molecules (it does not matter which two since they all are identical). Since the intermolecular potential $U(r_{12})$ depends only on the coordinates of the two molecules, integration over the remaining coordinates can be carried out, producing V^{N-2}. Because $N(N-1) \to N^2$ for $N \to \infty$ and $\ln(1+x) = x$ for $x \to 0$, we derive the following expression for the Helmholtz free energy $H_c(\beta; V; N)$:

$$H_c(\beta; V; N) = H_{id,c}(\beta; V; N) - \frac{N^2}{2\beta V^2} \int \int [e^{-\beta U(r_{12})} - 1] dV_1 dV_2, \qquad (1.97)$$

where $dV_1 dV_2$ is the product of the differentials of the coordinates of the two molecules. The pair potential energy $U(r_{12})$ is a function only of the distance between the two molecules, the difference of their coordinates. Thus, if the coordinates of the two molecules are expressed in terms of the coordinates of their centers of mass and relative coordinates, the potential $U(r_{12})$ will be dependent only on the latter. Therefore, by carrying out integration over the coordinates of center of mass in equation (1.97), one can rewrite it in the form:

$$H_c(\beta; V; N) = H_{id,c}(\beta; V; N) + \frac{N^2}{\beta V} B'(\beta), B'(\beta)$$

$$= B(T) = \frac{1}{2} \int \left[1 - \exp\left(\frac{U(r)}{k_B T}\right) \right] dV, \qquad (1.98)$$

where dV is the product of the differentials of the relative coordinates.

The equation of state for a gas of interacting molecules under all assumptions mentioned above can be obtained by substituting expression (1.98) for the Helmholtz free energy $H_c(\beta; V; N)$ into general result (1.71):

$$P = -\left(\frac{\partial H_c(\beta; V; N)}{\partial V}\right)_T = k_B T n [1 + B(T)n], k_B T = \frac{1}{\beta}, n = \frac{N}{V}, \qquad (1.99)$$

where the function $B(T)$ is known as the *second virial coefficient*. The first part of equation (1.99) $P_{id} = k_B T n$ is the equation of state for an ideal gas (see equation (1.71)). Therefore, the second virial coefficient allows one to take into account intermolecular forces in an integral way, that is, through integrals over the system volume with integrands dependent on intermolecular potentials. It is also important to note that the entire derivation procedure for a monatomic gas is also applicable for polyatomic gases.

Let us calculate the second virial coefficient $B(T)$ for two of the most popular pair intermolecular potentials.

1.4.4.1. *Calculation for the S Potential (see equation (1.88))*

In this case the integral in expression (1.98) for $B(T)$ gives:

$$B(T) = B_S(T) = \frac{1}{2} \int_0^\infty \left[1 - \exp\left(-\frac{U_S(R)}{k_B T} \right) \right] 4\pi R^2 \mathrm{d}R \approx \frac{a}{k_B T} - b,$$

$$a = \frac{12}{\nu - 3} \epsilon v_0, \, b = 4v_0,$$

(1.100)

where $v_0 = 4\pi D^3/3$ is the molecular volume. Substituting this result for $B_S(T)$ into expression (1.99) for the equation of state, we obtain:

$$\left(P + an^2 \right) \left(\frac{1}{n} - b \right) = k_B T.$$

(1.101)

This equation is also known as the *van der Waals equation of state*.

1.4.4.2. *Calculation for the LJ Potential (see equation (1.90))*

In this case the integral in equation (1.98) for $B(T)$ gives:

$$B(T) = B_{LJ}(T) = \frac{1}{2} \int_0^\infty \left[1 - \exp\left(-\frac{U_{LJ}(R)}{k_B T} \right) \right] 4\pi R^2 \mathrm{d}R$$

$$= \frac{\pi}{6} D^3 \sum_{k=0}^\infty \frac{2^k}{k!} \Gamma\left(\frac{k}{2} - \frac{1}{4} \right) \left(\frac{4\epsilon}{k_B T} \right)^{[(k/2)+(1/4)]},$$

(1.102)

where $\Gamma(\ldots)$ is the gamma function. The temperature $T_B \approx 3.44\epsilon/k_B$ corresponding to $B_{LJ}(T_B) = 0$ is known as the *Boyle temperature*. The isotherm corresponding to this temperature has a horizontal tangent as $P \to 0$ in coordinates $(PV/T, P)$, in other words, separate isotherms with positive and negative initial slopes.

The equation of state (1.99) derived in this section contains the first two terms in the expansion of the pressure P in powers of $1/V$. The generalized equation of state should have the following form:

$$P = k_B Tn[1 + B(T)n + C(T)n^2 + \ldots],$$

(1.103)

where the higher-order virial coefficients such as $C(T)$, take into account intermolecular interactions between groups of three, four, and higher groups of molecules. The calculation of these coefficients is beyond the scope of this chapter and can be found in several references (Isihara 1971; Maitland et al. 1987; Landau and Lifshitz 1988).

1.5. The Monte Carlo Method

1.5.1. *Origin and Fundamentals of the Monte Carlo Method*

The *Monte Carlo method* was developed by von Neumann, Ulam and Metropolis at the end of World War II to study the diffusion of neutrons in fissionable material. The name "Monte Carlo" was chosen because of the extensive use of random numbers. However, statisticians had used random sampling procedures long before this time. The English statistician W. S. Gossett, known under the name "Student," estimated the correlation coefficient in his *t*-distribution with the help of random sampling experiments and Lord Kelvin's assistant generated 5000 random trajectories to study the elastic collision of particles with shaped walls. The novel contribution of von Neumann and Ulam (1945) was in the development of understanding that some deterministic mathematical problems could be reformulated in terms of probabilistic analogs which then could be solved by applying the random sampling procedure.

The best and simplest illustration of the Monte Carlo technique is due to its application to integration. In this illustration, the integral of interest $F(x_1, x_2)$ is presented as the following arithmetic mean:

$$F(x_1, x_2) = \int_{x_1}^{x_2} dx\, f(x) \approx \frac{x_2 - x_1}{N_{\text{trials}}} \sum_{i=1}^{N_{\text{trials}}} f(\xi_i), \qquad (1.104)$$

where ξ_i ($i = 1, \ldots, N_{\text{trials}}$) are N_{trials} independent random numbers. These numbers are uniformly distributed in the interval $[x_1, x_2]$ (their probability density function $\rho(x)$ is $1/(x_2 - x_1)$). Let us apply formula (1.104) to the estimation of $\pi = 3.1415927$, both analytically and numerically. For this purpose one might consider the equation of a circle $f(x) = (1 - x^2)^{-1/2}$ in the first quadrant $[x_1 = 0, x_2 = 1]$. Each integration in equation (1.104) gives π, whereas a typical Monte Carlo numerical experiment carried out in correspondence with equation (1.104) gives 3.14169 after $N_{\text{trials}} = 10^7$ trials, showing that the result obtained utilizing the Monte Carlo technique differs from the correct result in the fourth decimal place. This illustrates that, for the simple one-dimensional integration, the Monte Carlo technique is not competitive with straightforward numerical methods such as Simpson's rule (Korn and Korn 1961), since the Simpson-rule estimate of π with only 10^4 function evaluations is 3.141593, which differs from the correct result in the seventh decimal place. For the multidimensional integrals of statistical mechanics, however, the random sample method, based on a suitable choice of the probability density function $\rho(x)$, is the only sensible solution. In order to explain this, we will consider the evaluation of the configuration integral $Z_{\text{c,conf}}(\beta; V; N)$ (see equation (1.94)) for a system containing $N = 100$ molecules in a cube of side L. Even a crude implementation of Simpson's rule might require ten function evaluations for each of the 300 coordinates in order to span the interval $[-L/2, L/2]$. This will result in the impractical number of 10^{300} function evaluations. Moreover, an overwhelming proportion of the contributions will be close to zero since the Boltzmann factor $\exp[-U(r)/k_B T]$ is extremely small whenever molecules overlap significantly, even zero in the case $U(r) = U_{HS}(r)$. The alternative Monte Carlo technique applied for the evaluation of the configuration integral may be formulated in the following way:

1. Pick a point at random in the 300-dimensional coordinate space, also known as the *configuration space*, by generating 300 independent random numbers from the interval $[-L/2, L/2]$. These numbers, being taken in triplets, will specify the coordinates of each molecule.
2. Calculate the potential energy $U(r)$, and hence the Boltzmann factor, for a given configuration (set of 300 coordinates).

This procedure is repeated for many trials and the configuration integral is estimated by using the following expression:

$$Z_{c,\text{conf}}(\beta; V; N) \approx \frac{V^N}{N_{\text{trials}}} \sum_{i=1}^{N_{\text{trials}}} e^{-\beta U(\xi_i)}. \qquad (1.105)$$

In principle, the number of trials N_{trials} may be increased until $Z_{c,\text{conf}}(\beta; V; N)$ is estimated to the desired accuracy. One would not expect to conduct 10^{300} function evaluations, as for the Simpson rule, but again a large number of trials would give a very small contribution to the mean. An accurate estimation of $Z_{c,\text{conf}}(\beta; V; N)$ for a dense liquid using a *uniform sample mean method* is beyond the capabilities of current computers. However, at realistic liquid densities the problem can be solved using a sample mean integration where the random coordinates are chosen from a nonuniform distribution. This variant of the Monte Carlo method is known as the *importance sampling method*.

1.5.2. *Theoretical Background of the Importance Sampling Method*

As has been concluded in Section 1.5.1, the efficiency of the Monte Carlo method requires the use of the importance sampling method introduced by Hammersley and Handscomb (1964). This method implies that molecular configurations should be sampled in such a way that the regions of configuration space that make the largest contributions to configuration integral $Z_{c,\text{conf}}(\beta; V; N)$ are the regions that are sampled most frequently.

The canonical ensemble average and its regular Monte Carlo estimate of any function $F(q) = F(q_1, \ldots, q_N)$ of the N molecule coordinates can be given in terms of the following expressions:

$$\langle F(q) \rangle = \frac{\int F(q) e^{-\beta U(q)} \prod_{i=1}^{N} dV_i}{\int e^{-\beta U(q)} \prod_{i=1}^{N} dV_i} \approx \frac{\sum_{i=1}^{N_{\text{trials}}} F(\xi_i) e^{-\beta U(\xi_i)}}{\sum_{i=1}^{N_{\text{trials}}} e^{-\beta U(\xi_i)}}, \qquad (1.106)$$

where $\xi_i = \xi_{i,1}, \ldots, \xi_{i,N} (i = 1, \ldots, N_{\text{trials}})$. The bias introduced in the random sampling inherent in the regular Monte Carlo method can be eliminated by attaching an appropriate weight to each configuration. If $w(\xi_i)$ is the probability of choosing a configuration ξ_i, then equation (1.106) has to be replaced by the following:

$$\langle F(q) \rangle \approx \frac{\sum\limits_{i=1}^{N_{\text{trials}}} w(\xi_i) F(\xi_i) e^{-\beta U(\xi_i)}}{\sum\limits_{i=1}^{N_{\text{trials}}} w(\xi_i) e^{-\beta U(\xi_i)}}. \tag{1.107}$$

A natural choice is to sample over the Boltzmann distribution. This can be achieved by setting:

$$w(\xi_i) = e^{-\beta U(\xi_i)}. \tag{1.108}$$

In this case, expression (1.107) for $\langle F(q) \rangle$ reduces to the following:

$$\langle F(q) \rangle \approx \frac{1}{N_{\text{trials}}} \sum\limits_{i=1}^{N_{\text{trials}}} F(\xi_i). \tag{1.109}$$

The canonical ensemble average is therefore obtained as an unweighted average over configurations in the sample.

The problem of deriving a scheme for sampling over configuration space according to a specific probability distribution is most easily formulated and studied in terms of the *theory of stochastic processes* (Doob 1953; Wax 1954; Cox and Miller 1965). Suppose we have a sequence of random variables, which in our case can be associated with a set of all possible coordinates of the molecules where the set elements represent different molecular configurations. Instead of speaking of the variable value at a particular point in the sequence, one may say that the system occupies a particular state at that point. If the probability of finding the system in state n at time t is determined solely by the state m, which was occupied by the system at the previous time instant $t_{\text{pr}} = t - 1$, then the sequence of states is said to form a *Markov chain*. The concept of time, introduced here, is merely for descriptive purposes; there is no connection with any molecular timescale.

Let $p_n(t)$ be the probability that the state n is occupied at time instant t:

$$\sum\limits_n p_n(t) = 1. \tag{1.110}$$

Then, the Markovian character of the random process can be expressed by means of the relationship:

$$p_n(t) = \sum\limits_m A_{n,m} p_m(t - 1), \tag{1.111}$$

where the *transition probability* $A_{n,m} = \text{prob}(n; t|m; t-1)$ is assumed to be independent of t. If we regard the probability distribution $\{p_n(t)\}$ as a column vector $p(t)$ and the quantities $\{A_{n,m}\}$ as a square *transition matrix* A, equation (1.111) can be rewritten in a more compact form:

$$p(t) = Ap(t-1), \tag{1.112}$$

where we adopt the usual convention that $A_{n,m}$ is the element in the nth row and mth column of the matrix A. The generalization of equation (1.112) for the probability distribution $p(t)$ at time t in terms of an arbitrary initial distribution $p(0)$ is obvious:

$$p(t) = A^t p(0), \tag{1.113}$$

where A^t is the t-fold product of A with itself. The elements $A_{n,m}^t$ of the matrix A^t are called the *"multistep" transition probabilities* of order t.

Since the elements of matrix A are interpreted as transitional probabilities, they have to satisfy the following conditions:

$$0 \le A_{n,m} \le 1, \sum_n A_{n,m} = 1. \tag{1.114}$$

Any matrix having properties given by relations (1.114) is called a *stochastic matrix*. It is easy to demonstrate that the matrix A^t is also stochastic since $A_{n,m}^t$ is merely the probability of reaching state n from state m in t steps rather than in one. In the case when the following limits:

$$\Psi_n = \lim_{t\to\infty} A_{n,m}^t \tag{1.115}$$

exist and are the same for all m states, the Markov chain is said to be *ergodic*. This usage of the term "ergodic" is closely related to the meaning of ergodicity in statistical mechanics discussed in Section 1.2.1. From relationship (1.113) as well as from the requirement that the vector column $p(0)$ be normalized, it follows that the vector column $\Psi = \{\Psi_n\}$ represents a unique asymptotic probability distribution:

$$\Psi = \lim_{t\to\infty} p(t). \tag{1.116}$$

Once the limiting distribution Ψ has been reached, it persists because the transition matrix A maps the limiting distribution into itself:

$$A\Psi = \Psi. \tag{1.117}$$

Equation (1.117), known as the *steady-state condition*, states that the limiting distribution Ψ is a right eigenvector of transition matrix A corresponding to the unit eigenvalue. Solution of the system of linear equations (1.117) under conditions (1.114) can be greatly simplified if one looks for a transition matrix A to which the *concept of microscopic reversibility* is applicable:

$$\Psi_n A_{n,m} = \Psi_m A_{m,n}. \tag{1.118}$$

In this case the steady-state condition (1.117) is satisfied automatically. The concept of microscopic reversibility, given by equation (1.118), is relevant to the concept of importance sampling because, in the case of sampling over the Boltzmann distribution, the limiting distribution Ψ is known:

$$\Psi_n = \frac{e^{-\beta U(n)}}{\displaystyle\sum_n e^{-\beta U(n)}}. \tag{1.119}$$

If the Monte Carlo method is employed, the sequence of configurations is generated in the following way:

1. It is assumed that at the given time instant t the system of molecules is in a state m.
2. An attempt is made to move to a trial state n, adjacent to (not very different from) the initial state m. Usually this trial molecular configuration state is generated by displacing one molecule of the system within small, prescribed limits.
3. The conditional probability $B_{n,m}$ of choosing the state n as the trial state is introduced. The quantities $B_{n,m}$ are treated as elements of a stochastic matrix B with $B_{n,m} = 0$ for all pairs of states that are not adjacent.
4. It is assumed that matrix B is symmetric, $B_{n,m} = B_{m,n}$.
5. The transitional matrix A is built in correspondence with the prescription that:
   $$A_{n,m} = B_{n,m} \text{ if } \Psi_n \geq \Psi_m \text{ for } n \neq m,$$
   $$A_{n,m} = B_{n,m}\Psi_n/\Psi_m \text{ if } \Psi_n \leq \Psi_m \text{ for } n \neq m,$$
   $$A_{n,m} = 1 - \sum_{m \neq n} A_{n,m}.$$

This procedure of the introduction of transition matrix A was first proposed by Metropolis et al. (1953). It still remains the most commonly used prescription for the introduction of matrix A.

1.5.3. *Implementation of the Importance Sampling Method*

Let us consider the case of a canonical ensemble and assume that its molecules interact via spherically symmetrical potentials. Suppose that at time $t = 0$ the system is in state m. The Monte Carlo process is initiated by selecting a molecule $n(n = 1, \ldots, N)$ and displacing it from its original position r_n to a trial position $r_n + \Delta r_n$. The displacement Δr_n is chosen at random. Any trial molecular configuration k that can be accessed from the initial configuration m by virtue of random infinitesimal displacement Δr_n is treated as a state adjacent to m in the sense that $B_{k,m} > 0$. The next step is to determine whether the trial configuration k is to be accepted or rejected. Let $\Delta U = U(k) - U(m)$ be an increase of the total potential energy resulting from the trial move. According to the Metropolis

criterion (5) introduced in Section 1.5.2, a trial configuration is accepted unconditionally if ΔU is negative. However, it is accepted only with the probability
$\exp[-\beta \Delta U]$ if ΔU is positive. In the latter case a random number ξ from the
interval $[0, \lambda]$ is generated, where λ is a parameter of the calculation. If
$\xi \le \exp[-\beta \Delta U]$, the trial configuration is accepted; otherwise it is rejected.
Therefore, there is a low probability of acceptance of a move that leads to a
large increase of potential energy. Acceptance of a move means that the system
passes from the state m at time instant $t = 0$ to the state k at time instant $t = 1$,
while rejection means that the system remains in the state m at time instant $t = 1$.
The entire cycle introduced is then repeated many times. Typically a number of
trials between 10^5 and 10^6 is sufficient for the determination of thermodynamic
properties such as energy and pressure.

The importance sampling method becomes only slightly more complicated
when the interaction between molecules is dependent on their mutual orientation. In such cases molecule n is subjected not only to a random displacement but
also to a random reorientation.

A number of variations in the importance sampling procedure are possible.
For example, the choice of molecule to be displaced may be made either at
random, with uniform probability, or in a serial fashion. One can also choose
to displace two molecules simultaneously or to alternate trial displacements with
trial rotations. The important question in the importance sampling procedure
concerns the choice of the calculation parameter λ. If λ is large and the system
has a liquid-like density ($\Delta U \gg 1$), a very high proportion of trials will be
rejected. A high rate of acceptance can be insured by making λ sufficiently
small. However, this will lead to a larger number of configurations to ensure
an adequate sampling of the relevant region of configuration space. The parameter λ is normally set to a value at which approximately a half of all trial
configurations are accepted. In practice, this means that λ should be of order
of one-tenth of the nearest-neighbor distance.

1.5.4. *Generation of Random Numbers*

The problem of random number generation in the Monte Carlo method is
usually overlooked in textbooks on molecular simulation. However, it is of
extreme importance and can sometimes undermine the entire simulation results.

Any modern programming language such as FORTRAN, C or C$+ +$, allows
generation of random numbers. The function generating these numbers is usually
called "rnd" or "random". In Fig. 1.6 we reproduce results obtained by utilizing
the FORTRAN random number generator for the square of *mean displacement*
$\langle R^2(N_{\text{trial}}) = X^2(N_{\text{trial}}) + Y^2(N_{\text{trial}}) \rangle$ of a two-dimensional random walk as a function of the number of steps N_{trial} in the generation of the random trajectories:

$$\langle R^2(N_{\text{trial}}) \rangle = \frac{1}{N_{\text{trial}}} \sum_{i=1}^{N_{\text{trial}}} R_i^2, \tag{1.120}$$

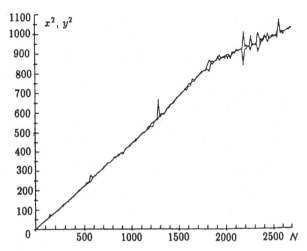

Fig. 1.6. Dependence of the mean square of displacement in random walk on the number N_{trial}.

where $R_i = (X_i^2 + Y_i^2)^{1/2}$ is the displacement corresponding to the i trajectory. The following two major facts related to $\langle R^2(N_{\text{trial}})\rangle$ are known from the random walk theory:

- Since the X and Y directions of any random step in the trajectory generation are of the same probability, one has to obtain the same results for the mean displacements along these directions. As shown in Fig. 1.6, these displacements do practically coincide.
- $\langle R^2(N_{\text{trial}})\rangle \propto N_{\text{trial}}$. As shown in Fig. 1.6 this proportionality law is not satisfied in the upper tail of the plotting!

In order to understand the problem one has to examine the generation of random numbers, because the general concept of Monte Carlo method is correct. The problem of generation of random numbers constitutes the essence of the Monte Carlo method. It is independent of sampling method and is present in any sampling method. The overall truthfulness of the Monte Carlo method is determined by the quality of generation of random numbers. Detailed discussion of the problem of generation of random numbers can be found in recent monographs by Knuth (1969), Kalos and Whitlock (1986), and Ahrens and Dieter (1990). A brief introduction to the problem will be presented here.

It is impossible to generate absolutely random numbers. Therefore, any procedure for their generation produces so-called *pseudorandom numbers*. The most widely used random number generator, suggested by Lehmer (1951), is based on the following recursive procedure:

$$\xi_i = a\xi_{i-1} + b \bmod(M, K), \tag{1.121}$$

where ξ_i $(i = 1, \ldots, N_{trial})$ are the pseudorandom numbers, a and b are some integers and mod(M, K) is the remainder of dividing integer M by integer K. This procedure produces pseudorandom numbers uniformly distributed in the interval $[0, 1]$. The statistics of the generated random numbers crucially depends on the choice of integers a, b and M, K that determine the set of pseudorandom numbers $\{\xi_i\}$ completely. Any wrong choice of these integer parameters leads to unavoidable mistakes in results obtained by the Monte Carlo method. Several different choices of these parameters are known in the literature (Knuth 1969; Kalos and Whitlock 1986; Ahrens and Dieter 1990).

There have been numerous tests of the major generators of pseudorandom numbers provided by different programming languages. All of them have demonstrated that the generated pseudorandom numbers were substantially correlated (dependent – not random – on each other). Sobol (1973, 1989) demonstrated that the correlations can lead to substantial errors. For example, it was demonstrated that trustable results (1–3 percent of error) of multivariate integration by the Monte Carlo method can be obtained only under the following conditions:

- The dimensionality of the integration is less than or equal to 12.
- The integrand belongs to a class of smooth functions.
- The region of integration is a convex hypercube.

There is a fast and simple test for the quality of any generator of pseudorandom numbers. This test consists in filling a d-dimensional manifold with a set of points, the coordinates of which are randomly generated. The results of such a filling for a two-dimensional hypercube (surface) produced by the FORTRAN generator of pseudo random numbers are shown in Fig. 1.7. The results show that the generator works unsatisfactorily since:

- Many regions of the surface are left completely unfilled by the generator.
- Pair correlations between points generated are clearly observable.

In the case of a three-dimensional hypercube, the results of this test are even worse.

Based on this problem, the general outlook for the usage of Monte Carlo method in multivariate integration in statistical physics is quite gloomy since the dimensionality of the integrals involved is around 1000. Therefore, any time the Monte Carlo method is used to perform such high-dimension multivariate integration, it is necessary to verify the generator of random numbers. This verification can be achieved by applying the Monte Carlo method to evaluation of some quantities known in advance from analytical or other considerations.

1.6. Simulation of Molecular Mechanics and Dynamics

During the past three decades computer experiments have come to play a major role in liquid-state physics. Their importance rests on the fact that they provide essentially accurate, quasiexperimental data on well defined models. As there is

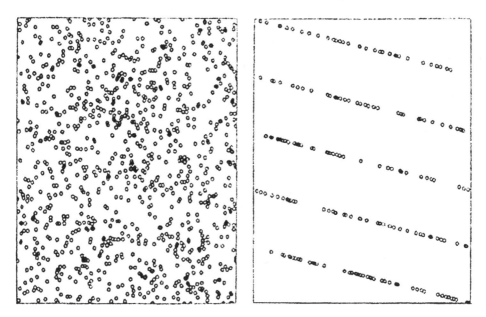

Fig. 1.7. Results of testing a generator of random numbers in a two-dimensional square lattice. On the right-hand side circles designate lattice sites chosen at random as prescribed by the generator. On the left-hand side the pair correlations are given.

no uncertainty about the form of interaction potential in these models, theoretical results can be tested unambiguously in a manner that is generally impossible with data obtained in experiments performed on real liquids. It is also possible to obtain information on quantities of theoretical importance that are not really measurable in the laboratory.

The behavior of solids, liquids, and dense gases can be simulated at the molecular level in one of two ways: by "molecular mechanics" or by "molecular dynamics," which are discussed in the following sections.

1.6.1. *Simulation of Molecular Mechanics*

Simulation of molecular mechanics emerged as the first and simplest application of the Monte Carlo method in statistical mechanics. Its usage implies numerical evaluation of system equilibrium characteristics within a specified ensemble. For example, the introduction of the Monte Carlo method (Section 1.5) was based on evaluation of the equilibrium characteristics of a canonical ensemble, also known as an NVT ensemble. This ensemble, discussed in Sections 1.2.3.2 and 1.3.11, is just a collection of system states characterized by the same values of N, V, and T. In this section we will demonstrate applications of the molecular mechanics simulation, based on the Monte Carlo method, for the NPT and grand canonical ensembles.

1.6.1.1. Simulation of Molecular Mechanics within the NPT Ensemble

In order to adjust the Monte Carlo method for the *NPT* ensemble, a few minor changes are required. The *NPT* ensemble average of a function $F(q; V) = F(q_1, \ldots, q_N; V)$ of the N molecular coordinates and volume V can be given in terms of the expression:

$$\langle F(q; V) \rangle = \frac{\int\limits_0^\infty \int F(q; V) e^{-\beta[PV+U(q)]} dV \prod\limits_{n=1}^N dV_n}{\int\limits_0^\infty \int e^{-\beta[PV+U(q)]} dV \prod\limits_{n=1}^N dV_n}. \tag{1.122}$$

In the calculations based on the Monte Carlo method all molecules are confined within a cubic cell of length L (see Section 1.5.1), which also can be considered as a random variable. Therefore, it is convenient to introduce the set $\{s\}$ of new scaled coordinates $s_1, \ldots, s_N$:

$$s_n = L^{-1} q_n, n = 1, \ldots, N. \tag{1.123}$$

Expression (1.122) for $\langle F(q; V) \rangle$ in terms of the scaled coordinates acquires the following form:

$$\langle F(q; V) \rangle = \frac{\int\limits_0^\infty \int V^N F(s; V) e^{-\beta[PV+U(s;L)]} dV \prod\limits_{n=1}^N d\Omega_n}{\int\limits_0^\infty \int V^N e^{-\beta[PV+U(s;L)]} dV \prod\limits_{n=1}^N d\Omega_n}, \tag{1.124}$$

where Ω_n $(n = 1, \ldots, N)$ are unit cubes corresponding to the molecular scaled coordinates. Equation (1.124) represents an average over the $(3N + 1)$-dimensional space of the variables $(s_1, \ldots, s_N; V)$ with the probability density function proportional to the pseudo-Boltzmann factor:

$$\rho(N, P, T) \sim e^{-\beta([PV+U(s;L)]+N \ln V)}. \tag{1.125}$$

In the implementation of the Monte Carlo method for the *NPT* ensemble, a trial configuration k is constructed from the current configuration m by randomly displacing a selected molecule and randomly changing the length L of the system cell within some predetermined limits. The trial configuration is unconditionally accepted if $\rho(N, P, T; k) \geq \rho(N, P, T; m)$ and is accepted with probability $\rho(N, P, T; k)/\rho(N, P, T; m)$ if $\rho(N, P, T; k) < \rho(N, P, T; m)$.

1.6.1.2. *Simulation of Molecular Mechanics within the Grand Canonical Ensemble*

The further extension of the Monte Carlo method for simulation of molecular mechanics takes into account the grand canonical ensemble (Section 1.2.3). In the grand canonical ensemble the average of any function $F(q) = F(q_1, \ldots, q_N)$ of the N molecular coordinates is given by the following expression:

$$\langle F(q) \rangle - \frac{1}{Z_{\text{gc}}(\beta; V; z)} \sum_{N=0}^{\infty} \frac{z^N}{h^{sN} N!} \int F(q) e^{-\beta U(q)} \prod_{n=1}^{N} \mathrm{d}V_n, \qquad (1.126)$$

where $z = \exp(\beta\mu)$ is the fugacity and s is a degree of freedom of each molecule (do not confuse s with the scaled coordinates $s_1, \ldots, s_N$). As usual, it is convenient to introduce the scaled coordinates in correspondence with equation (1.123). Then expression (1.126) for $\langle F(q) \rangle$ takes the form:

$$\langle F(q) \rangle = \frac{1}{Z_{\text{gc}}(\beta; V; z)} \sum_{N=0}^{\infty} e^{N\beta\mu^* - \ln(N!)} \int F(s) e^{-\beta U(s)} \prod_{n=1}^{N} \mathrm{d}\Omega_n, \qquad (1.127)$$

where $\mu^* = \mu - sN\ln(h)$ and $F(s) = F(s_1, \ldots, s_N)$, $U(s) = U(s_1, \ldots, s_N)$. Therefore, the probability density function $\rho(V, T)$, analogous to the one for the NPT ensemble, is proportional to the pseudo-Boltzmann factor taken in the form:

$$\rho(P, T) \sim \exp^{[N\beta\mu^* - \ln(N!) - \beta U(s)]}. \qquad (1.128)$$

The fixed parameters of the calculation are μ^*, V, and T. The trial configurations are generated in a two-step process. The first step involves the displacement of a selected molecule and then proceeds exactly the same as in the case of canonical ensemble: the displacement is accepted or rejected according to criteria discussed in Section 1.5.3. The second step involves an attempt either to insert or to remove a molecule at a randomly chosen point in the space of the system cell (Valleau and Cohen 1980). The decision whether to attempt an insertion or a removal is also made randomly with equal probabilities. In either case the trail configuration is accepted if the pseudo-Boltzmann factor (equation (1.128)) increases. If it decreases, the change is accepted with a probability equal to either:

$$\frac{\rho(N+1, P, T)}{\rho(N, P, T)} = \frac{1}{N+1} \exp^{-\beta\mu^* - \beta[U(s_1, \ldots, s_{N+1}) - U(s_1, \ldots, s_N)]} \qquad (1.129)$$

(the case of molecular insertion)

or:

$$\frac{\rho(N-1, P, T)}{\rho(N, P, T)} = N \exp^{-\beta\mu^* - \beta[U(s_1, \ldots, s_{N-1}) - U(s_1, \ldots, s_N)]} \qquad (1.130)$$

(case of molecular removal).

The Monte Carlo method for the grand canonical ensemble, as outlined above, works very well at low and intermediate densities. Its usefulness at higher densities is questionable because the probability of inserting or removing a molecule become very small at these densities. The method has, however, been used successfully in the study of a variety of interfacial phenomena with both atomic and molecular liquids (Nicholson and Parsonage 1982). The main advantage of the calculations based on the grand canonical ensemble is that they provide a direct means of determining the chemical potential and hence the free energy and entropy of the system under study.

1.6.2. *Simulation of Molecular Dynamics*

Simulation of molecular dynamics allows computation of equilibrium correlation functions that determine the transport coefficients of the system under study. In conventional molecular-dynamics calculations a system containing N particles (atoms, molecules, or ions) is placed within a cell of fixed volume, usually cubic in shape. A set of velocities, usually drawn from the Maxwell–Boltzmann distribution (see equation (1.22)), is also selected and assigned in such a way that the net molecular momentum is equal to zero. The subsequent trajectories of the molecules are then calculated by integration of the classical equations of motion. The molecules are assumed to be in interaction with each other via some prescribed force law and the bulk of computation is concerned with the calculation of forces acting on each molecule at every step. The resulting thermodynamic and other properties are obtained as time averages over the dynamic history of the system. Apart from the choice of initial conditions, simulation of molecular dynamics is, in principle, entirely deterministic in nature.

In most applications of the molecular mechanics simulation the goal is to determine the ensemble averages. In Section 1.6.1 the canonical, NPT, and grand canonical ensemble averages have been discussed. In the molecular dynamics simulation, apart from certain specialized calculations, the system is not subjected to any external field and its total energy is independent of time. Therefore, the dynamical states that the simulation generates represent a sample from a microcanonical ensemble (see Section 1.2.2) and the solution of the equations of motion is carried out in such a way that the corresponding phase-space sampling is performed over a surface of constant energy. The molecular dynamics simulation has a great advantage since it allows the study of time-dependent phenomena. If static properties alone are required, the molecular mechanics simulation is often simpler and more useful to use, because temperature and volume (canonical ensemble) or temperature and pressure (NPT ensemble) are more convenient parameters to keep fixed than volume and total energy.

There are a number of difficulties that are common to both simulations. The most obvious difficulty is based on the fact that on a macroscopic scale the size of sample that can be studied economically is extremely small: typically $N \sim 1000$ or less. Therefore, in order to minimize surface effects and thereby to simulate more closely the behavior of an infinite system, it is customary to use a periodic

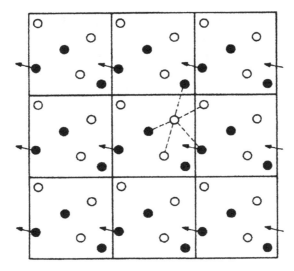

Fig. 1.8. Periodic boundary conditions used in computer simulation experiments.

boundary condition. The way in which the periodic boundary condition works is illustrated for a two-dimensional system in Fig. 1.8. The system molecules lie in the central cell, which is surrounded on all sides by images of itself: each image cell contains N molecules in the same relative positions as in the central cell. When a molecule enters or leaves through one face of a cell, the move is balanced by an image of that molecule leaving or entering, respectively, through an opposite face. A key question is whether the properties of a small periodic sample are truly representative of the macroscopic system that the model simulates. There is no easy answer to this question: it depends very much on the form of the potential and on the properties under investigation. The only safe course in cases of doubt is to make calculations for a range of different N and to look for signs and systematic trends in the results. Experience gained in simulation allows the following conclusions:

- The bulk properties are only weakly dependent on sample size for $N \geq 100$.
- The errors that do exist, relative to the $N \to \infty$ limit, are no larger than the statistical uncertainties.

Nonetheless, the restriction on sample size does have other important disadvantages. In particular, it is impossible to study spatial fluctuations and correlations having a wavelength larger than L (the length of the cell).

The method of molecular dynamics simulation was invented by Alder and Wainwright (1957, 1959) for the study of condensed phases. It was first applied to systems of molecules with hard cores: hard disks, hard spheres, and square-well potentials. These systems, which are characterized by discontinuous potentials and purely impulse forces, are the simplest ones for simulation. To demonstrate this, let us consider, for example, the motion of hard-sphere molecules, the velocities of which change only at collisions. Between collisions these molecules move along straight lines at constant velocities. Therefore, the time evolution of the system of hard-sphere molecules can be treated as a sequence of strictly binary, elastic collisions. Let v_1 and v_2 be the velocities of two hard spheres before a collision and let v_1' and v_2' be their velocities afterwards. Since the molecules are of the same mass, conservation of energy and momentum require that:

$$\text{energy:} \quad |v_1|^2 + |v_2|^2 = |v_1'|^2 + |v_2'|^2,$$
$$\text{momentum:} \quad v_1 + v_2 = v_1' + v_2'. \tag{1.131}$$

Thus, we obtain that: $\Delta v_1 = v_1' - v_1 = -(v_2' - v_2) = -\Delta v_2$. The projection of the relative velocity onto the vector joining the hard-sphere centers is reversed in an elastic collision:

$$v'_{12} \cdot r_{12} = (v_2' - v_1')(r_2 - r_1) = -(v_2 - v_1)(r_2 - r_1) = -v_{12} \cdot r_{12},$$

but the orthogonal components are unaltered. Thus, the change in velocity at a collision is:

$$\Delta v_1 = -\Delta v_2 = \left(\frac{b_{12}r_{12}}{D^2} \right)_{\text{contact}}, \quad b_{12} = v_{12} \cdot r_{12}, \tag{1.132}$$

where D is the hard-sphere diameter and the term on the right-hand side must be evaluated at contact. Therefore, the algorithm for the calculation of molecular trajectories consists of the following steps:

1. advancing the coordinates of all molecules until a collision occurs somewhere in the system;
2. calculating the changes in velocities of the colliding molecules according to equation (1.132);
3. repetition of steps 1 and 2 for as many collisions as are necessary to yield results of sufficient statistical reliability for the system under study.

The natural unit of time for the hard sphere molecules is the *mean collision time* τ_{coll}, the mean time between collisions suffered by a given molecule. The nonideal contribution to the pressure (see equation (1.99)) is proportional to the rate of collisions between molecules. If, therefore, we take the quantity $\beta P/n - 1 = nB(T)$ in the low-density limit by setting $B(T) = B_{\text{HS}}(T) = (2\pi D^3/3)g(D)$ in equation (1.98) we can obtain a relationship between a *collision rate* $\Gamma_{\text{coll}} = 1/\tau_{\text{coll}}$ and the collision rate Γ_0 of the dilute gas in the form:

$$\frac{\Gamma_{\text{coll}}}{\Gamma_0} = \frac{\dfrac{\beta P}{n} - 1}{\dfrac{2}{3}\pi n D^3} = \frac{B_{\text{HS}}(T)}{\dfrac{2}{3}\pi D^3} = g(D). \tag{1.133}$$

The low-density, or Boltzmann, collision rate Γ_0 is due to Moore and Pearson (1981):

$$\Gamma_0 = 4nD^2 \left(\frac{\pi k_B T}{m}\right)^{1/2}. \tag{1.134}$$

It is important to evaluate these expressions for conditions appropriate for real liquids. For this purpose, let us consider argon atoms which can be represented as hard spheres of diameter $D = 3.4$ Å. The liquid argon has a triple point at $T = 84$ K and $V = 28.4$ cm^3 mol^{-1}. The Boltzmann collision rate evaluated at this point is $\Gamma_0 = 2.3 \times 10^{12}$ s^{-1}, whereas the computer simulation results for hard spheres give that $g(D) = 4.4$ at the corresponding value of the reduced density nD^3. We, therefore, estimate the mean time between "collisions" in liquid argon to be $\tau_{coll} = 1/\Gamma_{coll} = 1/g(D)\Gamma_0 \approx 10^{-13}$ s.

The extension of the method of molecular dynamics simulation to the case where molecules of the system interact via continuous potentials was first described by Rahman (1964a, 1964b, 1966). Rahman's earliest calculations were made for the Lennard-Jones potential and the so-called exp -6 potential characterized by exponential repulsion and attraction varying as R^{-6}. In each case the parameters of the potential were chosen to simulate liquid argon. The more systematic study of the properties of Lennard-Jones liquids were later reported by Verlet (1967, 1968), Verlet and Levesque (1967), Levesque and Verlet (1970) and Levesque, Verlet and Kürkijarvi (1973). Since those pioneering investigations, the method of molecular dynamics for continuous potentials has been applied to many complex systems, including liquid metals, molten salts, etc. Reviews of modeling methods based on computer simulations for liquids and polymers have been given by Allen and Tildesley (1987) and Binder (1995).

In the case of continuous potentials, the trajectories of molecules, unlike those of hard spheres, can no longer be calculated exactly. For the sake of simplicity, let us consider the case of spherically symmetrical potentials. The system of equations of motion for N identical molecules is given by $3N$ coupled, second-order differential equations:

$$m\frac{d^2 r_n}{dt^2} = \nabla_n U(r_1, \ldots, r_N), n = 1, \ldots, N. \tag{1.135}$$

These equations are usually solved numerically by finite difference methods, which assume that if the coordinates of molecule n at time instant t are $r_n(t)$ then the coordinates at time instants $t \pm \Delta t$ can be given by the following Taylor expansion about $r_n(t)$:

$$r_n(t \pm \Delta t) = r_n(t) \pm \Delta t\frac{dr_n(t)}{dt} + \frac{(\Delta t)^2}{2!}\frac{d^2 r_n(t)}{dt^2} \pm \frac{(\Delta t)^3}{3!}\frac{d^3 r_n(t)}{dt^3} + O[(\Delta t)^4]. \tag{1.136}$$

The *central-difference prediction* for $r_n(t + \Delta t)$ is obtained by taking the addition alternative in expression (1.136):

$$r_n(t + \Delta t) \approx -r_n(t - \Delta t) + 2r_n(t) + \frac{(\Delta t)^2}{m} F_n(t), \quad F_n(t) = m\frac{d^2 r_n(t)}{dt^2}, \qquad (1.137)$$

where $F_n(t)$ is the total force acting on molecule n at time t. The error in predicted coordinates is of order $(\Delta t)^4$. This error is unavoidable but is small for $(\Delta t)/t \ll 1$. In the case of the subtraction option in expression (1.136), one obtains an estimate for the velocity $dr_n(t)/dt$ of molecule n at time t:

$$\frac{dr_n(t)}{dt} \approx \frac{1}{2\Delta t}[r_n(t + \Delta t) + r_n(t - \Delta t)]. \qquad (1.138)$$

The error in predicted velocity is of order $(\Delta t)^2$. In the approach outlined in equations (1.136) to (1.138) the time step (Δt) is constant. This contrasts with the previous case of the discontinuous hard-sphere potential, in which the analogue of (Δt), namely the interval between successive collisions, is a quantity varying throughout the calculation procedure.

Let us now apply result (1.137) to the case of Lennard-Jones molecules. If we introduce the Lennard-Jones parameter D (see equation (1.90)) as the unit of length, then equation (1.137) can be rewritten in the form:

$$r_n(t + \Delta t) \approx -r_n(t - \Delta t) + 2r_n(t) - (\Delta t)^2 \frac{48\epsilon}{mD^2} \sum_{m=1, m\neq n}^{N} \left(\frac{r_{nm}}{|r_{nm}|}\right)(|r_{nm}|^{-13} - \tfrac{1}{2}|r_{nm}|^{-7}).$$

$$(1.139)$$

The quantity $\tau_0 = 1/\Gamma_0 = [mD^2/(48\epsilon)]^{1/2}$ appears in equation (1.139) as a natural unit of time. Inserting parameters appropriate to liquid argon $(D = 3.4 \text{ Å}, \epsilon/k_B = 119.8 \text{ K}$ and $m = 6.63 \times 10^{-23}$ g, we obtain $\tau_0 = 3.1 \times 10^{-13}$ s. In the method of molecular dynamics, the choice of (Δt) is usually made on the basis of conservation of total energy. However, in the use of this method, errors in the integration scheme cause the total energy to fluctuate about its mean value. Fluctuations of the order of 10^{-2} percent are usually considered acceptable. In the case of liquid argon this level of accuracy in energy conservation can be achieved for $(\Delta t) \approx 10^{-14}$ s. If the result $\tau_{coll} \approx 10^{-13}$ s, obtained for the case of discontinuous hard-sphere potential, is taken into account, it is an obvious conclusion that the criterion based on energy conservation leads to the physically reasonable result that the time step (Δt) should be roughly an order of magnitude less than the mean collision time τ_{coll}. As the time step is increased, the fluctuations in total energy become larger, leading eventually to an overall drift in energy.

The simulation scheme of the molecular dynamics method based on continuous potentials is given by the sequence of the following procedures:

1. placing molecules at arbitrary chosen positions $r_n(0)$ ($n = 1, \ldots, N$);
2. specifying a set of coordinates $r_n(-\Delta t)$ at an earlier time instant $(-\Delta t)$ by ascribing to each molecule a random velocity drawn from the Maxwell–Boltzmann distribution (see equation (1.22)) in such a way that the total linear momentum of the molecules is zero;
3. organizing the subsequent calculation as a loop over time; at each step the time is increased by (Δt), the total force acting on each molecule is computed, and the molecules are advanced to their new positions.

In the early stages of calculation it is normal for the temperature, which is calculated from the mean kinetic energy, to drift away from the value at which it was originally set. Therefore, an occasional rescaling of the molecular velocities is necessary. Once equilibrium is reached, the system is allowed to evolve undisturbed with both potential and kinetic energies fluctuating around steady mean values. A run of the order of 10^3 time steps, excluding the time needed for equilibration, is generally sufficient to yield the mean potential energy with an uncertainty of about 1 percent and $\beta P/n$ to within ± 0.1.

References

Adler, B. J. and Wainwright, T. E. (1957). *Journal of Chemical Physics*, **27**, 1208–9.

Adler, B. J. and Wainwright, T. E. (1959). *Journal of Chemical Physics*, **31**, 459–66.

Ahrens, J. H. and Dieter, U. (1990). *Pseudo-Random Numbers*. New York: John Wiley & Sons.

Allen, M. P. and Tildesley, D. J. (1987). *Computer Simulation of Liquids*. Oxford: Clarendon Press.

Binder, K. (ed.) (1995). *Monte Carlo and Molecular Dynamics Simulations in Polymer Science*. Oxford: Oxford University Press.

Burnett, D. (1935). *Proceedings of the London Mathematical Society*, **36**, 385–97.

Cox, D. R. and Miller, H. D. (1965). *The Theory of Stochastic Processes*. Bristol: Chapman and Hall.

Ditchburn, R. W. (1952). *Light*. London: Blackie.

Doob, J. L. (1953). *Stochastic Processes*. New York: John Wiley & Sons.

Drude, P. K. L. (1933). *The Theory of Optics*. London: Longman.

Haase, R. (1969). *Thermodynamics of Irreversible Processes*. New York: Addison-Wesley.

Hammersley, J. M. and Handscomb, D. C. (1964). *Monte Carlo Methods*. New York: John Wiley & Sons.

Isihara, A. (1971). *Statistical Physics*. New York: Academic Press.

Kalos, M. N. and Whitlock P. A. (1986). *Monte Carlo Method*, Vol. 1. New York: John Wiley & Sons.

Knuth, D. (1969). *The Art of Computer Programming*, Vol. 2. New York: Addison-Wesley.

Korn, G. A. and Korn, T. M. (1961). *Mathematical Handbook for Scientists and Engineers*. New York: McGraw-Hill.

Landau, L. D. and Lifshitz, E. M. (1980). *Quantum Mechanics. Non-Relativistic Theory*. Oxford: Pergamon Press.

Landau, L. D. and Lifshitz, E. M. (1988). *Statistical Physics*, Part 1. Oxford: Pergamon Press.

Lehmer, D. H. (1951). *Proceedings of the 2nd Symposium on Large-Scale Digital Computing Machinery*, Vol. 142. Cambridge, MA: Harvard University Press.

Lennard-Jones, J. E. (1924). *Proceedings of the Royal Society*, **A106**, 473–82.

Levesque, D. and Verlet, L. (1970). *Physical Review*, **A2**, 2514–33.

Levesque, D., Verlet, L. and Kürkijarvi, J. (1973). *Physical Review*, **A7**, 1690–1700.

Maitland, G. C., Rigby, M. and Wakeham, W. A. (1987). *Intermolecular Forces. Their Origin and Determination*. Oxford: Clarendon Press.

Mayer, J. E. and Goeppert-Mayer, M. (1977). *Statistical Mechanics*. New York: John Wiley & Sons.

Metropolis, M., Rosenbluth, A. W., Rosenbluth, M. N., Teller, A. N. and Teller, E. (1953). *Journal of Chemical Physics*, **21**, 1087–92.

Moore, J. W. and Pearson, R. G. (1981). *Kinetics and Mechanics*. New York: John Wiley & Sons.

Nicholson, D. and Parsonage, N. G. (1982). *Computer Simulation and the Statistical Mechanics of Absorption*. London: Academic Press.

Onsager, L. (1944). *Physical Review*, **65**, 117–26.

Rahman, A. (1964a). *Physical Review*, **136**, A405–A411.

Rahman, A. (1964b). *Physical Review Letters*, **12**, 575–7.

Rahman, A. (1966). *Journal of Chemical Physics*, **45**, 2585–92.

Sobol, I. M. (1973). *Numerical Monte Carlo Methods* (in Russian). Moscow: Nauka.

Sobol, I. M. (1989). *USSR Computational Mathematics and Physics*, **29**, 935–42.

Sutherland, W. (1893). *Philosophical Magazine*, **36** (5th ser.), 507–17.

Valleau, J. P. and Cohen, L. K. (1980). *Journal of Chemical Physics*, **72**, 5935–41.

Verlet, L. (1967). *Physical Review*, **159**, 98–103.

Verlet, L. (1968). *Physical Review*, **165**, 201–14.

Verlet, L. and Levesque, D. (1967). *Physica*, **36**, 254–68.

von Neumann, J. and Ulam, S. (1945). *Bulletin of the American Mathematical Society*, **51**, 9–13.

Wax, N. (ed.) (1954). *Selected Papers on Noise and Stochastic Processes*. New York: Dover.

2

Crystallization Basics

ALLAN S. MYERSON

2.1. Crystals

Crystals are solids in which the atoms are arranged in a periodic repeating pattern that extends in three dimensions. When crystals are grown slowly and carefully they are normally bounded by plane faces that can be seen with the naked eye. This can be seen in the beautiful samples of many minerals that are exhibited in museums and stores that specialize in minerals and fossils or simply by looking at some table salt with a magnifying glass. The plane faces that are evident in the minerals (or the salt) are evidence of a regular internal structure. However, not all materials that are crystalline are so easy to distinguish. Many materials such as steel, concrete, bone, and teeth are made up of small crystals. Other materials such as wood, silk, hair, and many solid polymers are partially crystalline or have crystalline regions. The crystalline state of solids is the state of minimum free energy and is therefore their natural state. When a material is taken to a low temperature the molecules will always try to arrange themselves in a periodic structure. When they are unable to do so, because of the viscosity of the system or because the materials were cooled rapidly, noncrystalline solids can result. Solids such as glasses have no crystalline regions and no long-range order and are called amorphous solids. Many materials can form solids that are crystalline or amorphous, depending on the conditions of growth. Crystals of a particular chemical species can sometimes form into more than one possible three-dimensional arrangement of the atoms. Crystals can also be grown with different external planes, depending on the conditions of growth. The field of study that is devoted to the examination and analysis of crystals and their structure is called *crystallography* and is described in detail in many textbooks (Bunn 1961; Cullity 1978; McKie and McKie 1986). The purpose of this section is to present the basic concepts of crystallography.

Molecular Modeling Applications in Crystallization, edited by Allan S. Myerson. Printed in the United States of America. Copyright © 1999 Cambridge University Press. All Rights reserved. ISBN 0 521 55297 4.

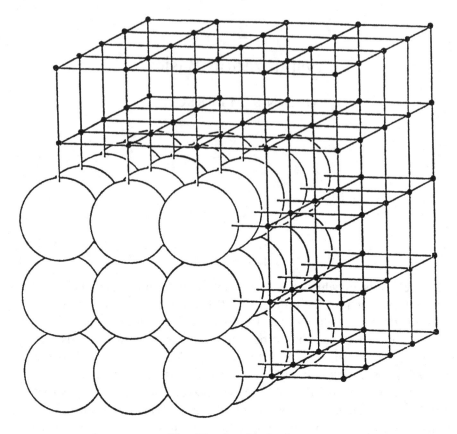

Fig. 2.1. A cubic lattice.

2.1.1. *Crystal Lattices*

The concept of a lattice is fundamental to the understanding of crystal structure and to crystallography. A lattice can be thought of as a repeating pattern in three dimensions. A two-dimensional example of a repeating pattern is a wallpaper pattern, in which a small unit is repeated in all directions. If we think of a crystal as made up of atoms (which we will imagine as spheres) arranged periodically in three dimensions and then join the centers of these atoms by lines, the crystal would then look like a periodic three-dimensional network with points in place of the atoms. This is illustrated in Fig. 2.1. A lattice is a set of points arranged so that each point has identical surroundings. In addition, a lattice can be characterized by three spatial dimensions: *a*, *b*, and *c*, and three angles: α, β, and γ. These dimensions and angles are known as *lattice parameters*. An example of a lattice is given in Fig. 2.2. Looking at the figure, we can see that the lattice is made up of repeating units that can be characterized by the lattice parameters and can be used to extend the lattice as far as desired in all dimensions. In Fig. 2.3 we have taken the smallest three-dimensional unit of the lattice that contains all the information necessary to replicate the lattice to any size. This repeat unit is called the *unit cell*.

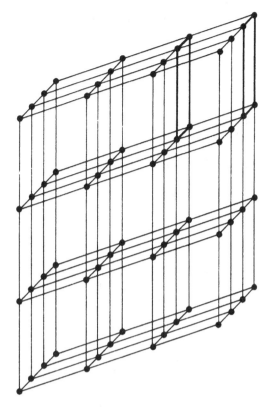

Fig. 2.2. A point lattice.

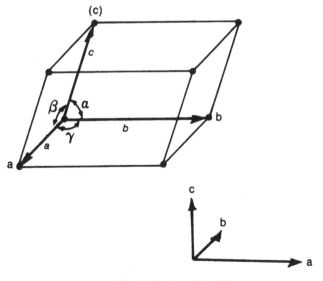

Fig. 2.3. A unit cell.

Table 2.1. *Crystal systems and Bravais lattices; data adapted from Cullity (1978)*

System	Axial lengths and angles	Bravais lattice
Cubic	Three equal axes at right angles $a = b = c, \alpha = \beta = \gamma = 90°$	Simple Body-centered Face-centered
Tetragonal	Three axes at right angles, two equal $a = b \neq c, \alpha = \beta = \gamma = 90°$	Simple Body-centered
Orthorhombic	Three unequal axes at right angles $a \neq b \neq c, \alpha = \beta = \gamma = 90°$	Simple Body-centered Base-centered Face-centered
Rhombohedral[a]	Three equal axes equally inclined $a = b = c, \alpha = \beta = \gamma = 90°$	Simple
Hexagonal	Two equal coplanar axes at 120°, third axis at right angles $a = b \neq c, \alpha = \beta = 90°, \gamma = 120°$	Simple
Monoclinic	Three unequal axes, one pair not at right angles $a \neq b \neq c, \alpha = \gamma = 90° \neq \beta$	Simple Base-centered
Triclinic	Three unequal axes, unequally inclined and none at right angles $a \neq b \neq c, \alpha \neq \beta \neq \gamma \neq 90°$	Simple

[a]Also called trigonal.

In 1848 A. Bravais and M.L Frankenheim mathematically investigated the number and types of three-dimensional lattices that could exist. They found that there are only fourteen possible point lattices that can be constructed and these are known as the *Bravais lattices*. All crystals have a three-dimensional structure that is categorized by one of the Bravais lattices.

The Bravais lattices can be arranged by categories according to the shapes (or by employing the lattice parameters). This results in the lattices being divided into seven categories, which are known as *crystal systems* and are shown in Table 2.1. All 14 of the Bravais lattices are shown in Fig. 2.4. Lattices are characterized by their lattice parameters. For example, the cubic lattices have equal lengths ($a = b = c$) and angles equal to 90°. The four orthorhombic lattices have three unequal axes ($a \neq b \neq c$) with all axes at right angles.

The simplest unit cell in each system is called the *primitive (P) lattice*. Primitive lattices have points in only the corners of the lattice and therefore have only one lattice point per unit cell. This is because each lattice point on a corner is shared with eight other unit cells so that one-eighth belongs to a single cell. Each crystal

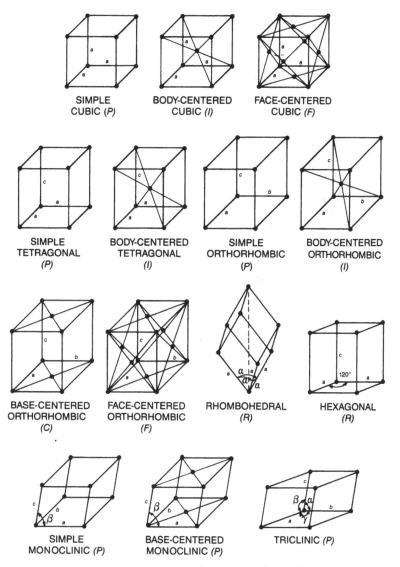

Fig. 2.4. The Bravais lattices (P = primitive lattice, F = face-centered lattice, I = body-centered lattice, and C = end-centered lattice).

system has a P lattice. The P lattice of the trigonal system is a rhombohedron and is normally referred to as the R lattice. There are three other lattice types, F (face-centered), I (body-centered) and C (end-centered). The F lattice has lattice points on all the corners and in the center of each face and is found in the cubic and orthorhombic systems. The I lattice has lattice points on the corners and in the center of the unit cell and is found in the cubic, tetragonal, and orthorhombic systems. The C lattice has lattice points in the corners and in the center of one set of parallel faces and occurs in the orthorhombic and monoclinic systems.

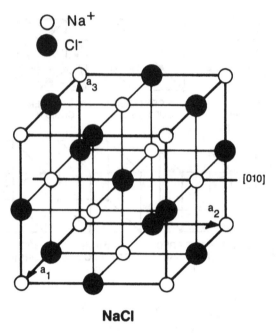

NaCl

Fig. 2.5. The structure of sodium chloride (NaCl).

2.1.2. *Crystal Symmetry*

When atoms of the same element in the unit cell have an identical atomic environment except for the orientation of the environment, they are said to be related
by symmetry. Let us look at a common atomic arrangement, which is the structure of sodium chloride and other simple ionic binary compounds that exhibit a
high degree of symmetry. The atomic arrangement of sodium chloride is shown
in Fig. 2.5. Looking at the figure, we can see that the unit cell is cubic. If we take
the corner of the cell to be the center of a sodium ion, there are also sodium ions
at the center of each face (since the lattice is face-centered cubic) and the chlorine
ions are halfway along the edges and also in the center of the unit cell. One type
of symmetry operation is rotation. If we define an axis perpendicular to any face
of the crystal and passing through the center and then rotate the cube about that
axis through 90°, the cube will look exactly the same. This will be true each time
we rotate the cube through 90° until the cube has been rotated through one
complete revolution (360°). Since the cube presents the same appearance four
times during the rotation, it is said to have an axis of fourfold symmetry. A cube
has three fourfold axes of symmetry, all at right angles to each other. Cubes also
have six axes of twofold symmetry and four axes of threefold symmetry. The
symmetry axes for sodium chloride are shown in Fig. 2.6. All of the axes shown
are symmetry elements of the atomic arrangement. In general an *n*-fold rotation
axis of symmetry is defined as a line about which rotation of $360°/n$ results in the
same appearance. The crystal therefore superimposes on itself *n* times during the
complete rotation.

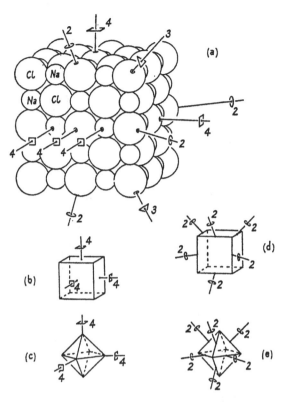

Fig. 2.6. (*a*) The atomic arrangement in sodium chloride and some of its axes of symmetry. (*b*) and (*c*) The fourfold axes of the cube and octahedron. (*d*) and (*e*) The twofold axes of the cube and octahedron. (Reprinted from *Chemical Crystallography* by C.W. Bunn (2nd ed., OUP, 1961) by permission of Oxford University Press.)

If the sodium chloride crystal is divided into two halves by a plane passing through the center of the crystal the two halves will be mirror images of each other. This type of plane of symmetry is called a mirror plane and is illustrated in Fig. 2.7. Sodium chloride has a set of three mirror planes, perpendicular to each other and parallel to each of the three sets of cube faces, and a set of six planes that bisect the angles between the first set.

Another type of symmetry axis involves the combination of rotation about a line with inversion through a point. The result is called an inversion axis of symmetry and is referred to as an inverse *n*-fold axis of symmetry. If we chose the origin as the point of inversion, then the coordinates of every point x, y, z, become $\bar{x}$, $\bar{y}$, $\bar{z}$. If the crystal has a structure such that every atom with coordinates x, y, z is duplicated by an identical atom at coordinates $\bar{x}$, $\bar{y}$, $\bar{z}$, the crystal structure is said to have a center of symmetry at the origin and is usually referred to as being *centrosymmetric*. Sodium chloride has a centrosymmetric structure. A comparison of a centrosymmetric and a non-centrosymmetric structure is shown in Fig. 2.8. There are also other types of inversion

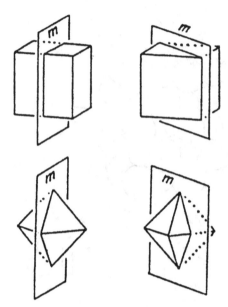

Fig. 2.7. Planes of symmetry in a cube and octahedron. (Reprinted from *Chemical Crystallography* by C.W. Bunn (2nd ed., OUP, 1961) by permission of Oxford University Press.)

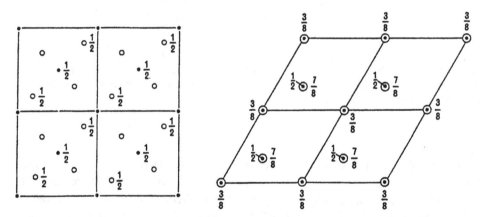

Fig. 2.8. Centrosymmetric and non-centrosymmetric structures. (*a*) Rutile, TiO_2, with titanium atoms shown as solid circles and oxygen atoms as open circles. (*b*) Wurtzite, ZnS, with zinc atoms shown as solid circles and sulfur atoms as open circles. Both structures are projected on the *xy* plane. (Reproduced with permission from McKie and McKie (1986).)

axes of symmetry that correspond to the inversion of the higher-fold axes of symmetry (such as the threefold and fourfold axes).

The symmetry elements all have simple symbols that are used to describe them. The rotation axes of symmetry are represented by the numbers 2, 3, 4, and 6, and the inversion axes are represented by the same numbers with a bar over them (representing negative numbers). If a crystal is asymmetric, it is repre-

Table 2.2. *Symmetry elements; data adapted from Cullity (1978)*

System	Minimum symmetry elements
Cubic	Four threefold rotation axes
Tetragonal	One fourfold rotation (or rotation-inversion) axis
Rhombohedral	Three perpendicular twofold rotation (or rotation-inversion) axes
Hexagonal	One sixfold rotation (or rotation-inversion) axis
Monoclinic	One twofold rotation (or rotation-inversion) axis
Triclinic	None

sented by a one (since it must go through a complete rotation to superimpose upon itself). If the crystal has a center of symmetry it is symbolized by $\bar{1}$, while a plane of symmetry is given the symbol *m* (for mirror).

2.1.3. *Crystallographic Point Groups*

The symmetry operations described in the previous section can be combined into thirty-two classes, which are called *point groups*. A crystallographic point group is defined as a group of symmetry elements that can operate on a three-dimensional lattice so as to leave one point unmoved. More information on the point groups and their derivation can be found in several sources (McKie and McKie 1986; Bunn 1961). The crystallographic point groups are normally divided into *crystal systems*, which have some common symmetry. These systems and the characteristic symmetry elements are given in Table 2.2.

The symbol for any of the 32 crystallographic point groups is obtained by first identifying the character of the principal plane of rotation or rotation inversion (*n* or $\bar{n}$). A plane of symmetry normal to the principal axis is written as n/m. The principal axis in this case must be a rotation axis. If a plane of symmetry contains the principal axis, it is written *nm* or $\bar{n}m$. An axis of twofold rotation (often called a diad axis) normal to the principal axis is written as *n2*. The 32 crystal classes are illustrated in Fig. 2.9 using stereograms. There are pairs of stereograms for each point group (except for triclinic), the left one representing the position of a plane *hkl* by its pole, and the right one showing the symmetry element of the point group. The figure is organized so that the first row represents the classes that have only principal rotation axes and the second row those that have corresponding inversion axes. In the third row a plane of symmetry perpendicular to a principal rotation axis is added. In the fourth row the plane of symmetry is parallel to the principal rotation axis and in the fifth row the plane of symmetry is parallel to a principal inversion axis. In the sixth row a secondary twofold axis is added to a principal rotation axis and in the last row there are

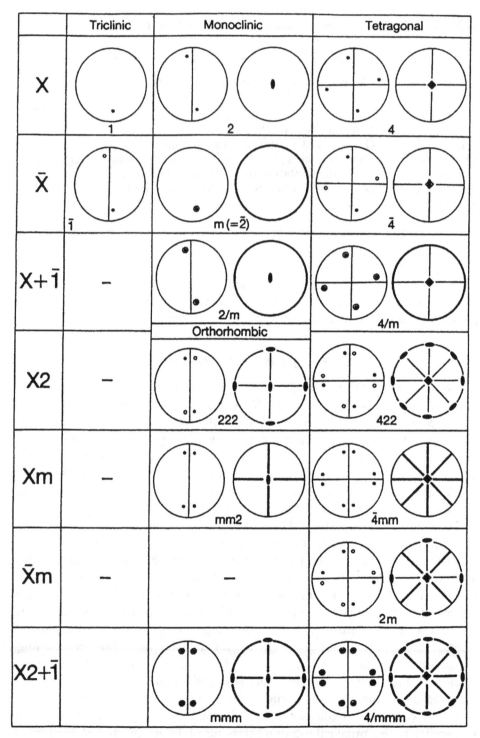

Fig. 2.9. The thirty-two crystallographic point groups. Each pair of stereograms shows, on the left, the poles of a general form and, on the right, the symmetry elements of the point group. Planes of symmetry are indicated by bold lines. (Reproduced with permission from McKie and McKie (1986).)

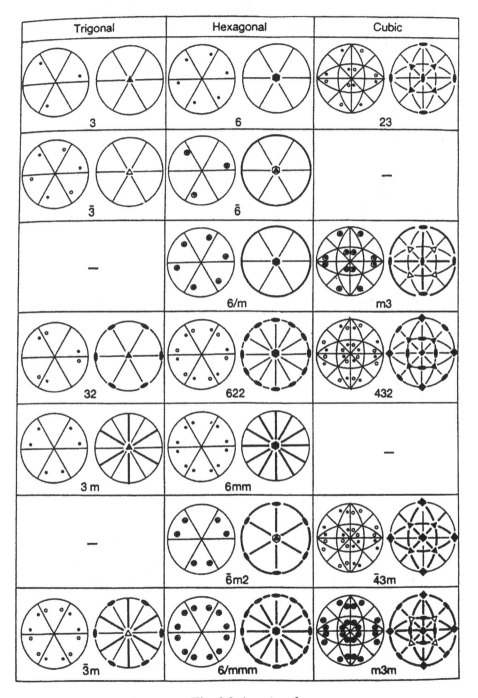

Fig. 2.9 *(continued)*

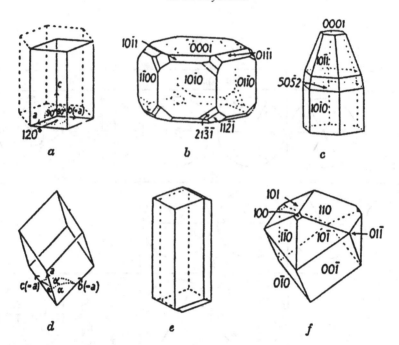

Fig. 2.10. Hexagonal and trigonal systems. (*a*) Hexagonal-type unit cell. (*b*) Apatite, $3Ca_3(PO_4)_2.CaF_2$. Class $6/m$. (*c*) Hydrocinchonine sulfate hydrate, $(C_{19}H_{24}ON_2)_2.H_2SO_4.11H_2O$. Class $6m$. (*d*) Rhombohedral-type unit cell. (*e*) A habit of calcite, $CaCO_3$. Class $3m$. (*f*) $KBrO_3$. Class $3m$. (Reprinted from *Chemical Crystallography* by C.W. Bunn (2nd ed., OUP, 1961) by permission of Oxford University Press.)

planes of symmetry both parallel and perpendicular to the principal rotation axis. The columns in the figure represent each of the crystal systems. Examples of crystals from several of the systems are given in Figs. 2.10–2.13.

2.1.4. *Miller Indices and Lattice Planes*

If we take any point on a lattice and consider it the origin, we may define a vector from the origin in terms of three coordinates. If we started with a cubic cell and defined a vector going from the origin and intersecting point 1,1,1, the line would go in the positive direction along the body diagonal of the cube and would intersect the point 2,2,2 and all multiples. This direction is represented by [1,1,1], where the numbers are called indices of direction. Negative numbers are indicated by bars. Indices are normally given as the smallest set of integers that can be obtained by division or by clearing of fractions.

The representation of planes in a lattice makes use of a convention known as Miller indices. In this convention each plane is represented by three numbers (*hkl*), which are defined as the reciprocals of the intercepts made by the plane with the three crystal axes. If a plane is parallel to a given axis its Miller index is zero. Negative indices are denoted by placing bars above the number. Miller

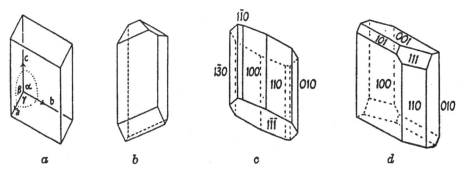

Fig. 2.11. Triclinic system. (*a*) Unit cell type. (*b*) Strontium hydrogen tartrate hydrate $SrH_2(C_4H_4O_6)_2.4H_2O$. Class 1. (*c*) $CuSO_4.5H_2O$. Class 1. (*d*) 1,4-dinitro 2,5-dibromo-benzene. Class 1. (Reprinted from *Chemical Crystallography* by C.W. Bunn (2nd ed., OUP, 1961) by permission of Oxford University Press.)

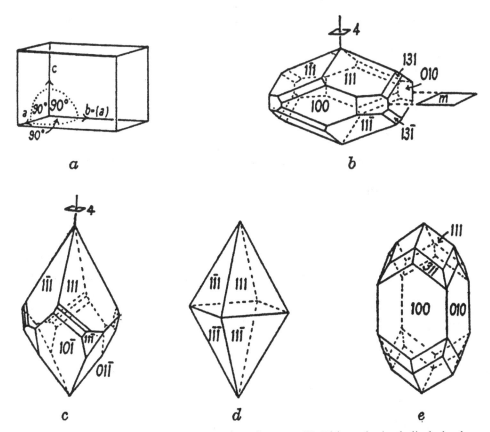

Fig. 2.12. Tetragonal system. (*a*) Unit cell type. (*b*) Phloroglucinol diethyl ether. Class 4/*m*. (*c*) Wulfenite, $PbMoO_4$. Class 4. (*d*) Anatase, TiO_2. Class 4/*mmm*. (*e*). Zircon, $ZrSiO_4$. Class 4/*mmm*. (Reprinted from *Chemical Crystallography* by C.W. Bunn (2nd ed., OUP, 1961) by permission of Oxford University Press.)

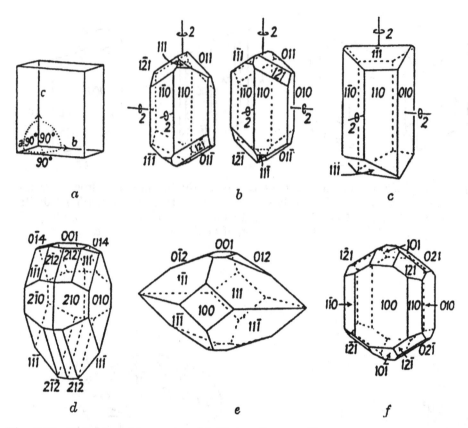

Fig. 2.13. Orthorhombic system. (a) Unit cell type. (b) (H.COO)$_2$Sr.2H$_2$O. Class 222. Left- and right-handed crystals. (c) 1-Bromo,2-hydroxynaphthalene. Class 222. (d) Picric acid. Class *mm*. (e) Oxalic acid. Class *mmm*. (f) C$_2$Br$_6$. Class *mmm*. (Reprinted from *Chemical Crystallography* by C.W. Bunn (2nd ed., OUP, 1961) by permission of Oxford University Press.)

indices refer to the plane specified by the indices and all planes parallel to it. The distance between parallel planes is called the interplanar spacing (sometimes called the *d* spacing) and is designated by the symbol d_{hkl}. If we wish to specify all planes that are equivalent, we put the indices in braces: for example, {100} represents all the cube faces. Examples of Miller indices in the cubic system are shown in Fig. 2.14. Miller indices are used to designate the faces present in grown crystals and are shown in Figs. 2.10 to 2.13.

2.1.5. *Translational Symmetry Elements*

In our discussion of symmetry up to this point, we have dealt with symmetry operations (rotation, reflection, and inversion) that, when applied to a lattice, result in a return to the initial position. There is also a different type of symmetry operation, which is called *translational symmetry*. In translational symmetry simultaneous use of reflection and translation or rotation and translation will

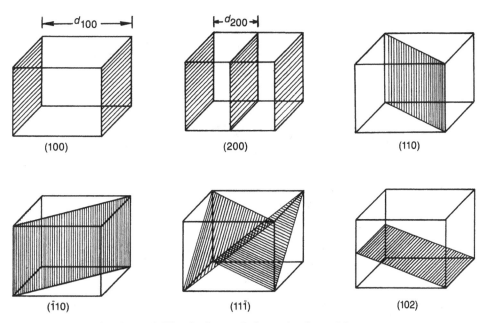

Fig. 2.14. Miller indices of planes in the cubic system.

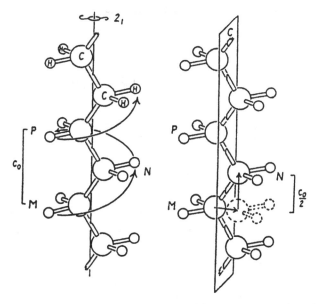

Fig. 2.15. Left: twofold screw axis. Right: glide plane. (Reprinted from *Chemical Crystallography* by C.W. Bunn (2nd ed., OUP, 1961) by permission of Oxford University Press.)

result in a displacement of the original position to a new position at an atom corresponding to the next lattice point. The former is called a glide plane and the latter a screw axis. These are illustrated in Fig. 2.15 with isolated molecules of polyethylene. Polyethylene is a polymer molecule with a long chain length made

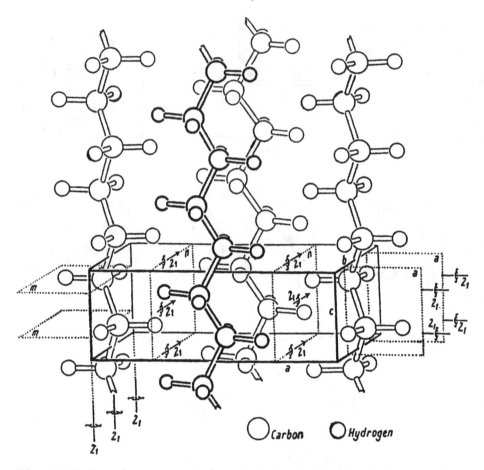

Fig. 2.16. Crystal structure of polyethylene. (Reprinted from *Chemical Crystallography* by C.W. Bunn (2nd ed., OUP, 1961) by permission of Oxford University Press.)

up of repeating units of two carbon and four hydrogen atoms in a straight line. The repeat units are labeled so that the first unit is $M-N$, the second unit $P-C$. If we choose any atom in the molecule in one unit and go to the equivalent molecule in the next unit there is a fixed distance which is labeled in the figure, c_0. Looking at the figure, imagine that we rotate the molecule around the axis by 180° and simultaneously move along the axis a distance $\frac{1}{2}c_0$. We have just moved point M to position N. If we repeat the process a second time point M is at position P, which is a position exactly equivalent to M, but just moved up the axis. This is called a screw axis and is given the symbol 2_1. Another way to accomplish the movement of point M to point N and then point P is to imagine the reflection of M in a plane perpendicular to the carbon atoms. The reflected group is then moved a distance of $\frac{1}{2}c_0$, which moves it to position N. If this is repeated a second time, point M moves to point P. This is called a glide plane and is given the symbol c, which indicates the direction of the translation. Glide planes and screw axes are further illustrated in Fig. 2.16, which shows the crystal structure of polyethylene. The figure shows that, in addition to the twofold screw axis possessed by the single molecule, the crystal also has a twofold screw axis relating the molecules to each other. There are three sets of these twofold screw axes, each one being parallel to a unit cell axis. Figure 2.16 also shows glide planes perpendicular to the *b* axis with a translation of $\frac{1}{2}a$.

2.1.6. *Space Groups*

In Section 2.1.3, we demonstrated that using concepts of symmetry and the Bravais lattices resulted in 32 crystallographic point groups. If we add the concept of the screw axis and the glide plane, the number of symmetry elements that can be derived increases to 230, which are known as the crystallographic *space groups*. The derivation of the 230 space groups and a complete listing can be found in a number of references (Buerger 1956; Hilton 1963; *International Tables for Crystallography* 1983). Space-group symbols start with a capital letter to indicate the lattice type, followed by a symbol for the principal axis. If there is a plane of symmetry or a glide plane perpendicular to the principal axis, the symbols are combined in the form $P2/m$ and $P2_1/c$ respectively. This is then followed by symbols for the symmetries of the secondary axis and planes of symmetry or glide planes parallel to these axes. For example, the space group *Pmcn* belongs to the point group *mmm* (orthorhombic), the lattice type is primitive (*P*) orthorhombic, a mirror plane is perpendicular to *x*, a *c*-glide plane perpendicular to *y*, and a diagonal glide plane perpendicular to *z*. The symbol *Pmmm* represents another space group that belongs to the orthorhombic point group and *P* lattice type and also includes three twofold axes of rotation; however, because these are implicit in the existence of the three mirror planes, they are not written, being unnecessary for unique characterization of the space group.

Space groups can be represented graphically by diagrams that show the distribution of symmetry elements in the unit cell and the equivalent positions that would be generated by operation of the symmetry elements on an initial position *x,y,z*. This is illustrated in Fig. 2.17 for the space group *Cmca*.

2.1.7. *Crystal Structure and Bonding*

In our discussions to this point we have looked at crystals from a purely geometric point of view and have presented the geometric structures that are used to represent the actual structure of real crystals. Real crystals, however, involve molecules that interact with each other to form chemical bonds of different kinds. Chemical bonds are the result of atoms being close enough together for their outer electron shells to interact. Formation of a chemical bond in a molecule and in a crystal must result in a decrease in the system energy. Chemical bonds are usually classified as ionic, covalent, metallic, van der Waals, and hydrogen bonds, with the first three types being stronger than the last two.

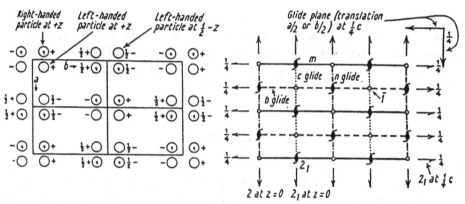

Fig. 2.17. Symmetry elements and equivalent positions in space group *Cmca*. (Reprinted from *Chemical Crystallography* by C.W. Bunn (2nd ed., OUP, 1961) by permission of Oxford University Press.)

Ionic bonds are the result of an electron distribution shift of outer electrons of neighboring atoms, resulting in charged ions that are attracted electrostatically. In covalent bonds, the shared outer electrons are concentrated in orbitals that are spatially fixed relative to the bonded atoms. In metallic bonds, the outer electrons are not associated with a particular atom but are distributed through the crystal lattice. Because these three types of bonds are strong, crystals that are ionic, covalent, or metallic display better mechanical properties (such as hardness) and have higher melting points than crystals made up of weaker bonds. Many common inorganic materials (such as sodium chloride) form ionic crystals. Obviously the definition of a metal indicates that it will form metallic crystals. An example of a covalent crystal is diamond.

Organic molecules crystallize into crystals that are known as *molecular crystals*, in which the molecules are held together by weak attractive forces called van der Waals forces; these can be the result of dipole–dipole interactions (if the molecule has a constant dipole) or the result of induced dipoles caused by molecular interactions. The major component of the intermolecular forces is the dispersion interactions between neutral atoms (or molecules) caused by rapidly formed temporary dipoles that result from the movement of electrons of neighboring atoms. A final type of weak bonding is formed between hydrogen atoms that are associated with atoms of nitrogen, oxygen, fluorine, chlorine, or sulfur. More information on chemical bonding can be found in Vainshtein, Fridkin, and Indenbom (1982) and Kitaigorodsky (1973).

When atoms form bonds with each other to form a crystal, the atoms are arranged at definite distances from one another. This distance depends on the type of bond and the atoms involved and is essentially independent of the crystal structure. The bond distance corresponds to a minimum in the potential-energy curve for the two atoms. This is illustrated in Fig. 2.18. It is also possible to assign sizes (radii) to individual atoms in crystals, depending on the type of bond they have, so that the sum of these values is equal to the distance between pairs

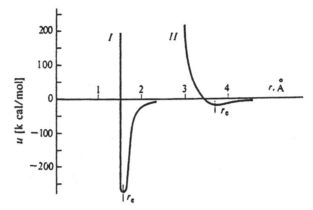

Fig. 2.18. The potential energy of interaction. I. A covalent, ionic and metallic bond. II. A van der Waals bond. The equilibrium distance between the atoms is r_c. (Data from Vainshtein et al. (1982).)

of molecules. These radii are known as the crystallochemical radii. Experimental work over many years has shown that, once determined for a particular type of atom and bond, these radii are a fixed property and can be used as a predictive tool in determining crystal structures.

2.1.8. X-Ray Diffraction

A major tool used to identify crystals and to determine crystal structure is *x-ray diffraction*. As we have seen in the previous sections, crystals are made up of atoms in a regular repeating pattern with a regular spacing. When x-rays of wavelengths of 0.5–2.5 Å are directed at a crystalline solid, the result is an observable diffraction pattern that occurs because the distances between atoms in a crystal are of a length similar to the x-ray wavelength, resulting in coherent scattering of the x-rays producing lines on a photographic film. The relationship between wavelength of the x-rays and the spacing between atoms in a crystal is known as Bragg's law:

$$\lambda = 2d \sin \theta, \tag{2.1}$$

where λ is the wavelength of the incident x-rays, d is the interplanar spacing in the crystal, and θ the angle of the incident x-rays on the crystal.

Bragg's law shows that, if x-rays of known wavelength are used and the incident angle of the radiation is measured, determination of the interplanar spacing of a crystal is possible. This is the foundation of x-ray diffraction methods that are used to analyze or determine the structure of crystals. Several different experimental methods that make use of x-ray diffraction and Bragg's law have been developed and used, depending on the type of sample that is available and the information desired.

The most commonly used x-ray method is *powder diffraction*, in which a fine powder of the crystal is used. The powder method relies on the fact that the array of tiny crystals randomly arranged will present all possible lattice planes for reflection of the incident monochromatic x-ray beam. In x-ray powder photography, the whole diffraction pattern is recorded simultaneously on a photographic film. In x-ray powder diffractometry, the diffraction pattern is counted by a device that plots the intensity against the Bragg angle. The powder pattern obtained for a particular substance acts as a signature for that substance so that a powder diffraction pattern can be used for identification, chemical analysis, and determination of degree of crystallinity. The powder x-ray patterns of more than 30,000 materials can be found in a reference known as the Powder Diffraction File (Joint Committee on Powder Diffraction Standards 1990) arranged so as to aid in searches based on measured patterns. In addition, other databases of complete crystal structures exist for both inorganic and organic compounds (Cambridge Database). Computer programs exist that allow computation of the powder pattern from a known crystal structure and thus powder patterns for virtually any material with a known crystal structure can be obtained. More information on powder diffraction methods can be found in Cullity (1978), Bunn (1961), and McKie and McKie (1986).

When a material does not have a known crystal structure, it must be determined employing a method which is commonly called single-crystal x-ray diffraction. A single crystal of good quality, measuring at least 1 mm in the smallest dimension, must be prepared. The crystal is then mounted with one of its axes normal to a monochromatic beam of x-rays and rotated in a particular direction. The crystal is surrounded by cylindrical film with the axis of the film being the same as the axis of rotation of the crystal. By repeating this process of rotation in a number of directions, the rotating-crystal method can be used to determine an unknown crystal structure. This structure determination includes determination of the lattice type, lattice parameters, space group, and atomic positions.

In the *Laue method* a single crystal is also used and is oriented at a fixed angle to the x-ray beam. The beam, however, contains the entire spectrum produced by the x-ray tube so that the angle θ in Bragg's law is fixed, but a variety of wavelengths are impinging on the crystal. Each set of planes that will satisfy Bragg's law with a particular wavelength will form a pattern known as a *Laue pattern*. The Laue method is used as a way to determine crystal orientation and to determine crystal quality. More information on the Laue method and other x-ray methods can be found in several references (Cullity 1978; Bunn 1961; Bertin 1975). X-ray diffraction methods are central to the analysis and understanding of crystal structures, as well as an important analytical tool for the identification of crystals of pure materials, mixed crystals, and semicrystalline materials.

2.1.9. Polymorphism

The crystal structure of a material determined by x-ray diffraction gives a complete picture of the arrangement of the atoms (or molecules) of the chemical species in the crystalline state. It is possible, however, for a given chemical species to have the ability to crystallize into more than one distinct structure. This ability is called *polymorphism* (or *allotropism* if the species is an element). Different polymorphs of the same material can display significant changes in their properties as well as in their structure. A dramatic example is carbon, which can crystallize as graphite or as diamond. Diamond is a cubic crystal, whereas graphite is hexagonal. In addition, properties such as hardness, density, shape, vapor pressure, dissolution rate, and electrical conductivity are all quite different for these two solids. These major differences in the properties of two polymorphs are not unique to carbon and can occur in all materials that display polymorphism.

Polymorphism is quite common in the elements and in inorganic and organic species. Many of the early identifications of polymorphs were minerals, such as calcium carbonate, which has three polymorphs (calcite, aragonite, and vaterite) and zinc sulfide, which has three polymorphs (wurtzite, sphalerite, and matraite). Some well known species have large numbers of polymorphs, for example water, which has eight different solid forms of ice. Organic molecular crystals often have multiple polymorphs that can be of great significance in the pharmaceutical, dye, and explosives industries. Polymorphic forms of a number of common materials are given in Table 2.3. The large number of materials that have polymorphs led to the statement of McCrone (1965) that "every compound has different polymorphic forms and that, in general, the number of forms known for a given compound is proportional to the time and money spent in research on that compound."

Under a given set of conditions, one polymorph is the thermodynamically stable form. This does not mean, however, that other polymorphs cannot exist or form at these conditions, only that one polymorph is stable and other polymorphs present can transform to the stable form. That is because the rate of transition of a polymorph depends on the type of structural changes that are involved. An example of this can be seen in heating (or cooling) a crystalline material with multiple polymorphs. As the temperature changes, the material will eventually enter a region where another polymorph is the stable form. The transformation of one polymorph to another, however, will occur at some rate that may be rapid or very slow. Transformations can be categorized by the types of structural changes involved, which can roughly be related to the rate of transformation (McCrone, 1965). For example, a transformation in which the lattice network is bent but not broken can be rapid. This type of transformation is known as *displacive transformation of secondary coordination*. Another type of rapid transformation can involve the breakage of weaker bonds in the crystal structure with the stronger bonds remaining in place. This is then followed by the rotation of parts of the molecule about the structure and the formation of new bonds. This type of transformation is a *rotational disorder transformation*. Slow transformations usually involve the breakage of the lattice network and major changes in the structure or type of bonding. The diamond-to-graphite transition is an example in which the nature of the bond type changes from metallic to

Table 2.3. *Polymorphic forms of some common substances; data from Verma and Krishna (1966)*

Element or compound	Chemical composition	Known polymorphic forms[a]
Cesium chloride	CsCl	Cubic (CsCl type) (*s*), $d = 3.64$ Cubic (NaCl type) (*m*), $d = 3.64$
Calcium carbonate	$CaCO_3$	Calcite (*s*), rhombohedral, uniaxial, $d = 2.71$ Aragonite (*m*), orthorhombic, biaxial, $d = 2.94$
Carbon	C	Diamond (*m*), cubic, very hard, $d = 3.5$, covalent tetrahedral binding, poor conductor Graphite (*s*), Hexagonal, soft, $d = 2.2$ layer structure, good conductor
Iron	Fe	γ-iron (*m*), face-centered cubic α-iron (*s*) and δ-iron (*m*), body-centered cubic
Mercuric iodide	HgI_2	Red (*s*), tetragonal Yellow (*m*), orthorhombic
Phosphorus	P_4	White phosphorus (*m*), $d = 1.8$, melts 44°C Violet phosphorus (*s*), $d = 2.35$; melts around 600°C
Silica	SiO_2	Quartz (*s*) (α and β forms), $d = 2.655$ Tridymite (*m*) (α and β forms), $d = 2.27$ Cristobalite (*m*) (α and β forms), $d = 2.30$
Sulfur	S	α, orthorhombic (*s*), $d = 2.05$, melts 113°C β, monoclinic (*m*), $d = 1.93$, melts 120°C
Tin	Sn	White tin, tetragonal, $d = 7.286$, stable above 180°C Gray tin, cubic (diamond type), $d = 5.80$, metastable above 180°C
Zinc sulfide	SnS	Wurtzite (*m*), hexagonal Sphalerite (*s*), cubic (diamond type)

[a] d = density (g/cm^3); *m* = metastable and *s* = stable.

Table 2.4. *Transformations of silica; data from McCrone (1965)*

Reconstructive	Displacive
Cristobalite	High cristobalite
	↕ 200–270°C
1470°C	Low cristobalite
	High tridymite
	↕ 160°C
Tridymite	Middle tridymite
	↕ 105°C
867°C	Low tridymite
	High quartz
Quartz	↕ 573°C
	Low quartz

covalent. Polymorphic transformations can also be classified as first- or second-order transitions. In a first-order transition, the free energy of the two forms become equal at a definite transition temperature and the physical properties of the crystal undergo significant changes upon transition. In a second-order transition, there is a relatively small change in the crystal lattice and the two polymorphic forms will be similar. There is no abrupt transition point in a second-order transition, although the heat capacity rises to a maximum at a second-order transition point. Examples of phase transitions as a function of temperature for silicon dioxide, which has seven crystals forms, are shown in Table 2.4.

When a material is crystallized from solution, the transition between polymorphs can occur at a much higher rate because the transition is mediated by the solution phase. Polymorphs of a given material will have different solubilities at a given temperature, with the more stable material having a lower solubility (and a higher melting point) than the less stable polymorph. If two polymorphs are in a saturated solution, the less stable polymorph will dissolve and the more stable polymorph will grow until the transition is complete. The rate of the transition is a function of the difference in the solubility of the two forms and the overall degree of solubility of the compounds in solution. This transition, of course, requires that some amount of the stable polymorph be present. This means that the stable polymorph must nucleate at least one crystal for the transition to begin. If a slurry of solution and crystals of a polymorph stable at a high temperature is cooled to a lower temperature, where another polymorph is the stable phase, the transition of the crystals already present will depend on the

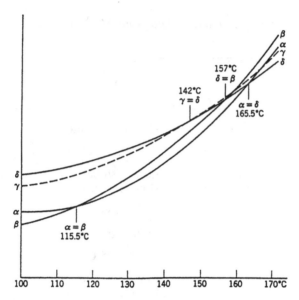

Fig. 2.19. Relative solubility curves for the four polymorphic forms of HMX (Data from McCrone (1965).)

presence of nuclei of the new stable phase. The more of these nuclei present, the faster the transition will occur.

It is often possible to crystallize a metastable polymorph by applying a large supersaturation (for example rapid cooling) so that crystals of the metastable form appear before crystals of the stable form. When this occurs, if the solid is removed from the solution rapidly and dried, it is possible to obtain samples of a metastable polymorph that will not easily transform to the stable phase unless heated. If the crystals of the metastable polymorph are left in solution for any length of time, they will likely transform to the stable form by going through the solution phase. If seeds of the stable polymorph are added to the solution, the transition will occur more rapidly. At first-order phase-transition temperatures, the solubility of the two forms will be equal and both can exist. An example of solubility curves for a material with multiple polymorphs, showing the transition points is shown in Fig. 2.19.

When a material that displays polymorphism is crystallized, a metastable phase often appears first and then transforms into a stable form. This observation is summarized by Ostwald's step rule, which is also known as the *Law of Successive Reactions*. This law states that, in any process, the state that is initially obtained is not the stablest state but the least stable state that is closest in terms of free-energy change, to the original state. In a crystallization process, therefore, it is possible to envision the phase transformation first occurring into the least stable polymorph (or even an amorphous phase), which transforms through a series of stages to successively more stable forms until the equilibrium form is obtained. While Ostwald's law has been observed in a wide variety of systems, it is most likely to be seen in organic molecular crystals.

Fig. 2.20. Schematic diagram of two possible arrangements of molecules of the same chemical composition, but different conformation in the crystal. The broken line represents a unit cell.

Another interesting feature of organic molecular crystals is that the molecular conformation of a species can be different in two polymorphs of the same material. By molecular conformation, we are referring to the shape of the molecule. The same molecule can display different shapes (conformations) by rotations about single bonds for example. *Conformational polymorphism* is the existence of polymorphs of the same substance in which the molecules present are in different conformations. An example of this is shown in Fig. 2.20. A detailed discussion of this topic can be found in Bernstein (1987).

2.1.10. *Enantiomorphism*

The ability of a crystal to rotate the plane of polarized light passing through it in certain directions is known as *optical activity*. Optical activity in crystals depends on the point-group symmetry and can only be present in crystals that have no inversion axis of symmetry and no mirror plane. There are eleven crystal classes that meet this requirement. Crystals that fall in these classes are called *enantiomorphous*. In addition to these eleven classes, crystals of four other noncentrosymmetric classes are theoretically capable of optical activity. These fifteen crystal classes are listed in Table 2.5. While many crystals are theoretically

Table 2.5. *Enantiomorphous and optically active point groups*

Crystal system	Enantiomorphous classes	Optically active classes
Triclinic	1	1
Monoclinic	2	2, *m*
Orthorhombic	222	222, *mm*2
Trigonal	3, 32	3, 32
Tetragonal	4, 422	4, 4, 422, 42*m*
Hexagonal	6, 622	6, 622
Cubic	23, 432	23, 432

Table 2.6. *Examples of differences in biological activity of enantiomers; data from Leusen (1993)*

Compound name	Effect of S or S,S enantiomer[a]	Effect of R or R,R enantiomer
Asparagine	(S) tastes bitter	(R) tastes sweet
Chloramphenicol	(S,S) inactive	(R,R) antibacterial
Acylalanine analog CGA-29212	(S) herbicide	(R) fungicide
Dopa	(S) anti-Parkinsonism	(R) toxic
Ethambutol	(S,S) tuberculostatic	(R,R) causes blindness
Limonene	(S) lemon smell	(R) orange smell
Paclobutrazol	(S,S) plant growth regulator	(R,R) fungicide
Penicillamine	(S) against metal poisoning, benefits in cystinuria, and relieves rheumatoid arthritis	(R) very toxic
Propranolol	(S) β-blocker	(R) contraceptive
Thalidomide	(S) teratogenic	(R) cures morning sickness

[a]Absolute configuration is given in parentheses.

optically active, they often are not so to an observable degree. Observation of optical activity usually requires large, well formed crystals.

Certain molecules of a given substance can exist as mirror image-isomers of each other. Such molecules are often called *chiral* and also are referred to as enantiomers. Many substances display this property and it is particularly important in pharmaceutical, agricultural, and food-related applications because usually only one enantiomer is biologically active. Table 2.6 gives examples of cases where the biological activity of enantiomers is important. Mixtures of molecules which are enantiomers are referred to as racemic mixtures. These molecules can crystallize into mirror image crystals of each other and an example of this is shown in Figs. 2.21 and 2.22. These optically active crystals are called D (dextro) and L (levo), indicating the direction of rotation of the plane of polarized light (right for D and left for L).

The separation of racemic mixtures into pure enantiomers is an important practical problem, which has been of great interest for many years. Pasteur is recognized as doing the first separation of enantiomers from a racemic mixture in 1848. He was able to separate D and L sodium ammonium tartrate tetrahydrate by crystallizing the compound at a temperature below 27°C, where pure D and pure L crystals form with recognizably different shapes and therefore could be separated by hand. This is a method that will not work for most compounds. When a racemic mixture is crystallized, it often forms crystals that are composed

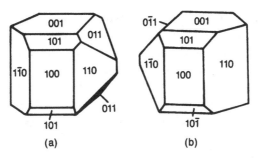

Fig. 2.21. The D and L crystals of tartaric acid.

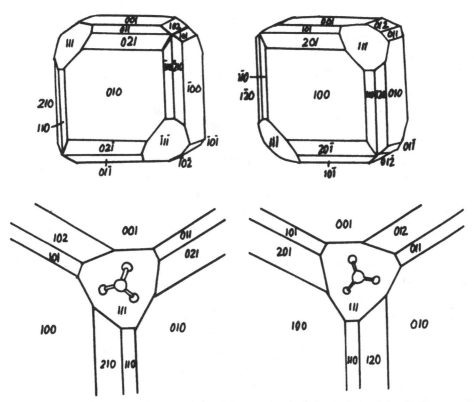

Fig. 2.22. L and D crystals of sodium chlorate (top). Orientation of the ClO_3 groups on (111) faces (bottom). (Reprinted from *Chemical Crystallography* by C.W. Bunn (2nd ed., OUP, 1961) by permission of Oxford University Press.)

of both D and L molecules in the crystal lattice (racemic crystals) or mixtures of individual D and L crystals (conglomerates), which cannot be distinguished by shape or size. There are four main techniques for the separation of racemic mixtures (Leusen 1993). The first two methods involve the use of a chemical that interacts specifically with one racemate, either through the rate of chemical reaction (kinetic resolution) or by absorption (chromatographic separation). The other two methods involve crystallization. In direct crystallization, one racemate is selectively crystallized from a racemic mixture in solution. The simplest situation is if the crystals are distinguishable by shape and can be separated on that basis. Other methods involve the seeding of the solution with one enantiomer or the use of chiral solvents or additives to aid in the selective crystallization of one enantiomer. The final method involves using an agent that will form a complex with each of the enantiomers, the two complexes having different solubilities, so that one complex can be selectively crystallized. This method is called resolution by diasteromeric salts.

2.1.11. *Isomorphism*

It is possible for different chemical species to form crystals in which the atomic arrangements are identical. For example, ammonium sulfate and potassium sulfate crystallize into identical structures as do sodium chloride and potassium chloride. The unit-cell dimensions of these crystals are very close to each other. In addition, ammonium and potassium ions are of similar size and chemical nature. Crystals with these properties are called *isomorphous*. Since the atomic structures of the ammonium and potassium sulfate are identical and the ions are of similar size, ammonium and potassium can replace each other in the crystal lattice forming a mixed crystal, which can have any proportion of the two materials. Such mixed crystals have unit-cell dimensions that are intermediate between those of the two pure components. These materials are sometimes referred to as *crystalline solid solutions*.

Not all isomorphous substances form mixed crystals. Calcite (calcium carbonate) and sodium nitrate have similar atomic arrangements, unit cells, and ions that are close in size, but they will not form a mixed crystal, probably because of the large difference in solubility of the two materials in water. Another method, however, in which mixtures of these two crystals can occur is by oriented overgrowth. Sodium nitrate will grow on the surface of a calcite crystal in a parallel orientation. It is possible for oriented overgrowth to occur in systems that are not isomorphous if the arrangement of atoms on a particular crystal face is similar in type and dimensions to the arrangement of atoms of a plane of the other material, even if the overall crystal structures are not similar. In this situation, one face acts as a template for the overgrowth of the other substance on a particular crystal face.

When different conformers of a molecule exist in the same crystal structure, the arrangement is known as *conformational isomorphism*. This is illustrated in Fig. 2.23. The figure shows that the conformers have a regular arrangement relative to each other. If the conformers are randomly arranged in the crystal lattice, the arrangement is known as *conformational synmorphism*.

2.1.12. *Twinning*

Certain substances commonly form crystals that are composed of two or more crystals joined together in a particular mutual orientation, which is constant for

Fig. 2.23. Schematic diagram of conformation isomorphism (left) and conformation synmorphism (right). The broken line represents a unit cell. (Data from Bernstein (1987).)

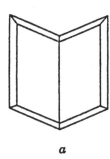 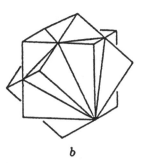

a b

Fig. 2.24. Examples of twinning. (*a*) Two individual crystals of gypsum, $CaSO_4.2H_2O$, joined at the (100) plane. (*b*) Interpenetration twin of fluorspar, CaF_2, with one crystal rotated 60° with respect to the other. (Reprinted from *Chemical Crystallography* by C.W. Bunn (2nd ed., OUP, 1961) by permission of Oxford University Press.)

that particular species. These types of crystals are called *twins* and their formation is known as *twinning*. Many substances commonly form twins, including gypsum (calcium sulfate dihydrate), calcite (calcium carbonate), copper, diamond, and fluorite (CaF_2).

There are three main classes of twin crystals, growth twins, deformation (or glide) twins, and transformation (or inversion) twins (McKie and McKie 1986). These names refer to the symmetry operation necessary to bring the lattice of one of the twin crystals into coincidence with that of the other twin crystal. The most common types of twin crystals have symmetry by rotation about an axis known as the twin axis or reflection in a plane known as the twin plane. In a rotation twin the rotation is usually through 180° (though other rotations are possible). A contact twin is a twin crystal in which the two component crystals are joined at a well defined plane. An example of this is shown in Figure 2.24 for gypsum twins that are joined at the (100) plane. Figure 2.24 also shows a different type of twin known as an interpenetrant twin, in which the interface between the component crystals is irregular and the two crystals can overgrow into each other. It is also possible for multiple twins to form in which the twin contains three or more component crystals; examples of such mimetic twins are shown in Fig. 2.25.

Another type of crystal formation is not a single crystal or twin, but an aggregate of many crystals radiating from a point. This type of aggregate crystal is called a spherulite. These occur in a wide variety of systems and are very common in polymer crystallization.

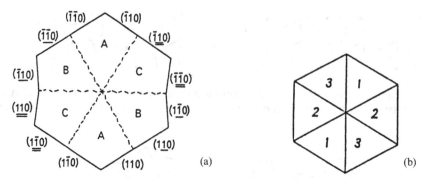

Fig. 2.25. Mimetic twinning: (*a*) Aragonite twins on (110) to produce interpenetrant triplets. (*b*) Ammonium sulfate. (Reproduced with permission from McKie and McKie (1986).)

2.1.13. *Crystal Lattice Defects*

In our discussion to this point, we have presented crystals as solids having a regular repeating structure with each atom in a precise location. Real crystals, however, even those grown slowly and carefully, will be very close to the ideal structure but will have deviations from the structure; these deviations are called defects. There are a number of different types of imperfections that can occur. If a foreign atom (or molecule) is present in the crystal lattice, this is known as chemical imperfection. The foreign atom can be present at a lattice site, having substituted for an atom in the structure, or it can be present by fitting between the atoms in the lattice. These are known as *substitutional impurities* and *interstitial impurities* respectively. Because these impurities do not fit exactly into the lattice structure, they will cause a displacement of neighboring atoms and cause a strain in the crystal. Another type of imperfection is due to vacancies in the crystal. A vacancy is a lattice point at which there is no atom.

The defects discussed so far are all examples of *point defects* as they involve only a single atom (or molecule) in the crystal and therefore occur at a point. Defects in one dimension are known as linear defects or dislocations. There are two types of dislocations, known as *edge dislocations* and *screw dislocations*. These types of dislocations disturb the long-range order in a crystal, distorting its entire structure. An example of an edge dislocation is shown in Fig. 2.26, which is a cross section. In the figure, half of a vertical row (the bottom half) in the middle of the lattice is missing in each plane of the lattice parallel to the page. Point A in the figure marks the dislocation. A line perpendicular to the page going through the crystal at point A represents the dislocation line. The figure shows that the lattice points are displaced in the region of the defect and that the displacement gets smaller as the distance from the dislocation increases. An example of a screw dislocation is shown in Fig. 2.27. The screw dislocation is very important to the growth of crystals and will be discussed in more detail in Section 2.2.4.

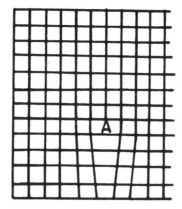

Fig. 2.26. An edge dislocation shown in two dimensions.

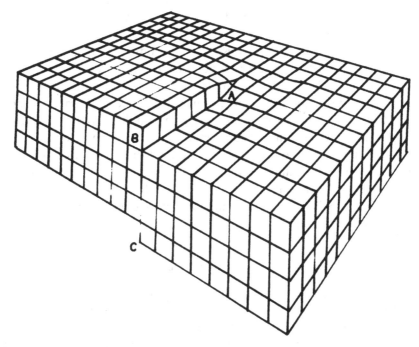

Fig. 2.27. A screw dislocation in a cubic crystal. The screw dislocation AD is parallel to BC (D is not visible).

2.2. Crystallization

Crystallization is a phase change that results in the formation of a crystalline solid. The most common type of crystallization is crystallization from solution, in which a material that is a crystalline solid at a temperature of interest is dissolved in a solvent. Crystallization is induced by changing the state of the system in some way that reduces the solubility of the crystallizing species. Crystallization from the melt refers to a crystallization in which the material to be crystallized is a solid at normal conditions and is heated until it melts and forms a liquid. The material is then cooled slowly until it crystallizes. The major difference between a melt and a solution is that in a melt there is no solvent. In both cases, however, there may be impurities in the system, the number of which will be reduced by the crystallization operation. Crystallization can also occur from the vapor phase. A material that has a significant vapor pressure is evaporated and then can be induced to crystallize if the state of the system is changed by lowering the equilibrium vapor pressure of the material.

Crystallization is used as a separation and purification technique for the production of a wide variety of materials. Crystallization from solution is commonly used in the chemical, petrochemical, pharmaceutical, and food industries as a final purification step. Melt crystallization is the method used to produce silicon wafers in the semiconductor industry and is also used to purify certain low-melting organic chemicals in industrial settings. As crystallization from the solution is by far the most common method used industrially, it is the method that will be discussed in detail.

The goal of all crystallization processes is to produce a crystal that meets certain standards of purity. However, in addition to purity a number of other properties of the crystals are also of importance. These properties include the average size and size distribution of the product, the shape of the product crystals, the control of polymorphism (if the system has multiple polymorphs), the ability to filter the material from the solution, the strength of the product crystal (does it break easily), and the properties of the final dry product crystals. The goal of this section is to introduce a number of basic and important concepts that are fundamental to crystallization from solution. More details on the topics mentioned above can be found in a number of references (Myerson 1993; Mullin 1993; Randolph and Larson 1988).

2.2.1. Solubility and Supersaturation

Solutions are made up of two or more components, of which one is known as the *solvent* and the others the *solute(s)*. When the solutes are solids, a solution is formed by adding these solid solutes to the liquid solvent, in which the solid dissolves. At a given temperature there is a maximum amount of solute that can dissolve in a given amount of solute. When this occurs, the solution is said to be *saturated*. The amount of solute required to make a saturated solution at a given temperature is called the *solubility*. The solubility of most materials is a function of temperature, generally increasing with increasing temperature (although there are some exceptions). The solubility of many inorganic materials in water is known with a fair degree of accuracy and can be found in the literature (Linke and Seidell 1958; Mullin 1993). The solubility of complex mixtures of electrolytes in water can also be determined with a good accuracy by using calculational techniques (Zemaitas et al. 1986). The solubility of materials in nonaqueous solvents is generally known with a great deal less accuracy. Few data are available for many substances in nonaqueous solvents and methods for the calculation of

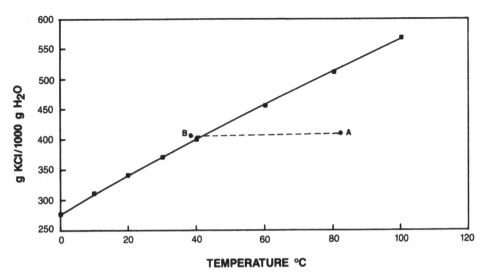

Fig. 2.28. Solubility of potassium chloride in aqueous solution. (Data from Linke and Seidell (1958, 1965).)

solubility of materials in these solvents have not yet achieved the degree of accuracy required to replace measurement. However, it is likely that computational methods similar to those discussed in the later chapters of this volume will allow calculation of solubilities in nonaqueous systems in the near future.

2.2.2. *Supersaturation*

When a solution of a solid solute dissolved in a liquid solvent is saturated, it is in thermodynamic equilibrium. In order for crystallization to occur, the state of the system must be shifted to a nonequilibrium state in which the concentration of the solute in the solution exceeds its equilibrium (saturated) concentration at the given solution conditions. Solutions that are in this nonequilibrium state are said to be *supersaturated*. The simplest method to create a supersaturated solution is by cooling. This is illustrated in Fig. 2.28. A solution is initially prepared at point A. If this solution is cooled it will be saturated when it intersects the saturation line. If it is cooled past the saturation line to point B it will be supersaturated. Just because this solution is supersaturated, however, does not mean that it will immediately crystallize. Supersaturated solutions are *metastable*. This means that there is a free energy barrier which must be overcome for the phase transition to be overcome. The thermodynamics of metastability and its relation to crystallization will be discussed in detail in Section 2.2.3.

The simplest and most common method of making a supersaturated solution is by cooling, but this is not the only method that can be used; there are five main methods:

1. Temperature change
2. Solvent evaporation

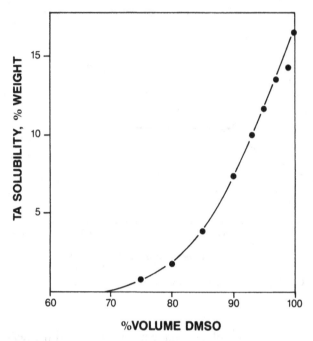

Fig. 2.29. Solubility of terephthalic acid in water–DMSO mixtures at 25°C. (Data from Saska (1984).)

3. Chemical reaction
4. Change in pH
5. Alteration in solvent composition.

As mentioned previously, temperature change is by far the most common method employed and, since most materials have an increase in solubility with increasing temperature, this temperature change is usually cooling. Solvent evaporation results in an increase in solute concentration as solvent is removed until the solution is supersaturated. Chemical reaction involves the addition of soluble reagents to the solution; these react and form a product of lower solubility and the process is often referred to as *precipitation*. Solubility can also be changed by changing the pH of the system or by altering the solvent composition. Altering the solvent composition usually involves the addition of a miscible solvent in which the solute has limited solubility. This is illustrated in Fig. 2.29, which shows the solubility of terephthalic acid (TPA) in DMSO–water mixtures at 25°C. TPA is very soluble in pure dimethylsulfoxide (DMSO) and essentially insoluble in water at this temperature. The addition of 30 percent water (by volume) reduces the TPA solubility from 16.5 wt percent to essentially zero.

Supersaturation is the fundamental driving force for crystallization and can be expressed in dimensionless form as:

$$\mu - \mu^*/RT = \ln a/a^* = \ln(\gamma c/\gamma^* c^*), \tag{2.2}$$

where μ is the chemical potential, c is the concentration, a is the activity, γ is the activity coefficient and * represents the property at saturation. In most situations, the activity coefficients are not known and the dimensionless chemical potential difference is approximated by a dimensionless concentration difference:

$$\sigma = (c - c^*)/c^*. \tag{2.3}$$

This substitution is only accurate when $\gamma/\gamma^* = 1$ or $\sigma \ll 1$ so that $\ln(\sigma + 1) = \sigma$. It has been shown that this is generally a poor approximation at $\sigma > 0.1$ (Kim and Myerson 1996), but it is still normally used because the needed thermodynamic data are usually unavailable. Supersaturation is also often expressed as a concentration difference, $\Delta c = c - c^*$ and as a ratio of concentrations $S = c/c^*$.

2.2.3. *Nucleation*

Crystallization from solution involves two distinct steps, *nucleation*, which is the birth of new crystals, and *crystal growth*, which is the growth of existing crystals to larger sizes. The properties of the crystals obtained (such as their size distribution and shape) depend on these two properties and their relationship to each other. As was discussed in Section 2.2.2, supersaturated solutions are metastable; that is, there is a free-energy barrier that must be passed in order for a solution to nucleate. Classical nucleation theory says that, when a solution enters the non-equilibrium supersaturated regions, the molecules of the solute begin to form aggregates (clusters), which are also referred to as concentration fluctuations. If it is assumed that the aggregates are spherical, an equation can be written that gives the change in Gibbs free energy required to form a cluster of a given radius:

$$\Delta F = 4\pi r^2 \sigma - (4\pi r^3/3V_{\rm m})RT \ln(1 + S), \tag{2.4}$$

where r is the cluster radius, σ is the solid–liquid interfacial tension, $V_{\rm m}$ is the specific volume of a solute molecule, and S is the supersaturation ratio.

The first term in equation (2.4) is the change in Gibbs free energy for forming the surface of the cluster, while the second term is the change in Gibbs free energy for forming the cluster volume. A plot of G as a function of cluster size is shown in Fig. 2.30. The plot shows that at smaller cluster sizes the slope of the free-energy curve is positive, meaning that growth of the cluster requires a positive change in the Gibbs free energy. Since all spontaneous processes require a negative change in the Gibbs free energy, this cluster will be likely to dissolve. Looking at larger sizes, we see that the slope of the curve is negative (so that the change in Gibbs free energy is negative), indicating that thermodynamically the clusters will want to grow. The point on the curve where the slope is horizontal is known as the critical size. Nucleation requires that a cluster of critical size forms in the solution. An equation for the critical size can be obtained by taking the derivative of equation (2.4) with respect to r and setting it equal to zero (since the derivative is zero at the critical size). This yields the expression,

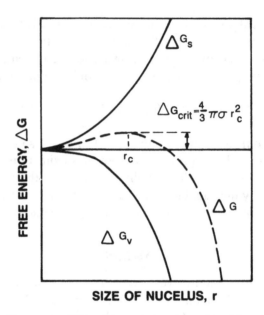

Fig. 2.30. Change in Gibbs free energy plotted against cluster radius.

$$r_c = 3V_m\sigma/RT\ln(1 + S). \tag{2.5}$$

Equation (2.5) shows that the critical size decreases as the supersaturation increases, thus increasing the likelihood that nucleation will occur.

Supersaturated solutions are metastable and the region in which nucleation and crystal growth occur is called the metastable region. This is a thermodynamically bounded region that is defined on one side by the solubility curve and on the other side by a curve known as the spinodal curve. The spinodal curve is defined as the locus of points at which the derivative of the chemical potential with respect to concentration equals zero. This is illustrated in Fig. 2.31. The region to the right of the spinodal curve on the figure is known as the unstable region. Phase transitions in this region do not occur by nucleation and growth, but by a mechanism called *spinodal decomposition*. Spinodal decomposition has never been observed in solutions made up of a solid solute and a liquid solvent because of the large width of the metastable zone that must be crossed without nucleation being induced. It is simply not possible to create that degree of supersaturation without nucleation occurring. However, in polymer–polymer systems and melt systems (particularly metals) that exhibit a critical solution point, it is possible to induce spinodal decomposition. This is because the metastable zone becomes very narrow near a critical point (see Fig. 2.31) and it is possible to go directly from the stable to the unstable region. Spinodal decomposition produces very different types of structures in these systems, since phase separation is spontaneous everywhere in the system and no critical size exists. More information on spinodal decomposition can be found in Cahn (1968) and Gunton, San Miguel, and Sahani (1983).

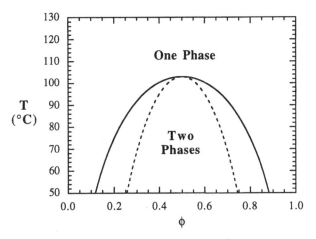

Fig. 2.31. Schematic phase diagram showing the binodal (equilibrium) curve (solid curve), the spinodal curve (dashed curve) and the critical point (where the curves meet).

The spinodal curve is rarely approached in practice because dust, dirt, and container walls usually aid in the nucleation process, such that an effective metastable zone exists for a solution of a given concentration, cooling rate, and vessel type and size. That is, a solution cooled at a given rate will eventually nucleate at a reproducible supersaturation. This effective metastable limit will increase with increasing cooling rate and will increase if the vessel is not stirred or if the volume is reduced.

Another property of supersaturated solutions that is related to the effective metastable zone width is the *nucleation induction time*. The nucleation induction time is the time required for a given supersaturated solution to nucleate if left at constant conditions. The higher the supersaturation, the shorter the nucleation induction time. This time can be thought of as the time required for a critical-sized cluster to form.

Nucleation of the type discussed in this section is known as primary nucleation and is driven by the thermodynamics of the phase-separation processes. Primary nucleation can be divided into homogeneous primary nucleation, which is the result of the formation of critical-sized clusters as described previously, and heterogeneous primary nucleation, which is nucleation in which the free-energy barrier (critical cluster size) has been lowered by the presence of dust, dirt, or the container walls. Secondary nucleation is nucleation that is due to the presence of existing crystals and to the interaction of these crystals with their environment, such as contact with the container walls or the stirrer and crystal collisions.

2.2.3.1. *Nucleation Theory*

The classical theory of nucleation (Volmer 1939; Nielsen 1964) was developed for the condensation of vapor droplets. This theory assumes that clusters are formed by an addition mechanism that continues until a critical-sized cluster is formed. The rate of nuclei formation based on this type of mechanism is given by:

$$B_0 = A \exp(-\Delta G_{cr}/kT), \tag{2.6}$$

where A is the preexponential factor and has a theoretical value of 10^{30} nuclei/ cm^3 s.

The change in Gibbs free energy required to form a nucleus can be written as the sum of the free-energy change required to form the surface of the cluster and the free-energy change required to form the area. This expression, in its most general form, is given below and makes no assumption about the shape of the cluster:

$$\Delta G = \Delta G_s + \Delta G_v = \beta L^2 \sigma + \alpha L^3 \Delta G_v, \tag{2.7}$$

where β and α are area and volume shape factors (based on the characteristic length L). These shape factors can be used to derive an expression for the free-energy change for a cluster of any shape. More information about shape factors can be found in Section 2.2.5. For spherical nuclei, $\beta = \pi$ and $\alpha = \pi/6$ (based on the diameter d of the nucleus) and the expression for the free-energy change becomes:

$$\Delta G = 4\pi r^2 \sigma + 4/3\pi r^3 \Delta G_v. \tag{2.8}$$

Minimizing equation (2.8) with respect to the cluster radius yields an expression for the critical size and the free-energy change at the critical size:

$$r_c = -2\sigma/\Delta G_v \tag{2.9}$$

$$\Delta G_{cr} = 4\pi r_c^2 \sigma/3. \tag{2.10}$$

The growth of clusters is governed by the Gibbs–Thomson equation:

$$\ln c/c^* = \ln S = 2\sigma v/kTr, \tag{2.11}$$

where c is the concentration of clusters of size r and v is the molecular volume. Thus, smaller clusters dissolve and larger clusters grow. Combining equations (2.10) and (2.11) at the critical size yields:

$$\Delta G_{cr} = (16\pi\sigma^3 v^2)/3(kT \ln S^2), \tag{2.12}$$

which can then be combined with equation (2.5) to give an expression for nucleation rate:

$$B_0 = A \exp[(-16\pi\sigma^3 v^2)/(3k^3 T^3 (\ln S^2))]. \tag{2.13}$$

This expression clearly indicates that the nucleation rate is a very strong function of supersaturation and, in fact, nucleation will not occur at any finite rate until a critical supersaturation is reached. Equation (2.13) is generally approximated as a power-law expression of the form:

$$B_0 = K_N S^n, \tag{2.14}$$

where K_N is an experimentally determined constant and n is the nucleation order.

Homogeneous nucleation is normally rarely observed in crystallization from solution except in cases of very high supersaturations obtained by chemical reaction or the addition of a miscible nonsolvent. In most solutions, nucleation occurs at much lower supersaturation than can be accounted for by equation (2.13). This is due to the presence of dust, dirt, and container walls in the system. Nucleation that occurs on these surfaces is called *heterogeneous nucleation* and is generally the nucleation mechanism that occurs in solutions that do not contain crystals. The work of nucleation on a substrate is less than that for homogeneous nucleation in the bulk solution. The ability of the nucleation substance to wet the foreign substrate determines the energy barrier to nucleation. If, for example, the wetting angle is 45°, the barrier to nucleation on the surface is an order of magnitude lower than for nucleation in the bulk. This implies that to achieve a high supersaturation without nucleation occurring, the solution must be very clean and the container must be made of a material that is not wetted by the crystallizing species.

Another type of nucleation is known as *secondary nucleation* and is nucleation due to the presence of crystals in the supersaturated solution. The existing crystals provide a source of nuclei through collisions with each other, the walls of the vessel, and the mixing device used. Secondary nucleation has been studied by a number of investigators, but the mechanisms and kinetics of the process remain poorly understood. More information on secondary nucleation can be found in several references (Mullin 1993; Myerson 1993; Randolph and Larson 1988).

In industrial crystallizers, most nuclei are generated by secondary nucleation and the nucleation rate is modeled by a modified version of the power law expression:

$$B = K_N W^i M_T^j S^n, \tag{2.15}$$

where M_T is the slurry density (mass of crystals per volume of solution) and W is the agitation rate, which can be expressed in rpm or as the tip speed of the impeller.

Additional information about nucleation and nucleation theory can be found in Chernov (1984).

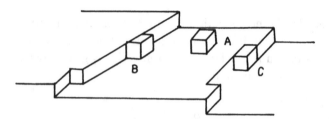

Fig. 2.32. Surface structure of a growing crystal.

2.2.4. *Crystal Growth*

When a nucleus is formed, it is the smallest sized crystalline entity that can exist under a given set of conditions. However, immediately after the formation of nuclei, they begin to grow larger through the addition of solute molecules to the crystal lattice. This part of the crystallization process is known as *crystal growth*. It is crystal growth, along with nucleation, that controls the final size distribution and shape of the crystals formed, as well as their purity and the type of defects present.

Factors that control the size and shape of crystals have long been of interest. Many of the early studies of crystals involved trying to understand the shape of minerals and of inorganic salts grown from solution in relation to the internal structure of the crystals and the thermodyamics of crystal growth. A discussion of crystal shape can be found in Section 2.2.5. Another area fundamental to crystal growth is the mechanism of the growth process and factors that control the rate of growth. Reviews of the theory of crystal growth can be found in a number of references (Ohara and Reid 1973; Strickland-Constable 1968; Nyvlt et al. 1985; Chernov 1984).

When the growth of a crystal is described, it is possible to discuss this growth in a macroscopic way by describing the overall change in some dimension of the crystal. This type of description gives no information about the mechanism of crystal growth, but is useful in describing the kinetics of the process and for use in developing models for the prediction of crystal size distribution. When crystal growth is viewed from a mechanistic point of view, it is necessary to look at the growth of a particular crystal face (plane). The crystal-growth process involves the incorporation of growth units into the crystal lattice. These growth units can be molecules, atoms, or ions, depending on the type of substance and type of crystal being grown. This growth unit must find its way to an appropriate site on the crystal surface where it can be incorporated into the crystal. The surface structure of a growing crystal is shown in Fig. 2.32. This figure shows three types of sites at which the growth unit can be incorporated into the crystal. The first type of site is a flat surface which is atomically smooth. If a growth unit (shown as A in the figure) attempts to become incorporated at this site, it can bond in only one place. Site B, which is known as a *step*, provides two places for the growth unit to bond to the crystal, while site C, which is known as a *kink* site, provides three places in which the growth unit can bond. From an energetic point of view, site C is the most favorable site followed by site B and then site A. This translates into a crystal growth mechanism by which molecules are absorbed on the surface and then diffuse along the surface until they are incorporated into the lattice at a step or kink site. The incorporation pattern indicates that the crystals grow by the spreading of steps. It does not provide any information on the source of these steps or on the rate-controlling factor in the crystal growth process. This type of information is the goal of crystal growth theory.

2.2.4.1. *Growth Theories for Perfect Crystals*

In the previous paragraphs we saw that crystals grow by the addition of molecules at kink sites and at steps. These steps will spread across the surface and complete a layer of growth on the face of interest, making the layer atomically flat. In order for this crystal face to continue growing a new layer must start. Starting a new layer requires *two-dimensional surface nucleation*, meaning that there is a Gibbs free-energy barrier to the formation of the new layer that must be overcome. Molecules will be continuously absorbed on the surface and then diffusing and being desorbed. The molecules will also collide with each other and form surface clusters. Applying nucleation theory in two dimensions results in an expression for the critical surface cluster size:

$$r_c = \sigma v_m / kt \ln S \tag{2.16}$$

where r_c is the radius of the critical cluster, σ the surface energy and v_m the volume of the molecule. In order to estimate the critical surface cluster size accurately, it is necessary to have an accurate value for the surface energy. Unfortunately, this is usually difficult with any degree of accuracy. The effect of variations in the estimated surface energy on the critical surface-cluster size is shown in Fig. 2.33, which also shows the effect of supersaturation on the critical radius.

A relationship that gives the surface nucleation rate can be derived using nucleation theory (Ohara and Reid 1973). A simplified version of the expression is:

$$I = C_1[\ln(S)^{1/2}] \exp[-C_2/T^2 \ln(S)] \tag{2.17}$$

$$C_1 = (2/\pi)n^2 v(v_m/h)^{1/2} \tag{2.18}$$

$$C_2 = \pi h \sigma^2 v_m / k^2, \tag{2.19}$$

where I is the two-dimensional nucleation rate, n is the equilibrium number of molecules in the solution, and v is the speed of surface absorbed molecules.

A number of theories of crystal growth that employ equation (2.17), along with an expression for the spreading velocity of the step have been developed.

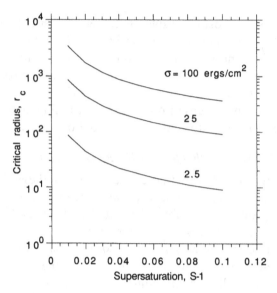

Fig. 2.33. Two-dimensional critical radii estimated from KCl at 300 K as a function of supersaturation. (Data from Ohara and Reid (1973).)

The result is a linear growth velocity for a crystal face as a function of temperature and supersaturation. These theories are generally known as *birth and spread* models and there a number of variations, which are described in several references (Ohara and Reid 1973; Myerson 1993). The major problem with all models of this type is that surface nucleation is the rate-determining step. A surface nucleus must form in order for crystal growth to occur. Since there is a free-energy barrier to the formation of a surface nucleus, which requires a significant supersaturation, these models all predict that crystals will not grow at low supersaturations. Unfortunately, this does not conform with experimental observations, in which crystals are seen to grow at finite rates at supersaturations of 1 percent or less. This observation was first explained by Frank (1949), who showed that, by considering the imperfections in crystals, it was possible to explain the growth of crystals without need of surface nucleation events.

2.2.4.2. *Growth of Imperfect Crystals*

In 1949 Frank put forward the idea that dislocations in crystals could be the source of new steps and allow crystals to grow continuously. Looking at a screw dislocation in Fig. 2.34, we see that molecules can be absorbed on the crystal surface and diffuse to the top step of the two planes of the screw dislocation. The surface then becomes a spiral staircase that is self-perpetuating. This theory was developed into a growth model by Burton, Cabrera, and Frank (1951) with surface diffusion assumed to be the rate-determining step; this is known as the BCF theory. Derivations of this theory can be found in several references (Nyvlt et al. 1985; Chernov 1984; Ohara and Reid 1973).

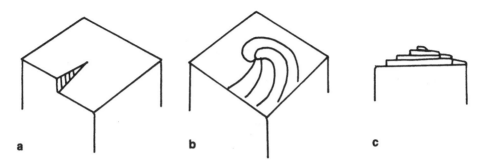

Fig. 2.34. Development of a growth spiral from a screw dislocation.

The derivation yields the following kinetic growth expression:

$$G = K_1 T(S - 1)\ln(S)\tanh[K_2/T\ln(S)]. \tag{2.20}$$

This equation reduces at low supersaturations to a form that states that crystal growth is proportional to supersaturation to the second power, while at higher supersaturations growth rate is a linear function of supersaturation. In addition, the equation states that growth rate is a continuously increasing function of supersaturation. The predictions of BCF theory match observations in many, if not most, cases.

The BCF model was derived for the case in which surface diffusion is the rate-limiting step in the crystal growth process. This is true in growth from the vapor phase. However, in solution growth, diffusion of the solute in the liquid phase can often be the rate-determining step. Chernov (1961) developed a model employing the screw dislocation of the BCF model with bulk diffusion in the liquid phase as the rate-limiting step. The resulting growth equation is:

$$G = (v_m \xi C^*(S - 1)^2 h)/S_{cr}\delta[1 + (\xi h/D)\ln\{S_{cr}\delta/((S - 1)h)\sinh((S - 1)/S_{cr})]\}, \tag{2.21}$$

where ξ is a coefficient that depends on kink density, δ the boundary layer thickness, and S_{cr} the critical transition supersaturation, which is a dimensionless group defined as:

$$S_{cr} = 4v_m\sigma/kT\delta. \tag{2.22}$$

When $S_{cr} \gg (S - 1)$, equation (2.21) reduces to:

$$G \propto (S - 1)^2/[1 + k\ln(\delta/h)], \tag{2.23}$$

where $k = \xi h/D$. The results predicted by equation (2.23) are similar to those predicted by the BCF model with the added condition that the growth rate is a function of the boundary-layer thickness, which is a function of the hydrody-

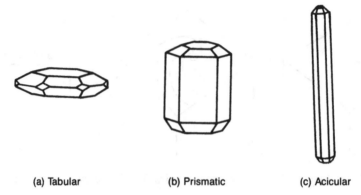

(a) Tabular   (b) Prismatic   (c) Acicular

Fig. 2.35. The external shapes of hexagonal crystals displaying the same faces.

namic conditions. Other more complex models based on the BCF theory, in which surface and bulk diffusion limitations are treated in series and in parallel, have also been developed (Gilmer, Ghez, and Cabrera 1971; Bennema 1969).

2.2.5. *The Shapes of Crystals*

As we have seen in previous sections, crystals are solids with the atoms, molecules, or ions in a regular repeating structure. In addition, these structures can be characterized by planes or faces which can grow. It is the crystal planes (or faces) that exist on a grown crystal, along with the relative areas of these faces, which determine the overall form of the crystal. This overall form is referred to as the *crystal habit, crystal shape* or *crystal morphology*. These terms refer to the external shape of the crystal and the planes present, without reference to the internal structure. The habit of a crystal refers to the overall shape of the crystal, without reference to the faces present; another term the *combination of crystallographic forms* (Hartman 1963) describes the faces exhibited by crystal. It is possible for crystals of the same substance to have the same combination of crystal forms (the same faces) and have very different habits. This is illustrated in Figs 2.35 and 2.36 for a hexagonal crystal and an orthorhombic crystal. In each case the three crystals display exactly the same faces, but their habits are different because of the variation of the relative areas of these faces on the grown crystal. It is possible for a crystal of the same material to have the same habit but a different combination of crystal faces. This is illustrated in Fig. 2.37, which shows two orthorhombic crystals that have the same habit but vary in the crystal faces present. This variation of crystal habit and crystallographic form is a result of the fact that crystals are the end result of a growth process that can occur by a number of different methods in different environments and at different rates. Crystals obtained experimentally are generally said to exhibit a crystallographic growth form and a growth habit. The growth form and habit are often compared to an idealized equilibrium form known as the *Wulff form*. The Wulff form is a minimum-energy form that satisfies the condition:

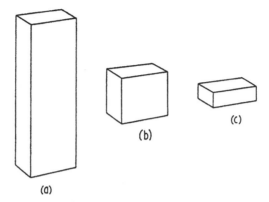

Fig. 2.36. Three orthorhombic crystals with the same faces but different habits.

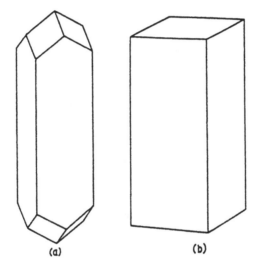

Fig. 2.37. Two orthorhombic crystals with the same habit, but a different combination of crystal faces. The crystal on the right has (110), (101) and (011) faces, whereas the crystal on the left has (100), (010) and (001) faces. (Data from Hartman (1963).)

$$\Sigma \sigma_i A_i = \text{minimum}, \tag{2.24}$$

where σ_i is the surface free energy of face I and A is its area.

The growth of crystals is a nonequilibrium process, so that the Wulff form is usually not produced. A crystal which is stored in a saturated solution for a long period of time will, however, transform into the equilibrium form with time through solvent-mediated dissolution and regrowth of the crystal faces. This is known as an aging process and has been reported by a number of investigators (Saska and Myerson 1987; Stranski and Honigmann 1950).

The characterization of a grown crystal consists of knowing the crystal faces present, the relative areas of these faces, the relative lengths of the three axes and

Table 2.7. *Shape factors for various geometries; data from Nyvlt et al. (1985)*

Body	α	β	$F = \beta/\alpha$
Sphere	0.524	3.142	6.00
Tetrahedron	0.182	2.309	12.7
Octahedron	0.471	3.464	7.35
Six-shaped prism	2.60	11.20	4.31
Cube	1.00	6.00	6.00
Prism	10.00	2.00	5.00
Platelet	0.20	2.80	14.00
Needle	10.00	42.00	4.20

the angles between the faces. Another common method of describing the shape makes use of a concept known as the *shape factor*. Shape factors are a simple way of describing the geometry of a crystal and relating that geometry to a characteristic dimension of the crystal. The *area* and *volume shape factors* are defined as:

$$V = \alpha L^3 \tag{2.25}$$

$$A = \beta L^2, \tag{2.26}$$

where L is a characteristic dimension of the crystal.

Shape factors for a number of common geometries are given in Table 2.7. Shape factors are employed in the analysis of crystal-size distribution data because measurement methods measure a characteristic dimension of a crystal and must be related to the actual crystal shape in order to produce an accurate representation of the distribution that can be used in modeling of industrial systems.

A number of different properties that can be of importance in different applications vary with the habit and form of the crystals. The dissolution rate of crystals will vary with the surface area of the crystal (which is a function of the habit), as well as with the faces present in the crystal. Just as different crystal faces can grow at different rates, so different crystal faces can also dissolve at different rates. Other properties of the dry crystalline powder depend on the habit and form. These include the bulk density (the volume mass that the crystals take up when randomly packed) and the flow properties of the solid. The rheological properties of a crystal–liquid suspension are also a function of crystal shape.

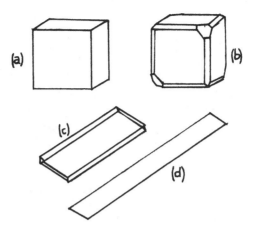

Fig. 2.38. Variation of crystal shape with conditions of growth. Sodium chlorate grown (*a*) rapidly and (*b*) slowly. Gypsum grown (*c*) slowly and (*d*) rapidly. (Reprinted from *Chemical Crystallography* by C.W. Bunn (2nd ed., OUP, 1961) by permission of Oxford University Press.)

Crystal habit and form can change dramatically as a function of the rate of crystal growth, the solvent used, and the impurities present in the system. Figure 2.38 shows the effect of growth rate on sodium chlorate and gypsum. In both cases the faces present in the crystals change, with fewer faces being present in the rapidly grown crystals. This can be explained by the fact that the smallest faces on a grown crystal are the fastest growing faces, while the largest faces are the slowest growing (since the growth rate of a face is the rate at which the face moves perpendicular to itself). The fast-growing faces grow so rapidly that they effectively disappear on the crystal. Impurities often change the habit and form of crystals through specific interactions with particular crystal faces, thus retarding the growth rate of these faces. This can result in faces appearing that are usually not present in crystals grown without the impurity or in modifications of the shape through changes in the relative areas of the faces. This can be illustrated by Fig. 2.35. If an impurity were specifically absorbed on the fast-growing faces of an acicular crystal (Fig. 2.35c), the growth rate of those faces would slow and the shape of the crystal might change to prismatic (Fig. 2.35b) or even tabular (Fig 2.35a). There are many additives that are used industrially to control the shape of crystals and also many references and reviews of the subject (Klug 1993; Buckley 1951).

Solvents can have an effect similar to that described for impurities. Changing solvents can therefore also result in a change in the crystal habit because the specific solvent–solute interactions will change with the change in solvent, as will specific absorption on crystal faces. This is illustrated in Fig. 2.39, which shows the result of taking a crystal that was grown in one solvent and continuing its growth in a different solvent. The role of the solvent in crystal growth and crystal shape can also be analyzed on thermodynamic grounds. This was done for melt growth by Jackson (1958), who defined a surface entropy factor that could be

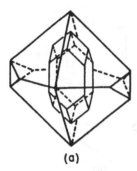 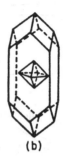

Fig. 2.39. (*a*) Crystal of anthranilic acid from aqueous solution, grown larger in ethyl alcohol solution. (*b*) Crystal from (*a*) transferred into acetic acid solution and allowed to grow. (Data from Hartman (1963).)

related to growth mechanism. This work was extended to solution growth (Davey, Mullin, and Whiting 1982; Davey, Milisavljevic, and Bourne 1988; Bourne 1980) and used to relate the likely growth mechanism to the solute employed. A review of these theories and their application can be found in Klug (1993).

2.2.6. *Crystal Structure and Crystal Habit*

The habit of grown crystals depends both on the structure of the crystal and on the conditions of growth. It has long been a problem of interest to relate the crystallographic structure of a material to its observed growth morphology. The first work in this area was done by Bravais in 1849, who said that the largest faces on a grown crystal should be those with the largest interplanar spacing. This was verified by Friedel in 1909 by the observation of mineral samples and is therefore known as the *Bravais–Friedel law*. This work was later extended by Donnay and Harker to take into account the influence of lattice centering, screw axes, and glide planes upon the spacing of lattice planes and is now generally referred to as the *Bravais–Friedel–Donnay–Harker law* (BFDH) and is quite useful in determining the important faces on a grown crystal or in making a first estimate of the crystal habit. Because the area of a plane on a grown crystal is roughly inverse to its linear growth velocity, the BFDH law is often stated as saying that the growth rate of a face is inversely proportional to the interplanar spacing of that face.

The BFDH law does not take into account the chemical nature of the crystals, that is, the nature of the bonding between atoms and molecules in the crystal. A theory that takes this bonding into account was developed by Hartman and Perdok (1955*a*, *b*, *c*) and is called *periodic bond chain theory*. This theory assumes that strong bonds should form more easily and faster than weaker bonds and that a crystal can only grow in a given direction when an uninterrupted chain of strong bonds that form during crystallization exists in the structure. This chain of strong bonds is called a *periodic bond chain* (PBC). This concept is illustrated

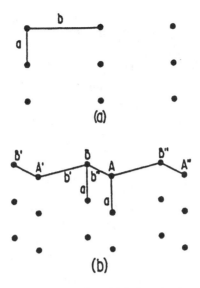

Fig. 2.40 (*a*) Two-dimensional crystal in which bond *a* is stronger than bond *b*. The crystal will be elongated in the direction of bond *a*. (*b*) Two-dimensional crystal in which bonds *a* and *b″* are of the same strength and bond *b′* is much weaker. The shape of the crystal is determined by bonds *a* and *b′*. The crystal will be elongated in the direction of bond *a*. (Data from Hartman (1963).)

in two dimensions in Fig. 2.40. When applied in three dimensions, the PBC theory divides crystal faces into three types:

- *F* faces (flat faces), which are parallel to at least two periodic bond chains;
- *S* faces (stepped faces), each of which is parallel to one periodic bond chain;
- *K* faces (kinked faces), which are not parallel to any periodic bond chain.

The attachment energy of a crystal face is the difference between the crystallization energy and the slice energy. For an *F* face the attachment energy can also be expressed as the bond energy of bonds belonging to the PBCs that are not parallel to the face. The attachment energy of an *S* face will require the formation of at least one more strong bond than an *F* face, and a *K* face will require at least one more strong bond than an *S* face. Hartman–Perdok theory indicates that the larger the attachment energy, the faster the growth rate of a face. Since the largest faces on grown crystals are the faces that grow the least, Hartman–Perdok theory indicates that *F* faces will be the important faces on a grown crystal, with *S* faces occurring rarely and *K* faces never occurring. This means that Hartman–Perdok theory provides a means to predict the morphology of a grown crystal from the attachment energy of the *F* faces. More information on Hartman–Perdok theory can be found in Hartman (1963) and in Chapter 3.

The use of attachment energy as a method for the calculation of crystal morphology was developed by Hartman and Bennema (1980) and has since become the standard method used to calculate the morphology of organic molecular crystals. The basic method used by most investigators, as well as the method used in software packages (CERIUS2 1998; Clydesdale, Roberts, and Docherty 1996), first identifies the important crystallographic faces of the crystal by employing the BFDH method. The attachment energy of these faces is then calculated using interatomic potential functions (see Chapters 1 and 3) and the morphology predicted by assuming that relative growth velocity of a face is directly proportional to the attachment energy. More information on this method can be found in Chapter 3.

References

Bennema, P. (1969). *Journal of Crystal Growth*, **5**, 29–43.

Bernstein, J. (1987). In *Organic Solid State Chemistry*, ed. G. R. Desiraju, pp. 471–517. Amsterdam: Elsevier.

Bertin, E. P. (1975). *Principles and Practice of X-Ray Spectrometric Analysis*. New York: Plenum Press.

Bourne, J. R. (1980). *American Institute of Chemical Engineers Symposium Series*, **76**(193), 59–64.

Buckley, H. E. (1951). *Crystal Growth*. New York: John Wiley & Sons.

Buerger, M. J. (1956). *Elementary Crystallography*. New York: John Wiley & Sons.

Bunn, C. W. (1961). *Chemical Crystallography*, 2nd ed. London: Oxford University Press.

Burton, W. K., Cabrera, N., and Frank, F. C. (1951). *Philosophical Transactions*, **A243**, 299–358.

Cahn, J. W. (1968). *Transactions of the Metallurgical Society of the AIME*, **242**, 166–90.

CERIUS2 Software (1998). Molecular Simulations Inc., San Diego, CA.

Chernov, A. A. (1961). *Soviet Physics*, usp. **4**, 116–25.

Chernov, A. A. (1984). *Modern Crystallography*. III. *Crystal Growth*. New York: Springer-Verlag.

Clydesdale, G., Roberts, K. J., and Docherty, R. (1996). In *Crystal Growth of Organic Materials*, eds. A. S. Myerson, D. Green, and P. Meenan, pp. 43–52. Washington, DC: American Chemical Society.

Cullity, B. D. (1978). *Elements of X-Ray Diffraction*, 2nd ed. Reading, MA: Addison-Wesley.

Davey, R. J., Mullin, J. W., and Whiting, M. J. L. (1982). *Journal of Crystal Growth*, **58**, 304–12.

Davey, R. J., Milisavljevic, B., and Bourne, J. R. (1988). *Journal of Physical Chemistry*, **92**, 2032–6.

Frank, F. C. (1949). *Symposium on Crystal Growth, Discussions of the Faraday Society*, No. 5, 48–60.

Gilmer, G. A., Ghez, R., and Cabrera, N. (1971). *Journal of Crystal Growth*, **8**, 79–93.

Gunton, J. D., San Miguel, M., and Sahani, P. S. (1983). In *Phase Transitions and Critical Phenomena*, eds. C. Daub and J. L. Lebowitz, Vol. 8. New York: Academic Press.

Hartman, P. (1963). In *Physics and Chemistry of the Organic Solid State*, eds. D. Fox, M. M. Labes, and A. Weissberger, pp. 370–409. New York: Interscience.

Hartman P. and Bennema, P. (1980). *Journal of Crystal Growth*, **49**, 145–56.

Hartman P. and Perdok, W. G. (1955a). *Acta Crystallographica*, **8**, 49–52.

Hartman P. and Perdok, W. G. (1955b). *Acta Crystallographica*, **8**, 521–4.

Hartman P. and Perdok, W. G. (1955c). *Acta Crystallographica*, **8**, 525–9.

Hilton, H. (1963). *Mathematical Crystallography*. New York: Dover.

International Tables for Crystallography (1983). Vol. A. Dordrecht: Reidel.

Jackson, K. A. (1958). In *Growth and Perfection of Crystals*, eds. R. H. Doremus, B. W. Roberts and B. Turnbull, pp. 319–24. New York: John Wiley & Sons.

Joint Committee on Powder Diffraction Standards (1990). 1601 Park Lane, Swarthmore, PA.

Kim, S. and Myerson, A. S. (1996). *Industrial and Engineering Chemistry Research*, **35**, 1078–84.

Kitaigorodsky, A. I. (1973). *Molecular Crystals and Molecules*. New York: Academic Press.

Klug, D. L. (1993). In *Handbook of Industrial Crystallization*, ed. A. S. Myerson. Stoneham, MA: Butterworth-Heinemann.

Leusen, F. J. J. (1993). PhD dissertation, Nijmegen University, Nijmegen, The Netherlands.

Linke, W. R. and Seidell, A. (1958, 1965). *Solubilites–Inorganic and Metal–Organic Compounds*, Vol. I (1958), Vol. II (1965). Washington, DC: American Chemical Society.

McCrone, M. C. (1965). In *Physics and Chemistry of the Organic Solid State*, eds. D. Fox, M. M. Labes, and A. Weissberger, pp. 726–67. New York: Interscience.

McKie, D. and McKie, C. (1986). *Essentials of Crystallography*. London: Blackwell Scientific Publications.

Mullin J. W. (1993). *Crystallization,* 3rd ed. London: Butterworth-Heinemann.

Myerson, A. S. (1993). *Handbook of Industrial Crystallization*. Stoneham, MA: Butterworth-Heinemann.

Nielsen, A. E. (1964). *Kinetics of Precipitation*. New York: Pergamon Press.

Nyvlt, J., Sohnel, O., Matuchova, M., and Broul, M. (1985). *The Kinetics of Industrial Crystallization*. Amsterdam: Elsevier.

Ohara, M. and Reid, R. C. (1973). *Modeling Crystal Growth from Solution*. Englewood Cliffs, NJ: Prentice-Hall.

Randolph, A. D. and Larson, M. A. (1988). *Theory of Particulate Processes*, 2nd ed. San Diego: Academic Press.

Saska, M. (1984). Ph.D. dissertation, Georgia Institute of Technology, Atlanta, GA.

Saska, M. and Myerson, A. S. (1987). *AIChE J*. **33**, 848–53.

Stranski, I. N. and Honigmann, B. (1950). *Zeitschrift für Physikalische Chemie (Leipzig)*, **194**, 180–9.

Strickland-Constable, R. F. (1968). *Kinetics and Mechanism of Crystallization*. New York: Academic Press.

Vainshtein, B. K., Fridkin, V. M., and Indenbom, V. L. (1982). *Modern Crystallography. II. Structure of Crystals*. New York: Springer-Verlag.

Verma, A. R. and Krishna, P. (1966). *Polymorphism and Polytropism in Crystals*. New York: John Wiley & Sons.

Volmer, M. (1939). *Kinetic der Phasenbildung*. Dresden: Steinkoff.

Zemaitis, J. F. Jr, Clark, D. M., Rafal, M., and Scrivner, N. C. (1986). *Handbook of Aqueous Electrolyte Thermodynamics*. New York: American Institute of Chemical Engineers.

3

The Study of Molecular Materials Using Computational Chemistry

ROBERT DOCHERTY AND PAUL MEENAN

3.1. Introduction

Many current industrial products involve the direct formation of particles or at some point in the manufacturing process exist in the solid state. Consequently, consideration of solid-state properties is an integral component of any process to produce particles that will have desirable properties for optimum processing and end use. The application of molecular modeling techniques to the study of solid-state chemistry is playing an increasingly significant role in helping to predict and optimize such properties and also to alleviate many of the problems associated with particle formation and processing. An example where solid-state modeling is having a major impact on processing is crystallization, where the primary engineering concern has been to maximize process yields. Traditionally, little consideration has been given to particle property optimization. Typically, whatever crystal produced has been accepted and subsequently processed to improve materials handling and the property characteristics desired by the end user.

Crystallization of the same molecule under different process conditions can produce particles of radically different crystal structure, size, shape, and properties. In addition to properties that can influence downstream processing, particle characteristics can also influence chemical properties such as activity, reactivity, selectivity, rate of dissolution, and bioefficacy. Differences in the crystal structure can lead to the formation of enantiomorphs or polymorphs (materials with the same chemical structure, but different crystal structures), where one form may have undesirable properties. Changes in polymorphic form can be either problematic or beneficial. In the pharmaceutical industry different polymorphic forms can produce different dissolution rates and hence bioavailability. With pigments and dyes, changes in polymorphic form can produce different shades, which can be commercially important (although control is obviously vital). In the area of electrophotography, certain polymorphs of phthalocyanines are photosensitive,

Molecular Modeling Applications in Crystallization, edited by Allan S. Myerson. Printed in the United States of America. Copyright © 1999 Cambridge University Press. All Rights reserved. ISBN 0 521 55297 4.

yet others are inactive. The key issue is, therefore, particle engineering – "tailoring" particles with controlled structure, shape and material properties to optimize both production efficiency and final product properties.

On a bulk macroscopic level, in order to optimize properties, the crystal shape, size, and size distribution must be controlled. Impurities have to be rejected to the desired degree of purity. Crystals present different chemistry at each distinct crystal face, so the overall chemical nature of the particulate material depends on the surface-area ratio of the various crystal faces. The level of attraction between the solvent medium and/or impurities and the growing crystal faces, as well as between the crystals themselves, must be controlled or at least understood, in order to prevent habit modification or aggregation/agglomeration.

Understanding the intermolecular microscopic forces that dictate these macroscopic factors is necessary if molecular modeling and computational methods are to explain successfully process occurrences and the final solid-state chemistry at a molecular level. The molecular and crystal chemistry of the material in question must be elucidated. The functional units present in the molecule dictate the crystal chemistry. Accurate descriptions of the charge distribution and the intermolecular forces are hence necessary if one is to perform realistic modeling simulations of crystal properties. Charge distributions can be examined using basic semi-empirical molecular-orbital methods or more detailed *ab initio* methods and intermolecular force functions based on molecular mechanics methods can be parameterized.

The advent of sophisticated modeling program packages in conjunction with cheaper high-performance computing hardware and the increasing availability of single-crystal structure information has enabled more realistic molecular and solid-state modeling of industrial processes. The application of modeling complements experimental studies and thus reduces the need for expensive, trial-and-error, labor-intensive experiments. An example would be the application of molecular modeling in conjunction with experimentation as a "screening" methodology to determine the influences of impurities and/or solvents on particle morphology. A logical progression is the application of molecular modeling to the *ab initio* design of additive molecules for a specific influence on a crystallizing system.

Given the importance of polymorphism, crystal engineering and solid-state chemistry in the production of specialty effect chemicals (pharmaceuticals, agrochemicals, dyestuffs, pigments, and optoelectronics), our knowledge of the solid-state structures of such materials is surprisingly limited. This is due, primarily, to difficulties in the preparation of single crystals of sufficient size and quality in the desired polymorph for conventional single-crystal studies. The application of molecular modeling techniques through lattice-energy minimization, in combination with high-resolution powder diffraction, provides a route to detailed and reliable structural information.

The aim of this chapter is to provide an overview of the theories and methodologies governing modeling of molecular structure, crystal structure, and the crystal chemistry of molecular materials. A particular emphasis will be placed on understanding the important intermolecular interactions within different chemical classes through lattice-energy calculations. The link between lattice-energy calculations, shape prediction, and crystal engineering will be discussed. As crystal-structural information is central to many of these calculations, a review of crystal-structure databases, where such information can be found, is included. Crystal-structure predictive techniques are initially outlined and will be explained in greater detail in subsequent chapters. The potential for predictive methods is widespread, but they remain restricted, with rigid structures being the most successful candidates. Structure determination from powder diffraction data has recently been the subject of much investigation and the use of x-ray diffraction removes some of the concern of global minimization and force-field accuracy that has hindered progress with predictive methods.

3.2. The Crystal Chemistry of Molecular Materials

The structures and crystal chemistry of molecular materials are often categorized according to the type of intermolecular forces present. The more important classes of intermolecular interactions include:

- simple van der Waals attractive and repulsive interactions;
- classical hydrogen bonding;
- electrostatic interactions between specific polar regions;
- nonclassical hydrogen bonding of the type X—H:::O, where $X = sp$, sp^2 or sp^3 carbons;
- short directional contacts epitomized by Cl--N and S--S interactions.

Molecules can essentially be regarded as impenetrable systems, the shape and volume characteristics of which are governed by the radii of the constituent atoms. The atomic radii are essentially exclusion zones around the atoms that no other atom may enter except under special circumstances (for example, bonding). Figure 3.1 shows the traditional stick and van der Waals (spacefill) representations of a typical donor–acceptor azobenzene.

The awkward shape of organic molecules results in only a limited number of low symmetry crystal systems being adopted by molecular materials (Kitaigordskii 1973). The general uneven shape (Fig. 3.1) tends to result in unequal unit cell parameters and so the vast majority of the structures reported (around 95 percent) prefer crystal systems with unequal unit cell parameters, that is, triclinic, monoclinic and orthorhombic. This is shown in Table 3.1, which is an analysis of the most recent Cambridge Crystallographic Database (Version 5.1, October 1995). The unusual shapes of molecules results in difficulties in packing the molecules together and illustrates the importance of molecular complementarity. The fitting of the bumps of one molecule within the gaps of the surrounding molecules is necessary to maximize the contacts between molecules. Organic molecules prefer to adopt space groups that have inversion centers, glide planes, and screw axes, as this allows the most efficient spatial packing of the protrusions of one molecule into the gaps left by the packing arrangements of its

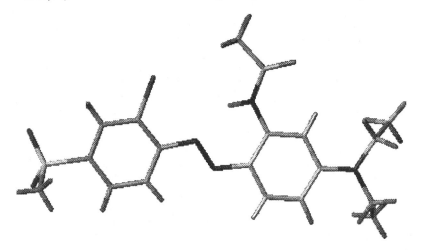

Fig. 3.1. Stick and spacefill representation of a donor–acceptor azobenzene color coded by atom type. The molecular structure is shown in Figure 3.2(*l*).

neighbors (Kitaigordskii 1973). The space group $P2_1/c$, which belongs to the monoclinic crystal system, is the single most important, with over 35 percent of molecules preferring to adopt arrangements that can be described by the symmetry elements present within this space group.

A useful parameter for judging the efficiency of a molecule for using space in a given solid state arrangement is the packing coefficient (C) (Kitaigordskii 1973):

$$C = \frac{ZV_{\mathrm{m}}}{V_{\mathrm{cell}}} \tag{3.1}$$

Table 3.1. *An analysis of the Cambridge Crystallographic Database (V.5.1, October 1995) by crystal system*

System	Axes	Angles	Number of entries	Percentage
Triclinic	a, b, c	α, β, γ	29,313	20.2
Monoclinic	a, b, c	$\alpha = \gamma = 90°; \beta$	76,628	52.7
Orthorhombic	a, b, c	$\alpha = \beta = \gamma = 90°$	31,483	21.7
Tetragonal	$a = b; c$	$\alpha = \beta = \gamma = 90°$	3,763	2.6
Hexagonal	$a = b; c$	$\alpha = \beta = 90°; \gamma = 120°$	1,811	1.2
Cubic	$a = b = c$	$\alpha = \beta = \gamma = 90°$	875	0.6
Rhombohedral	$a = b = c$	$\alpha = \beta = \gamma$	1,470	1.0

Table 3.2. *The packing coefficients of some organic molecules, as calculated using equation (3.1); the labels (a) to (l) correspond to Fig. 3.2*

Structure	Name	C
(a)	Benzene	0.68
(b)	Naphthalene	0.70
(c)	Anthracene	0.72
(d)	Perylene	0.80
(e)	Benzophenone	0.64
(f)	Benzoic acid	0.62
(g)	Urea	0.65
(h)	Indigo	0.68
(i)	1,4-Diketo-3,6-diphenylpyrrolo(e,4-c) pyrrole	0.66
(j)	N,N-Dimethylperylene-3,4:9,10-bis(dicarboxamide)	0.78
(k)	Phthalocyanine (β-polymorph)	0.73
(l)	6′-Acetamido-6-bromo-2-cyanodiethylamino-4-nitroazobenzene	0.66

where V_m is the molecular volume, V_{cell} is the unit cell volume, and Z is the number of molecules in the unit cell. This simple model assumes that the molecules within the crystal will attempt to pack in a manner such as to make the most of the space available. Table 3.2 shows the packing coefficient for a selection of organic molecules. The molecular structures are given in Fig. 3.2(a)–(l).

An inspection of Table 3.2 shows a rough correlation between the higher packing coefficient and increasing aromatic size (Wright 1987). The perylene pigment (*j*) has a packing coefficient of 0.78, slightly lower than perylene 0.8 (*d*). This is due to the slight disruptive effect on crystal packing of the methyl substituent in (*j*). The large flat "disk-like" aromatic molecules tend to stack face to face with lateral displacements to minimize the repulsive contribution to the π–π stacking energy (Hunter and Sanders 1990). Variations in the packing arrangements in these systems have been studied as a function of aromatic surface area and attempts to generate crystal structures from such information have recently been investigated (Gavezzotti 1991*a*). Once deviations from planarity are introduced, the effective packing ability falls. The azobenzene based dyestuff (Fig. 3.2(*l*)), which is shown in Fig. 3.1, is planar apart from the end groups and the packing coefficient falls to 0.68. For benzophenone (Fig. 3.2(*e*)), an aromatic ketone, the phenyl groups are twisted to 54° with respect to each other and the packing efficiency falls to 0.64.

One of the most interesting features in Table 3.2 is the low packing coefficient for hydrogen-bonded systems. Analysis of this table shows that these molecules, such as urea (0.65) and benzoic acid (0.62), tend to have packing coefficients less than that for benzene (0.68). This rather surprising feature is due to the open architecture of hydrogen-bonded structures, which is the result of a need to adopt particular arrangements to maximize the hydrogen-bonded network. This open nature of hydrogen-bonding motifs is shown in Fig. 3.3, which shows the one-dimensional packing motif present in benzamide (Fig. 3.2(*m*)). Figure 3.4 illustrates the two-dimensional hydrogen-bond motif observed for trimesic acid (Fig. 3.2(*n*)). The arrangements consist of six trimesic acid molecules forming a hexagon. The molecules are held together by carboxylic acid dimers. This hexagonal network results in large cavities that are filled by three perpendicular trimesic acid molecules, which form a similar pattern. Urea, which is one of few organic molecules to crystallize in the high-symmetry tetragonal crystal system, forms a three-dimensional hydrogen-bonding motif in which each urea molecule is hydrogen-bonded to six partners. This cluster of urea molecules (Fig. 3.5) is responsible for 85 percent of the total lattice energy (Docherty et al. 1993). Approaches for estimating the strengths of each of these intermolecular interactions will be described in the next section. The desire to achieve these complicated networks is not always consistent with the need to pack in the most efficient manner. Recognition of hydrogen-bonding patterns (Etter 1991) and their role in crystal engineering (Aakeroy and Seddon 1993) remains an area of active research. It is a key element in molecular solid-state chemistry, in the design of molecular aggregates (Fan et al. 1994), and in the understanding and construction of molecular recognition complexes for biologically interesting substrates (Chang and Hamilton 1988).

The role of special hydrogen bonds such as, C—H:::O═C interactions, remains the subject of some debate. The crystallographic evidence for their existence and classification as special intermolecular interactions was considered in detail (Taylor and Kennard 1982; Desiraju 1989) through an analysis of the Cambridge Crystallographic Database. These "special" hydrogen bonds are generally weaker than traditional hydrogen bonds and although spectroscopic,

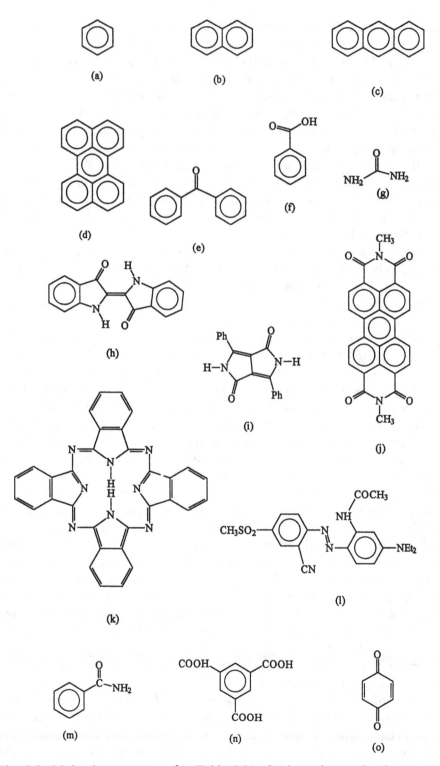

Fig. 3.2. Molecular structures for Table 3.2(*a–l*) plus other molecular structures used in Sections 3.2, 3.3, and 3.5.

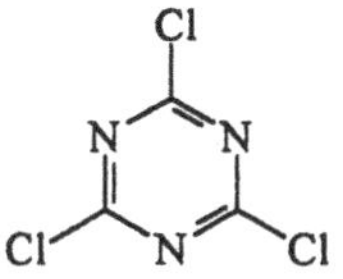

(p)

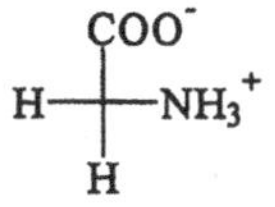

(q)

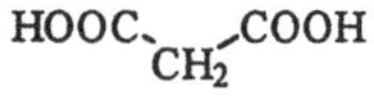

(r)

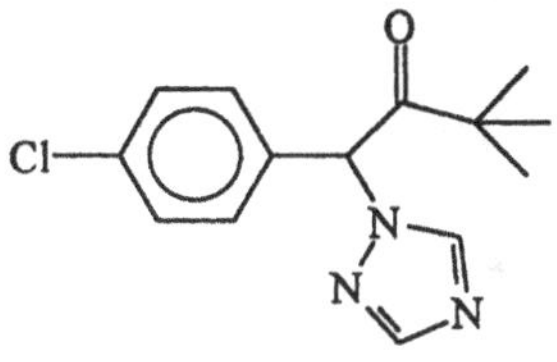

(s)

(t)

Fig. 3.2 (*continued*)

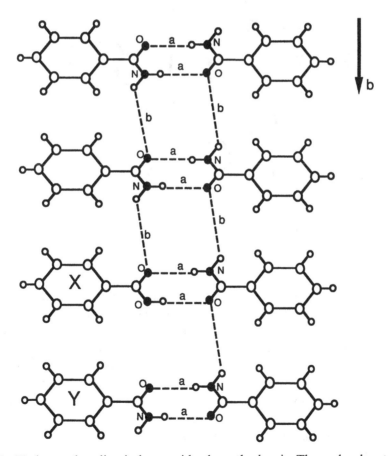

Fig. 3.3. Hydrogen bonding in benzamide along the *b*-axis. The molecular structure is shown in Fig. 3.2(*m*).

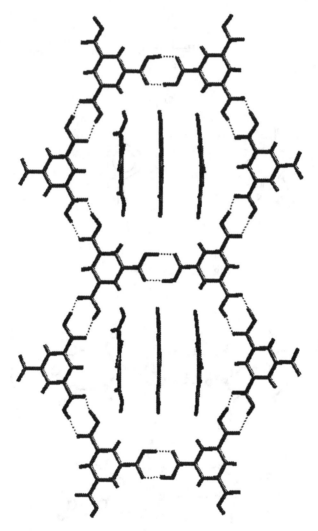

Fig. 3.4. Hydrogen bonding and packing of trimesic acid. Figure 3.2(*n*) shows the molecular structure.

crystallographic, and theoretical investigations continue to probe their magnitude and directionality, their overall role in determining structural patterns remains the subject of controversy. Figure 3.6 illustrates the layer structure found in benzo-quinone, which is held together through C—H:::O contacts. The layer consists of chains held together by C—H:::O $=$ C dimers at distances of 2.37 Å. These chains are held together in the layer by contacts at 2.3 Å. Both interactions are significant in that they are both shorter than the sum of the van der Waals radii (2.45 Å). The strength of these C—H:::O interactions seems to depend on the acidic nature of the H atom: alkyne > quinone > alkene > aromatic > aliphatic.

Similar doubts exist over the exact role of polar contacts and directional van der Waals interactions. Debate continues over whether these weak interac-

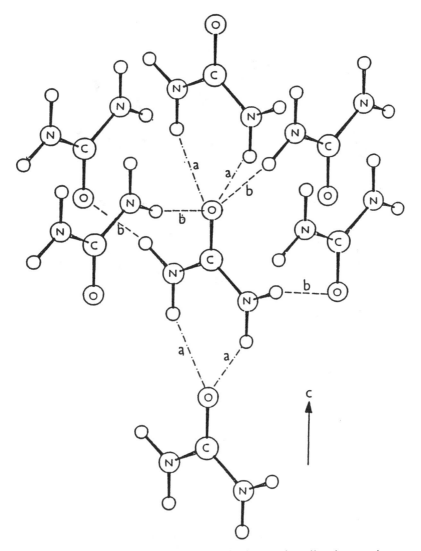

Fig. 3.5. Urea cluster illustrating hydrogen bonding interactions.

tions have a primary or secondary role in structure-arrangement determination. Short Cl--Cl and Cl--N contacts have been identified in a number of structures. In cyanuric chloride (Fig. 3.2(p)) Cl--N interactions hold the molecules in a layer (Fig. 3.7). Each cyanuric chloride molecule forms six interactions, three C—Cl--N and three N--Cl—C at around 3.14 Å. These interactions are anisotropic and consequently a monoclinic structure, rather than the hexagonal structure that one might expect, is observed (Desiraju 1989; Maginn et al. 1993).

   In general, the emphasis on characterizing the important intermolecular interactions has focused on the individual directional contacts, such as hydrogen bonds and special interactions. It is important to recognize the collective strength of a large number of nominally weak interactions. Essentially the needs of the

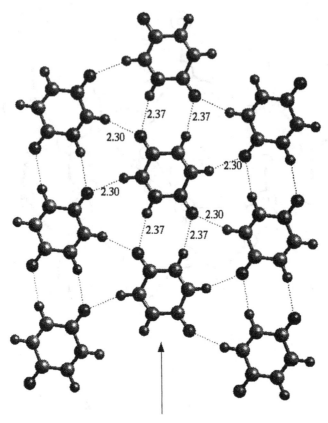

Fig. 3.6. The layer structure of benzoquinone (Fig. 3.2(*o*)), highlighting C—H:::O contacts.

many weak interactions can outweigh the needs of the individual stronger inter-actions. For example, π–π attractions consist of a number of interactions between the π cloud and the partial charges on the underlying molecular frame-work. In porphyrins and phthalocyanine pigment structures (Fig. 3.2(*k*)), the π–π interactions can be worth around −12 kcal/mol (Hunter and Sanders 1990), which is stronger than the average value for a conventional hydrogen bond.

It is probably safe to suggest that, in the majority of cases with molecular materials, it is the desire to pack effectively in the solid state that is the single biggest driving force towards selected structural arrangements. The notable exceptions will be in cases where the need to form complex hydrogen-bonding arrangements will compete with this need. Weaker interactions such as special hydrogen bonds and polar interactions are probably not primary movers in the arrangements adopted, but will tend to be optimized within a given efficient spatial arrangement. The exploitation of intermolecular interactions in the "non-covalent synthesis" of supramolecular assemblies has become the subject of a recent review (Desiraju 1995; Whitesides et al. 1995). The rules or "gram-mar" that govern the packing of molecules still remains the subject of study and

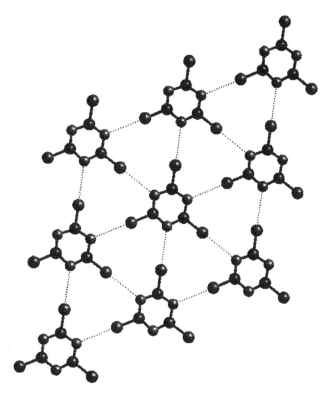

Fig. 3.7. View of the layer structure of cyanuric chloride (Fig. 3.2(*p*)), illustrating the Cl--N intermolecular interactions.

discussion (Brock and Dunitz 1994). Ultimately the structural arrangement adopted by a molecule will depend on the subtle balance of intermolecular interactions achievable through particular packing arrangements. Crystallization and the properties of the solid state are dependent on a molecular recognition process that occurs on a grand scale. Polymorphism and changes in properties are due to the recognition of different balances of these subtle inter-molecular interactions.

3.3. Lattice Energy Calculations

In order to understand the principles that govern the wide variety of solid-state properties and structures of organic materials, it is important to describe the interactions of molecules in specific orientations and directions. This refines our understanding from just considering shape, as in packing coefficients, to considering the functional units present in molecules. As a result of pioneering work (Williams 1966; Kitaigordskii, Mirskaya, and Tovbis 1968) in the use of atom–atom potentials, it is now possible to interpret packing effects in organic crystals in terms of interaction energies. The basic assumption of the atom–atom method is that the interaction between two molecules can be considered to consist simply of the sum of the interactions between the constituent atom pairs.

The lattice energy E_{latt}, often referred to as the crystal binding or cohesive energy, can, for molecular materials, be calculated by summing all the interactions between a central molecule and all the surrounding molecules. Each intermolecular interaction can be considered to consist of the sum of the constituent atom–atom interactions. If there are n atoms in the central molecule and n' atoms in each of the N surrounding molecules, then lattice energy can be calculated by the equation (3.2). In most cases n and n' will be equal, but in the case of molecular complexes they may differ. N is just the total number of atoms in the crystal:

$$E_{latt} = \sum_{k=1}^{N} \sum_{i=1}^{n} \sum_{j=1}^{n'} V_{kij}, \tag{3.2}$$

where V_{kij} is the interaction between atom i in the central molecule and atom j in the kth surrounding molecule. Each atom–atom interaction pair consists of a van der Waals attractive and repulsive interaction (V_{vdw}), an electrostatic interaction (V_{el}), and in some cases a hydrogen-bonding potential (V_{hb}). The van der Waals interactions can be described by a number of potential functions, all of which have the same basic form. The Lennard-Jones 6–12 potential function (Jones 1923) is one of the most common, consisting of an attractive and repulsive contribution as shown in equation (3.3). A and B are the atom–atom parameters for describing a particular atom–atom interaction and r is the interatomic distance:

$$V_{vdw} = -A/r^6 + B/r^{12}. \tag{3.3}$$

The electrostatic term can be described by assigning a fractional charge $\pm q$ to each atom, the sign determining whether the atom in the molecular arrangement has an excess or deficiency of electrons compared with the neutral atom. These charges are usually determined from molecular-orbital calculations. The electrostatic interaction is determined using equation (3.4) where q_i and q_j are the charges on atoms i and j, and D is the dielectric constant:

$$V_{el} = \frac{q_i q_j}{D r}. \tag{3.4}$$

Hydrogen-bonding interactions (V_{hb}) are essentially very special van der Waals interactions. Special modified versions of the Lennard-Jones potential functions have been employed (Momany et al. 1974) to describe the hydrogen-bond interactions in carboxylic acids and amides. The 10–12 potential is very similar in construction to equation (3.3) except that the attractive part is dependent on r^{10} rather than r^6. This gives a much steeper potential curve, which helps account for some of the important structural features of hydrogen bonds.

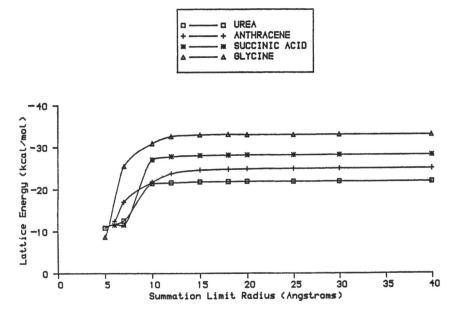

Fig. 3.8. The plot of lattice energy versus crystal radii (summation limit) for urea, anthracene, succinic acid and glycine.

The parameters A and B can be derived from fitting the chosen potential to observed properties, including crystal structures, heats of sublimation, and hardness measurements (Lifson, Hagler, and Dauber 1979). *Ab initio* quantum-chemistry calculations on interaction energies can also be used as a source of "experimental" data on which to fit these parameters.

Calculation of all the interactions between a central molecule and all the surrounding molecules in a real crystal is clearly a considerable task. Inspection of the potential functions shown in equations (3.3) and (3.4) shows that A, B, and the fractional charges for a particular atom pair are constant and so the interaction energy is dependent upon the separation distance r. Clearly, at large distances the interaction will be negligible. The van der Waals interactions will be short-range interactions, depending as they do upon inverse powers of r. The electrostatic interactions will act over a longer range, depending as they do upon $1/r$. Figure 3.8 shows the profiles of the calculated lattice energy as a function of summation limit for α-glycine, anthracene, β-succinic acid, and urea (Fig. 3.2 (q), (c), (r), (g)). These plots show the same general trend, upon increasing the size of the crystal there is an initial increase in the lattice energy. This is followed by the attaining of a plateau region beyond 20 Å. Further increases in the crystal size have no effect on the calculated lattice energy. Calculations are normally carried out at or beyond this plateau value.

The lattice energy is a crucial parameter to determine, as the calculated value can be compared with the experimental sublimation enthalpy, as a check of the description of the intermolecular interactions between the molecules by the force field in question. Table 3.3 contains a selection of calculated lattice energies and

Table 3.3. *Calculated and "experimental" lattice energies for a range of molecular materials; this is a subset of the data presented in Fig. 3.9. Experimental lattice energies derived from sublimation enthalpies*

| | Lattice energies | |
Material	Calculated (kcal/mol)	Experimental (kcal/mol)
n–Octadecane	−35.2	−37.8
Biphenyl	−21.6	−20.7
Naphthalene	−19.4	−18.6
Anthracene	−24.9	−26.2
Benzophenone	−24.5	−23.9
Trinitrotoluene	−25.1	−24.4
Glycine	−33.0	−33.8
L-Alanine	−33.3	−34.2
Benzoic acid	−20.4	−23.0
Urea	−22.7	−22.2
β-Succinic acid	−30.8	−30.1

experimental sublimation enthalpies from a dataset of over eighty compounds. Figure 3.9 shows a plot of calculated versus experimental lattice energies for the complete dataset and illustrates the excellent fit of theory and experiment. This selection includes not only a range of intermolecular interaction types (hydrocarbons through to amino acids), but also a selection of all the crystal classes, particularly the important monoclinic, triclinic, and orthorhombic classes. The average difference between calculated and experimental is less than 6 percent. The mean error is 1.5 kcal/mol and the maximum error 3.5 kcal/mol. The molecular classes reported include aliphatic hydrocarbons (Williams and Cox 1984), aromatic hydrocarbons (Docherty et al. 1991), alcohols (Royer et al. 1989), oxohydrocarbons (Gavezzotti 1991b), azahydrocarbons (Gavezzotti and Filippini 1992), carboxylic acids (Momany et al. 1974), polymers, and dyestuffs (Charlton, Docherty, and Hutchings 1995).

The calculated and experimental values show excellent agreement. The calculated value has the advantage that it can be broken down into the specific interactions along particular directions and further partitioned into the constituent atom–atom contributions. This is the key link between molecular and crystal structure. The urea structure (Fig. 3.5) consists of two types of hydrogen bonding with two interactions of type *a* and four interactions of type *b*. This is reflected in Table 3.4, which shows the breakdown of the lattice energy calculated for urea. The two interactions of type *a* (approximately −3.6 kcal/mol) involve two N—H:::O interactions along the **c** axis. The four interactions of type *b* (−3.0 kcal/mol) involve single amide hydrogen bonds. These interactions form the cluster

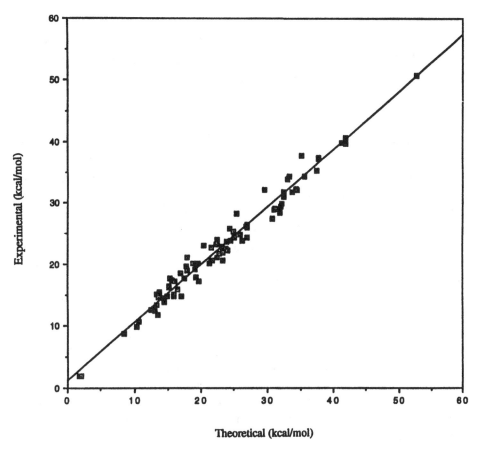

Fig. 3.9. Plot of the lattice energy versus sublimation enthalpy for a range of organic molecules.

Table 3.4. *The important intermolecular interactions in urea (see Fig. 3.5)*

Position (UVW)	Symmetry Z	Interaction (kcal/mol)	Label (Fig. 3.5)	Fraction of lattice energy (%)
001	1	−3.64	*a*	16.1
001	1	−3.64	*a*	16.1
100	2	−3.00	*b*	13.2
110	2	−3.00	*b*	13.2
000	2	−3.00	*b*	13.2
010	2	−3.00	*b*	13.2

Note: All calculations were carried out from a central molecule with (*UVW*) as (000) and *Z* = 1.

shown in Fig. 3.5 and constitute 85 percent of the total lattice energy. This allows a profile of the important intermolecular interactions to be built up within families of compounds and an understanding of the interactions that contribute to particular packing motifs to be established.

3.4. Databases

Single-crystal x-ray diffraction is the most useful experimental technique available for the determination of crystal structures. Improvements in automated data collection methods on computer-controlled diffractometers and the development of more robust refinement software has led to structure solution becoming more routine. Over the past ten years the total number of structures stored in the Cambridge Crystallographic Database has increased from 40,000 to over 145,000. There has also been a noticeable improvement in the quality and an increase in the complexity of the structures solved in recent times. The crystallographic discrepancy factor R (which is a measure of how well the postulated structure matches the experimentally recorded intensities) has fallen from an average of 12 percent in 1965 to 5 percent in recent years. The number of atoms in the asymmetric unit of structures being investigated has increased from about twenty to fifty in recent years (Allen, Kennard, and Taylor 1983).

A single-crystal structure can provide a vast amount of potentially useful information on both molecular and intermolecular arrangements. Bond distances, bond angles, molecular conformations, intra- and intermolecular nonbonded contacts, and preferred packing motifs can be examined within classes of compounds. The structures are obtained from data published in the literature (over 560 sources) or from data deposited directly by authors. Each entry is checked on submission for accuracy and any unusual features highlighted. This collection of data is a unique and valuable resource. It provides a comprehensive source of information on molecules and crystals, which is crucial for the study of molecule interactions within crystals, in drug/protein interactions, and in molecular recognition systems. In molecular modeling studies the database can be used for fragment construction as well as force-field generation and evaluation. The database is a vital resource used by both industry and academia.

The database stores three main categories of information for each entry, the bibliographic summary, the two-dimensional chemical structure, and the full three-dimensional structural details. The bibliographic summary includes chemical names, molecular formula, accuracy, basic structure-solution details, chemical classification code, and full literature citation. The two-dimensional chemical details include atom type and connectivity information. The full three-dimensional crystal-structure information includes the unit-cell dimensions, the spacegroup symmetry, and the individual atomic fractional coordinates. The crystallographic connectivity is also stored.

The database can be searched using molecular formula, journal name, year, author, property, and a number of other keywords. A two-dimensional representation of the molecule of interest or even a substructure may be used as a

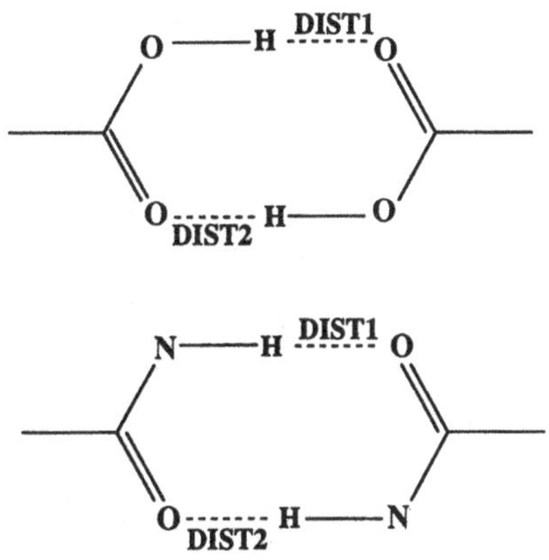

Fig. 3.10. Input query for carboxylic acid and amide dimer search of the Cambridge Crystallographic Database.

query. In recent years many advances have been made in the input and output interfaces to the database. In Version 4 of the database, a graphical input/output interface was introduced, which simplified the setting up and analysis of searches. A similarity searching technique was also introduced, which allowed fuzzy matches to structure queries to be identified. This allowed the examination of structures close to, but not exactly the same as, the query molecule. In Version 5, three-dimensional searching capability was introduced. A specific full three-dimensional query can now be inputed. This allows particular patterns to be examined both within and between molecules. Clearly, this is a powerful search facility that significantly increases the value of the information already within the database.

A number of surrounding programs have been developed for post-search analysis, including VISTA, GSTAT, and PLUTO. Drawing packages, which allow single entries to be displayed and the molecular structure and packing arrangements of hits to be examined, have been developed. The software also allows intermolecular contacts to be examined. Collections of hits may be tested for the statistical significance of selected molecular structures, nonbonded contacts, certain conformational arrangements, and packing motifs. Clearly these search facilities can be an enormous benefit in attempts to predict crystal structures systematically from molecular details, because they allow tentative links between molecular features and packing motifs to be established. These facilities have been used to analyze all the bond lengths for the accurate structures within the database, producing a compendium of bond distances (Allen et al. 1987), to examine the architecture of traditional hydrogen bonding (Taylor and Kennard 1984), to prove the existence of special hydrogen bonds such as $C{-}H{:::}O{=}C$ contacts (Taylor and Kennard 1982) and to examine packing trends within molecules in an attempt to predict crystal structures (Gavezzotti 1991a).

Figure 3.10 shows a query designed to investigate the structural features around the traditional centrosymmetric dimer structure for carboxylic acids

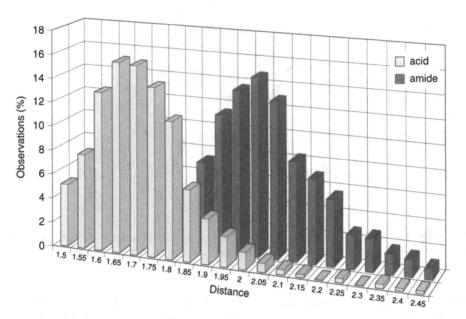

Fig. 3.11. Plots of the number of hits versus contact distance for the carboxylic acid and amide dimer query in Fig. 3.10.

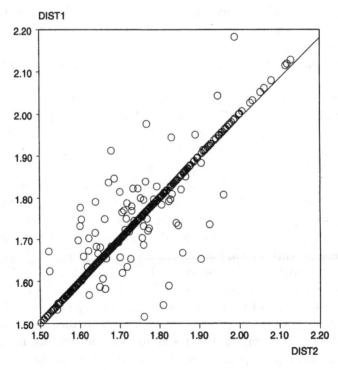

Fig. 3.12. Regression analysis of the contact distances DIST1 and DIST2, as part of carboxylic acid dimer query in Fig. 3.10.

and amides. The search was restricted to structures with a R factor of less than 0.07 and to systems with no metals present. The results are shown in Fig. 3.11 and Fig. 3.12. Figure 3.11 shows the percentage of observations at distances ranging from 1.5 to 2.5 Å. The plot shows that for carboxylic acids the mean distance occurs at about 1.7 Å. For amides, this mean is about 0.3 Å longer, occurring at about 2 Å. In Fig. 3.12 the centrosymmetric nature of the carboxylic acid dimer is examined. For all the carboxylic acids, DIST1 and DIST2 are plotted, with the resulting correlation coefficient of 0.94. The correlation coefficient is even higher if the corresponding oxygen–oxygen distances are used. This reflects the problem of locating the hydrogen positions in organic crystals and the potential for proton disorder in these carboxylic acid systems.

This approach has been used to examine the geometry of intermolecular hydrogen bonding (Gavezzotti and Filippini 1994). Mean hydrogen-bonding distances for acids (1.67 Å) and amides (1.95 Å) were found, in excellent agreement with the aforementioned search. The centrosymmetric dimer is the motif adopted by the majority of carboxylic acids and consequently the frequency of the two most closely packed space groups is enhanced. The frequency of $P2_1/c$ is 51 percent in the case of carboxylic acids which compares against 35 percent for all organic compounds.

3.5. Crystal Shape Prediction

3.5.1. *Crystal Shape Calculation*

A crystal consists of an ordered solid array of atoms or molecules exhibiting long-range order in three dimensions. Early crystallographers were fascinated by the flat, plane, and symmetrical faces in both natural and synthetic crystals. The field of morphological crystallography was initiated by showing that if cubes were stacked, various crystal shapes could be obtained. This basic building block of the crystals became known as the unit cell. The equilibrium crystal morphology was subsequently related to surface-energy considerations (Gibbs 1906; Wulff 1901).

Morphological simulations based on crystal lattice geometry were developed over a number of years (Bravais 1866; Friedel 1907; Donnay and Harker 1937). This work is often quoted collectively by modern workers as the BFDH law. It can be summarized as follows:

Taking into account morphologically submultiples of the interplanar spacing d_{hkl} due to space-group symmetry, the most important (MI) crystallographic forms will have the greatest interplanar spacings. If there are two forms: (h_1, k_1, l_1) and (h_2, k_2, l_2) and:

$$d(h_1, k_1, l_1) > d(h_2, k_2, l_2).$$

then

$$\text{MI}(h_1, k_1, l_1) > \text{MI}(h_2, k_2, l_2).$$

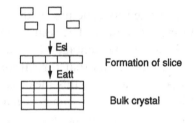

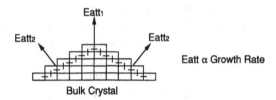

Fig. 3.13. Schematic illustrating the partition of slice and attachment energies and the relationship between attachment energy and growth rate at different faces.

This model can rapidly provide morphological data. However, the level of accuracy of the prediction is subject to the degree of anisotropy in the intermolecular bonding within the crystal structure.

The BFDH model was further refined (Hartman and Perdok 1955) by taking into account bonding within the crystal structure. Hartman and Perdok identified chains of bonds, known as Periodic Bond Chains (PBCs), within the crystal structure. The weakest bond within the PBCs is the rate-determining step and governs the rate of growth along the direction of the chain. Three categories of crystal faces could be defined depending on the geometry of the PBC chains.

A flat (*F*) face contains two or more PBCs in a layer d_{hkl}, a stepped (*S*) face contains one PBC in a layer d_{hkl} and a kinked (*K*) face contains no PBCs within a layer d_{hkl}. These categories grow by different mechanisms:

- *F* faces grow according to a layer mechanism, are slow growing, and hence are important faces on a morphology.
- *S* faces grow according to one-dimensional nucleation.
- *K* faces need no nucleation, grow fast, and are not normally found on crystals.

The fundamentals of slice and attachment energy are developed from this theory. The slice energy (E_{sl}) is defined as the energy released upon the formation of a slice of thickness d_{hkl}. The attachment energy (E_{att}) can be defined as the fraction of the lattice energy released upon attachment of a slice of thickness d_{hkl} to a crystal surface. The assumption is made that the attachment energy is proportional to the growth rate and that the larger the attachment energy (E_{att}), the faster the growth rate and hence the less important the corresponding form within the morphology. This assumption has been shown to be valid for a number of growth theories and is illustrated pictorially in Fig. 3.13. A unit coming onto the top flat face forms one bond to the surface. A unit attaching at the side face can form two interactions. Clearly E_{att1} will differ from E_{att2} and consequently these faces will have different morphological importance.

Given the necessary structural information (such as unit cell dimensions and atomic coordinates), a simulation of the crystal morphology can be undertaken. Attachment and slice energies can be determined from a PBC analysis, or directly from the crystal structure by partitioning the lattice energy into slice and attachment energies for specific crystal planes (Hartman and Bennema 1980; Berkovitch-Yellin 1985; Docherty and Roberts 1988). Predictions based on this methodology have shown excellent correlation to growth morphologies in many cases (Berkovitch-Yellin 1985; Docherty et al. 1991). This can be seen in Fig. 3.14 from the examples of two chemically very different materials, anthracene and β-succinic acid (Docherty and Roberts 1988).

These calculations have also been applied to industrially important materials, including the precursor in the manufacture of the agrochemical paclobutrazole (Fig. 3.2(*s*); Black et al. 1990). This material crystallizes in an orthorhombic space group $P2_12_12_1$ with four molecules per unit cell of dimensions $a = 5.799$ Å, $b = 13.552$ Å, $c = 19.654$ Å. The packing of the molecules is dominated by chains of C—H:::O interactions running in the **b** direction and a cascade of edge-to-face interactions running down the **a** direction. The predicted and experimental crystal morphologies are shown in Fig. 3.15. The morphology predicted is a barrel-like morphology, dominated by {002} and {011} front faces with capping {101} faces. The major difference between prediction and observation is the presence of small {012} and {110} faces.

The overall degree of correlation between prediction and experiment is dependent on both the underlying crystal chemistry under examination and the experimental conditions under which the growth morphology was obtained. Clearly the presence of impurities and solvent molecules can have a dramatic effect on the crystal shape. If the intermolecular bonding within the crystal structure is anisotropic, then normally the BFDH model will not simulate the morphology as well as a prediction based on attachment-energy calculations. This is clearly illustrated in the comparison of BFDH and attachment-energy models for β-succinic acid and anthracene (Fig. 3.14). For anthracene (isotropic interactions), the BFDH and attachment-energy models are in very good agreement with each other and the observed morphology. In succinic acid (directional interactions), the BFDH model shows significant differences from the attachment model and observed forms.

3.5.2. *Habit Modification Studies*

Habit modification occurs during crystallization when some factor is introduced that alters the relative crystal growth rates of different faces and hence the morphology obtained. Within a crystallizing system there are a number of factors that can be altered and may influence the resultant habit, for example choice of solvent, level of supersaturation, and the presence of impurities. Frequently, a

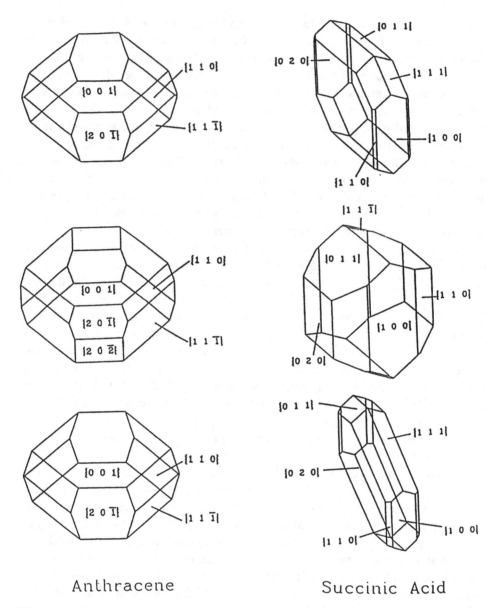

Anthracene   Succinic Acid

Fig. 3.14. Calculated and observed morphologies for anthracene and β-succinic acid. Observed (top), BFDH (middle), and attachment-energy (bottom) morphologies are shown.

habit shift occurs when a byproduct or a reactant is present to some extent and the resultant habit shift may be advantageous or disadvantageous. In many cases, the actual crystal habit obtained from a process does not have optimum particulate properties for end use. Additives may be employed to help modify the shape to optimize properties, such as ease of handling, filtration, and drying. For example, industrial crystallization of sodium chloride normally produces small,

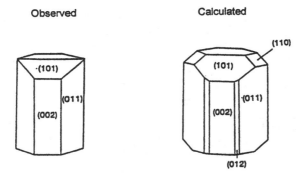

Fig. 3.15. Calculated and observed morphologies for an intermediate in the manufacture of the agrochemical paclobutrazole (Fig. 3.2(*s*)).

cubic crystals. Sodium cyanide is used as an anticaking additive and functions by changing the small cubic habit to large dendrites (Phoenix 1966). In the detergent industry, burkeite (sodium carbonate disodium sulfate) crystallizes with a blocky prismatic habit and is converted to dendritic crystals via the use of surfactants (Atkinson et al. 1987), ultimately forming a desirable, highly porous, high-surface-area powder.

There has been a drive within industry to modify particles during the formation stage to optimize habit and material properties, such as attrition resistance and particle size distribution, rather than the traditional approach of relying on postformation techniques to handle and process the crystals formed. This approach may be termed *crystal engineering*.

Traditionally, additives for habit modification are chosen after labor-intensive batch experiments. The results obtained are specific to defined and fixed conditions. These are difficult to apply to different process conditions and usually are not transferable to different chemical systems. However, experiments combined with solid-state modeling techniques can help elucidate a fundamental understanding of the mechanism of additive influence on crystal size and shape. Molecular modeling techniques can complement experimental studies by acting as a "screening tool," helping to assess rapidly the choice and potential design of molecules as crystal-habit-modifying additives.

The effects of impurities on the crystal shape of organic materials have been accounted for through elegant structural explanations (Lahav et al. 1985). The modification of crystal morphology using structurally similar molecules can be considered as the incorporation of either a *disrupter* or a *blocker* molecule into the surface of a growing crystal. Both use their molecular similarity to enter a surface site where the molecular difference is not recognized. A disrupter molecule will interrupt crystal packing through the lack of certain functional groups, which upsets the conventional intermolecular bonding pattern adopted by the host. A blocker molecule is usually bigger than the host system and uses its structural similarity to incorporate itself into selected surface sites. Then, as a result of steric hindrances, it physically inhibits the path of incoming host

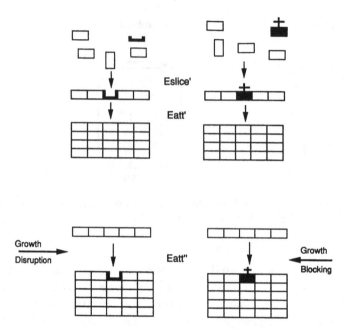

Fig. 3.16. Schematic illustrating the effect of disrupter and blocker type additives.

molecules to the growth surface. Such systems will obviously alter attachment energies and hence predicted crystal morphology. The computational methodology is to substitute the additive molecule for host molecules in the growth slice at each lattice site respectively, leading to different slice $E_{sl'}$ and attachment $E_{att'}$, $E_{att''}$ energies. This is summarized in Fig. 3.16 (Clydesdale, Roberts, and Docherty 1994a).

The effects of an impurity can be considered in two stages. The initial stage involves the calculation of a binding energy $E_{b'}$ between an additive and a surface site. This can be compared to the binding energy of the host molecule to the same site E_b. It then becomes possible to compare the change in binding energy ΔE_b between a host molecule and an additive molecule binding at the same surface site. This is essentially a measure of the ease of incorporation of the impurity. The effect of the impurity on subsequent growth can then be evaluated by determining the attachment energy for a surface containing the impurity $E_{att''}$.

The effect of a disrupter additive can be seen in the benzamide/benzoic acid host/additive system. Benzamide (Fig. 3.2(m)) crystallizes in a flat-plate-like morphology, dominated by {001} faces with {100} side faces and {011} top faces (Fig. 3.17). The calculated morphology is in excellent agreement with experiment. Calculations of the binding energies for benzoic acid indicate that preferential incorporation occurs in the {011} surfaces. There is very little change in the binding energy compared to the incorporation of a benzamide molecule itself. Calculated attachment energies show a reduction in the growth rate along the b-axis. Experimentally, benzoic acid additives tend to reduce growth along the b-axis. Figure 3.3 shows the hydrogen bonding along the b-axis for benzamide.

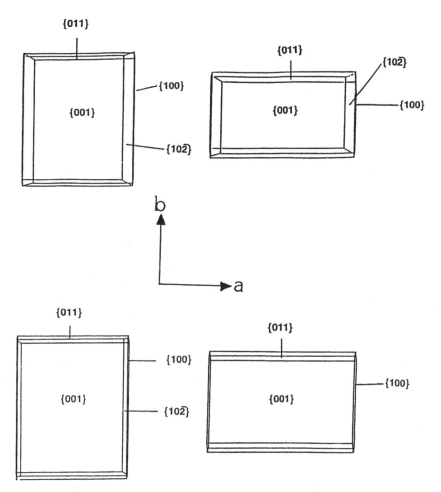

Fig. 3.17. Calculated and observed morphologies for pure benzamide (top and bottom – left) and benzamide doped with benzoic acid (top and bottom – right).

Benzoic acid can be incorporated into this direction as shown at point X. At position Y a benzamide attempts to form the proper intermolecular bonding pattern. It cannot achieve this, because where it would expect to find a N—H unit, to form an N—H:::O hydrogen bond, it finds a C=O. This results in a repulsive C=O--O=C interaction. Calculated and experimental morphologies in the presence of benzoic acid, showing good agreement, are given in Fig. 3.17.

The effects of blocker additives are more difficult to consider and the methodology to investigate the effects is still being developed (Clydesdale et al. 1994*b*). However, it is possible to quantify the effects of blocker-type additives and this can be highlighted by considering the case of benzophenone (Fig. 3.2(*e*)) grown from the melt and in the presence of toluene (Fig. 3.2(*t*); Roberts et al. 1994). Figure 3.18 shows the morphology of melt-grown benzophenone, which is dominated by {110} front faces and smaller {020}, {021}, {011}, and {111} side and top faces. Crystals grown in the presence of toluene are dominated by {021}

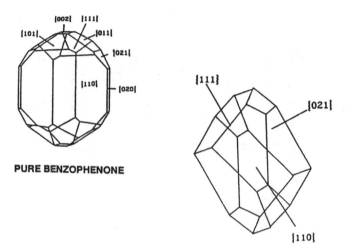

PURE BENZOPHENONE

BENZOPHENONE GROWN WITH TOLUENE

Fig. 3.18. Calculated and observed morphologies for melt-grown benzophenone (Fig. 3.2(*e*)) and benzophenone grown in the presence of toluene (Fig. 3.2(*t*)).

and {110} faces. The biggest difference in the morphologies is the massive increase in the morphological importance of the {021} faces. If toluene is considered as a blocker impurity, binding energy calculations indicate that it is most likely to be incorporated into the {021} and {111} faces. Attachment-energy calculations indicate that the toluene molecule in the {021} face will have a dramatic effect on the binding of future layers of benzophenone. The attachment energy for a new surface $E_{att'}$ is positive, which is an indication of the blocking ability of the toluene molecule. This is further highlighted in Fig. 3.19, which shows the packing of benzophenone with the {021} plane highlighted and a toluene molecule fitted on the surface.

3.6. Property Estimation Methods

The prediction of thermochemical data is also of considerable importance in the identification of commercially viable processes, in the formulation of many products such as dyes, drugs, and agrochemicals, and in plant design and operation. The sublimation enthalpy SE of a compound is the energy required to break up the solid state and convert the system into the gas phase. It is a property of particular interest in the crystal chemistry of molecular materials. Measurement of thermochemical data such as the SE can be a costly and time-consuming process and the prediction of solid-state properties before a compound is synthesized would obviously be of considerable use.

Recent applications of statistical methods (including neural networks) within the chemistry field have incorporated the determination of structure–activity relationships in drug design and the classification of spectra. Thermochemical

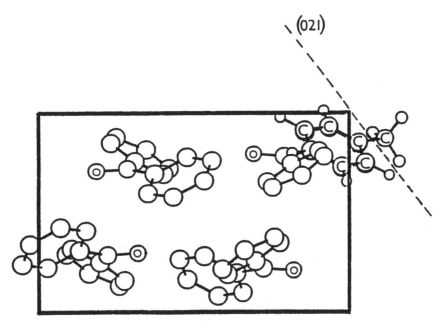

Fig. 3.19. The packing of benzophenone, highlighting the {021} face with a toluene additive presence.

data such as solubilities and boiling points have also been the subject of investigation. There are a number of reviews of the use of neural networks in chemistry (Lacy 1990; Zupan and Gasteiger 1993). It has been recently shown (Gavazzotti 1991*a, b*) that the lattice energy (and consequently the sublimation enthalpy) can be predicted based upon linear regression studies of molecular crystal descriptors. The heat of sublimation can be estimated from the packing potential energy PPE, which correlates with molecular descriptors such as molecular weight, van der Waals surface, volume, and molecular outer surface. It has also been shown that there is a correlation between the melting point of a substance and the magnitude of the intermolecular forces in terms of the enthalpy of sublimation (Westwell et al. 1995).

Recently, various quantitative techniques have been applied to model and predict the SE of a series of organic molecules (Charlton et al. 1995). These range from simple hydrocarbons, polycyclic aromatic compounds, carboxylic acids, amides, and amino acids to heterocycles and dyestuffs. The compounds under study in this work were limited to those containing only carbon, hydrogen, oxygen, and nitrogen, although the system could, in principle, be extended to other elements. A feed-forward Neural Network NN was trained to reproduce sublimation enthalpies. The results were compared both with calculated values from traditional molecular modeling techniques and with a multilinear regression analysis MLRA model.

The network used was a feed-forward network, trained by back propagation. This architecture has achieved greatest popularity in chemical studies. The net

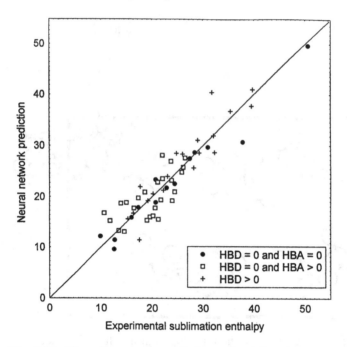

Fig. 3.20. Neural-network predictions of sublimation enthalpy.

has three layers (input, output, and one hidden layer), which is considered to be sufficient for practical structure–activity studies. It was decided that the ability to predict the sublimation enthalpy of a compound would be of most use if the input to the network was as simple as possible. The absence of any quantum-mechanically derived parameters (dipole moments, charges, etc.) from the model reflects this intention. With this in mind, seven parameters were selected that could be important factors in determining the SE. The results from the neural network are shown in Fig. 3.20 and Table 3.5. The parameters used as input were:

number of carbon atoms (C);
number of hydrogen atoms (H);
number of nitrogen atoms (N);
number of oxygen atoms (O);
number of π-atoms (PI);
number of hydrogen-bond donors (HBD);
number of hydrogen-bond acceptors (HBA).

To assess the NN model, a multilinear regression analysis MLRA and crystal-packing calculations (Section 3.3) were undertaken on the same dataset. In the MLRA analysis the independent variables used were exactly the same as in the NN analysis. As a pre-evaluation of the independent variables, the Pearson correlation matrix was determined. This revealed strong collinearities between a number of variables. Thus, it is statistically unacceptable to include all these

Table 3.5. *Statistical results of neural network MLRA and crystal packing calculations*

	Theory[a]	NN[b]	MLRA[c]
Maximum error	3.5	10.1	8.9
Mean error	1.4	2.5	1.8
r^2	0.97	0.87	0.92

[a]Theoretical prediction using crystal-packing calculations.
[b]Neural network prediction (Leave-one-out experiment).
[c]Multilinear regression analysis prediction.

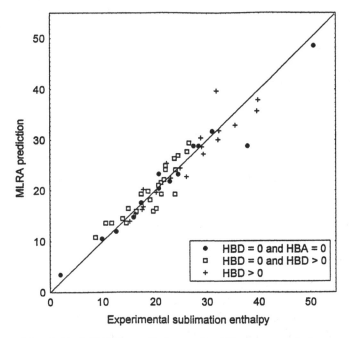

Fig. 3.21. MLRA predictions of sublimation enthalpy.

parameters in the MLRA. A check for multilinearities revealed the correlation between C and HBD and HBA to have an r^2 of 0.21. It was concluded that this combination would not result in coefficient bias. The sublimation enthalpy values were then regressed against these three parameters. It was concluded that a simple three-parameter MLRA model, as shown in equation (3.5), is the best that can be derived with the data used and the results are summarized in Table 3.5 and plotted on Fig. 3.21.

$$SE = 3.47 \pm 0.88 + 1.41 \pm 0.06C + 4.55 \pm 0.30HBD + 2.27 \pm 0.24HBA \quad (3.5)$$

It is clear from Table 3.5 that the NN and MLRA results are statistically less accurate than the crystal-packing calculations. Nevertheless, they show that, with the exception of a few outliers, they reproduce the experimental trends reasonably well. The NN predictions tend to be worse for larger values of SE, which is in line with the fact that there are only a few large molecules in the training set. Such molecules tend to have the largest values of SE and the largest values for the input parameters. Neural nets, although good at interpolation, are relatively poor at extrapolating beyond the maximum values within the training set. This could partly explain the poor performance for these molecules.

One of the main outliers is indigo, which is widely overestimated by the model. Indigo has two potential N—H hydrogen-bond donor sites, but they are involved in intramolecular hydrogen bonds and are less available for intermolecular hydrogen bonding. Thus, they contribute less to the sublimation enthalpy. In fact, the crystal structure of indigo indicates only weak intermolecular hydrogen bonds, with long hydrogen-bond distances of 2.1 Å. Altering the HBD parameter for indigo to zero to reflect the absence of intermolecular hydrogen bonding improves the model ($r^2 = 0.94$) and indigo falls nicely onto the regression line. Although neither neural network nor regression analysis reproduced the experimental results nearly as accurately as the theoretical crystal-packing calculations, both have yielded surprisingly good models. It is particularly interesting that the SE could be reproduced from the simple three parameter MLRA and this in itself could be a useful tool.

3.7. The Application of Computational Chemistry to Structure Prediction/ Determination

A traditional full single-crystal structure analysis can reveal a great deal of information on the structure of molecular materials. This information can be classified into three categories:

1. Unit cell details – the analysis will yield the dimensions of the unit cell (a, b, c) and the angles between the unit cell sides (α, β, γ).
2. Molecular information – the fractional coordinates of the atoms in the asymmetric unit (usually one molecule). These coordinates are the relative positions of the atoms in space and thus bond distances, angles, and torsional parameters responsible for the molecular conformations can be established.
3. Symmetry information – the number of asymmetric units (molecules) in the cell and the relationship between these. This gives the overall packing motif.

In order to produce information comparable with a conventional single-crystal x-ray diffraction study, a theoretical approach for the prediction of crystal structure must therefore be able to describe accurately both the molecular structure, including the low energy conformations, and the interactions between these molecules in a postulated packing arrangement. A systematic, efficient search/

structure generation algorithm is also needed to generate the possible structures. The prediction of solid-state packing arrangements from molecular structure has recently been reviewed (Docherty and Jones 1997).

3.7.1. *Molecular Structure Calculation*

In molecular modeling studies of organic materials it is often necessary to build molecules for which there is little or no experimental structural information. Models can be built using standard templates and fragments, but the linked fragments often do not accurately reflect the three-dimensional structure of the material of interest. One method to obtain an improved model of the system is to minimize the structure. Both molecular-orbital (semi-empirical and *ab initio*) and molecular mechanics approaches can be used. Semi-empirical molecular-orbital methods can be used for small and medium-sized molecules and are generally accepted as the most robust and reliable approach for small molecules (Stewart 1990; Dewar 1993; Grant and Richards 1995). Even with the increase in computer power available, *ab initio* molecular orbital calculations remain difficult to apply in a routine manner because they are computationally intensive for even relatively small molecules. Molecular mechanics provides a route to rapid, accurate, and reliable molecular structure information on a large range of systems. Molecular mechanics will be discussed in some detail in the following section because of the similarities in the description of non-bonded interactions with crystal-packing calculations. Consequently a route to a unified molecular-structure/crystal-packing calculation algorithm is possible.

3.7.1.1. *Molecular Mechanics*

The basis of the molecular mechanics method is the assumption that atoms in molecules have a natural arrangement (bond lengths and angles) that they adopt relative to each other in order to produce a geometry of minimum energy. The method is an empirical one based on parameterizing a force field by fitting experimental observables to a set of potential functions. The justification of the method is in its ability to reproduce experimental values for related molecules. The method assumes the transferability of the force field between related families of compounds. The interested reader is referred elsewhere (Allinger and Burket 1982) for a detailed perspective of the development of the molecular mechanics approach.

The molecular mechanics method is one in which the total energy for a molecule is dependent on how far the structural features of that molecule deviate from "natural" or "ideal" values that the atoms within the molecule should adopt. The molecular mechanics force field includes descriptions, in terms of the energy, of bond lengths, bond angles, torsion angles, non-bonded interactions, and electrostatic forces. The total energy E_m of the system is the sum of all the terms shown in equation (3.6) and can be considered as the difference between the real molecule and a hypothetical molecule where the atoms adopt their ideal geometric values:

$$E_m = E_b + E_\theta + E_\Phi + E_w + E_{vdw} + E_{el} + E_{b-\theta} + E_{hb}, \qquad (3.6)$$

where

E_b = the bond-stretching term;
E_θ = the angle-bending term;
E_Φ = the torsional energy;
E_w = the out-of-plane deformation energy;
E_{vdw} = the nonbonded van der Waals interaction;
E_{el} = the electrostatic interaction;
$E_{b-\theta}$ = the bond-stretching, angle-bending term;
E_{hb} = the hydrogen-bonding term.

Although the total energy E_m has no real physical meaning, the relative energies of different conformations can be considered as a measure of their comparative stability. Each of the terms in equation (3.6) attempts to relate the structural properties to an energy term. It is not proposed to consider all the terms in detail, as this has been done elsewhere (Allinger and Burket 1982). However, outlining one of the terms may further serve to illustrate the molecular mechanics methodology.

A molecule in the molecular mechanics approach can be considered as atoms joined together as springs that are attempting to restore the *natural* or *ideal* geometric features within the molecule. The stretching of a bond is often described by a harmonic potential, as shown in equation (3.7):

$$E_b = 1/2K(r - r_0)^2, \qquad (3.7)$$

E_b is the bond-stretching energy, K is the force constant, r is the observed or current bond distance, and r_0 is the ideal bond distance. At large deformations from the ideal values, deviations from the harmonic potential would occur. This is compensated for by including a cubic term in equation (3.7). This is a simple approximation, is computationally convenient, and gives the desired performance characteristics. The angle-bending term is similar in nature to the bond-stretching term described above. The nonbonded interactions including van der Waals, electrostatics, and hydrogen bonds are identical to those described in the crystal-packing calculations (Section 3.3).

In a typical minimization, three distinct stages can be identified, the preoptimization stage, the preliminary optimization, and final refinement. Preoptimization involves the correction of bonds that are too long and the removal of unacceptably short, nonbonded, contact distances. The short contacts are usually those with distances less than the sum of the van der Waals radii. The importance of the van der Waals radii in describing the molecular shape has already been highlighted in the discussions on molecular crystal chemistry in Section 3.2. In the preliminary optimization stage, nonbonded contacts and electrostatic interactions are only calculated below certain distances. The

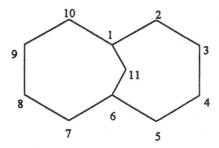

Fig. 3.22. The molecular structure of bicyclo (4.4.1) undecapentane.

Table 3.6. *A summary of the calculated and experimental structural features for bicyclo(4.4.1) undecapentane (Fig. 3.22) (Allinger and Burket 1982)*

Structural feature	Observed	Calculated
Bond 1–2 (Å)	1.409	1.406
Bond 2–3 (Å)	1.383	1.400
Bond 3–4 (Å)	1.414	1.423
Bond 1–11 (Å)	1.477	1.470
Angle 1–11–6 (°)	99.6	99.1
Angle 2–1–11 (°)	116.1	115.8
Angle 1–2–3 (°)	122.3	120.4

truncation distances applied during this stage help reduce the calculation time on larger systems. In the refinement stage all interactions are switched on.

Structural calculations have also been carried out and the experimental bond lengths have been found to be in excellent agreement with the calculated values. Figure 3.22 shows the molecular structure of bicyclo (4.2.1) undecapentane and Table 3.6 compares the calculated bond lengths and angles against the observed values (Allinger and Burket 1982). The calculated and experimental values show excellent agreement. In the validation of a general-purpose force field (Clark, Cramer, and Van Opdenbosch 1989) a comparison was made of structures, optimized by molecular mechanics, of seventy-six organic molecules and the average difference in bond lengths was found to be 0.011 Å and the difference in angles to be 0.13°. A similar evaluation (Mayo, Olafson, and Goddard 1990) for the Dreiding force field found average bond errors of 0.009 Å and angles of 0.56° and torsions of 0.23°. These average errors are very impressive, though it should be noted that in both these studies the maximum error for an individual angle or torsion can be considerably greater. Molecular properties, such as heats of formation, have also been reported and for a selection of 110 compounds the standard deviation was 2.1 kJ/mol (Allinger and Burket 1982).

3.7.1.2. Molecular Orbital Methods

It is not within the scope of this chapter to discuss the basic theory behind molecular-orbital calculations and the interested reader is referred to review articles (Dewar 1993; Stewart 1990; Grant and Richards 1995; Simons 1991). With the increase in computer power and the falling costs of such machines, it is now possible to carry out semi-empirical and even *ab initio* calculations on reasonably complex molecular structures. Such calculations are important in polymorph prediction studies and this will be discussed in subsequent chapters. Molecular-orbital calculations can yield details on a number of important molecular properties including:

- structural features,
- heat of formation,
- dipole moment,
- atomic charges,
- electrostatic potential,
- ionization potential,
- tautomer stability,
- spectral characteristics,
- reaction pathways.

Molecular-orbital methods can be employed across a wider range of molecular classes with a greater confidence than for molecular mechanics. The parameterization of molecular-mechanics force fields results in difficulties in examining molecules that differ from the classes of compounds used in deriving the force field. The greater the differences, the less the confidence can be placed in the results. The differences between molecular-orbital and molecular-mechanics calculations is that at one extreme an *ab initio* molecular-orbital calculation can, in principle, be carried out on any novel, unique molecular structure. No experimental information about the system need be known. In the case of molecular mechanics, the parameterization of the force field requires a great deal of experimental information on structurally similar molecules. Clearly, this means that molecular-mechanics calculations can only be applied with confidence to structures closely related to known molecular classes.

In Table 3.7 a summary is given of the errors found in the calculation of various properties, including heats of formations and dipole moments (Stewart 1990). In Table 3.8 a summary of the average errors in bond lengths is presented for compounds with a wide range of elements common to molecular materials and in addition some metals (Stewart 1990). The average error in bond lengths of 0.031 Å is greater than that for the molecular-mechanics studies described earlier, but the molecular-orbital calculations have been performed over a much larger dataset (ten times larger) and over a much more diverse range of chemistry.

Figure 3.23(*a*) shows the molecular structure of triethylammonium-6,7-benzo-3-*o*-hydroxyphenyl-1,4-diphenyl-2,8,9-trioxa-4-phospha-1-boratricyclo   (3.3.1)

Table 3.7. *A summary of the errors in calculated properties using the semi-empirical method AM1 (Stewart 1990)*

Property	Number of compounds	Error
Heat of formation	1116	14.8 kcal/mol
Bond lengths	587	0.031 Å
Ionization potential	319	0.7 eV
Dipole moments	185	0.4 Debye

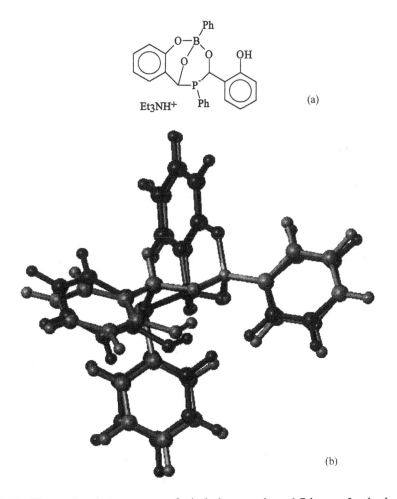

Fig. 3.23. The molecular structure of triethylammonium-6,7-benzo-3-*o*-hydroxyphenyl-1,4-diphenyl-2,8,9-trioxa-4–phospha-1-boratricyclo(3.3.1) nonane (CSD refcode JINSUU) (*a*) and the overlay between calculated (dark gray) and experimental structures (*b*) (light gray).

Table 3.8. *A summary of errors in bond lengths for some common elements as reported from calculations using the AM1 semi-empirical quantum-chemistry procedure (Stewart 1990)*

Element	Number of compounds	Error (Å)
Hydrogen	147	0.027
Carbon	311	0.028
Nitrogen	75	0.052
Oxygen	87	0.045
Fluorine	86	0.055
Phosphorus	37	0.079
Sulfur	66	0.087
Chlorine	52	0.063
Germanium	64	0.040
Mercury	19	0.065
Zinc	9	0.037

Table 3.9. *A comparison of the calculated structural features and the experimental data around the central boron atom in the compound triethylammonium-6,7-benzo-3-o-hydroxyphenyl-1,4-diphenyl-2,8,9-trioxa-4-phospha-1-boratricyclo(3.3.1) nonane (CSD refcode JINSUU) as shown in Fig. 3.23(a)*

Structural feature	Experimental	Calculated
Ph—B bond length (Å)	1.59	1.57
B—O bond lengths (Å)	1.47–1.53	1.48–1.49
O—C bond lengths (Å)	1.39–1.41	1.39–1.41
B—O—C angles (°)	112.8–113.9	112.9–113.6
O—B—O angles (°)	107.3–109.9	107.3–107.9
O:::O nonbonded (°)	2.41–2.43	2.40–2.41

nonane (CSD refcode JINSUU) (Litvinov 1990). Table 3.9 compares the calculated and experimental structure around the central boron atom. The analysis has concentrated around the central boron atom. The distances and angles show good agreement between the experimental and calculated structures. Figure 3.23(b) also shows the overlay between the main structural features of the molecular-orbital and the experimental crystal structure. In this case, the

RMS fit of the respective heavy atoms around the central boron (0.025) indicates that the molecular-orbital calculations are in excellent agreement with the experimental structure.

3.7.2. *Ab Initio Approach to Crystal Packing*

Ab initio prediction of crystal packing, based purely on the molecular structure, is the most attractive of all possible approaches. It involves calculating the arrangement for a collection of molecules that maximizes the intermolecular interactions. A methodology that would permit the prediction of the most stable solid-state arrangement for a molecule requires:

- A comprehensive search/structure generation algorithm, which will analyze all structural space, identifying not only local minima, but also the global minimum.
- A similarity search procedure, which will reduce the dataset by clustering similar structures, or rejecting poor ones.
- A refinement methodology based on lattice-energy minimization.

The lack of a methodology for predicting solid-state structural arrangements from molecular descriptors has been described as "one of the continuing scandals in the physical sciences" (Maddox 1988). Despite the inherent difficulties, prediction from first principles remains an admirable scientific goal and this is reflected in the recent elegant work of a number of authors (Gavezzotti 1991*a, b*; Holden, Du, and Ammon 1993; Karfunkel and Gdanitz 1992; Perlstein 1994). Progress has been hindered by problems both in locating the global minimum and in force-field accuracy, where the description of the electrostatic interactions continues to receive much consideration (Price 1996; Price and Stone 1984).

The potential of the *ab initio* approach was highlighted (Fanfani et al. 1975) when the packing of pyridazino-(4,5)-pyridazine (PP) and 1,4,5,8-tetramethoxy-pyridazino-(4,5)-pyridazine (TMPP), shown in Fig. 3.24, was determined by calculating the lattice energy as the three rotational parameters (ψ_1, ψ_2, ψ_3) were systematically altered. No translational degrees of freedom were permitted as the molecule centers lie on crystallographic centers of symmetry. The unit-cell dimensions and space-group symmetry, which had been determined by traditional diffraction methods were held constant during the calculations. The resultant minimum found from the calculations is in good agreement with the known experimental values, as shown in Table 3.10.

Without space-group and unit-cell information, the generation of possible structures becomes a considerably more complex problem. Gavezzotti and his coworkers have pioneered the concept of the "molecular nuclei." Given a molecular structure, clusters of molecules (nuclei) are built using various symmetry operators, including glide, inversion, and screw elements. These symmetry operations, accompanied by translations, account for around 80 percent of the space groups and more importantly all the principal space groups for molecular materials (Section 3.2). The relative stability of each of these nuclei is appraised through a calculation of its intermolecular energy. When these nuclei have

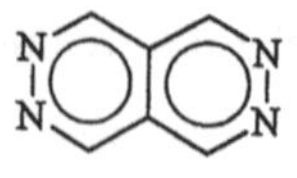

(PP)

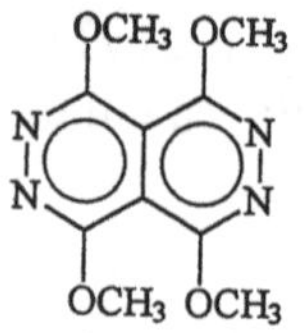

(TMPP)

Fig. 3.24. The molecular structures of pyridazino-(4,5)-pyridazine (PP) and 1,4,5,8-tetramethoxypyridazino-(4,5)-pyridazine (TMPP).

Table 3.10. *The calculated and experimental minima for the structures for pyridazino-(4,5)-pyridazine (PP) and 1,4,5,8-tetramethoxypyridazino-(4,5)-pyridazine (TMPP) as shown in Fig. 3.24 (Fanfani et al. 1975)*

	(PP)		(TMPP)	
Angle	Exp.	Calc.	Exp.	Calc.
Ψ_1	41.0	34.0	21.7	22.0
Ψ_2	61.0	62.0	64.4	65.0
Ψ_3	270	260	76.7	77.0

been selected, a full crystal structure can be generated using a systematic translational search. Some restrictions can be imposed on the possible cell lengths by using the molecular dimensions. The translational search is carried out in a systematic manner with lines, layers, and full three-dimensional structures being built. A further selection process is used based on packing coefficients (Section 3.2) and packing-density limits as determined from the Cambridge Database (Section 3.4) to filter out the most likely structures.

Refinement of the selected structures is carried out using lattice-energy minimization; calculation of the lattice energy has already been described in some detail in Section 3.3. During refinement the unit-cell dimensions ($a, b, c, \alpha, \beta, \gamma$) are allowed to alter, along with the molecular orientation and translational parameters, in order to minimize the lattice energy. The results of this method show considerable promise and it is best depicted by considering a published example (Gavezzotti 1991a). The molecule bicyclohepta(de,ij)naphthalene, which is often referred to as dipleiadiene (CSD refcode DIHDON), is shown in Fig. 3.25. The unit-cell dimensions from the predictions based on this approach are given in Table 3.11. The best predicted value shows excellent agreement with the experimental observations, the largest deviations being for the angle β. For the

Fig. 3.25. The molecular structure of bicyclo(de,ij)naphthalene (CSD refcode DIHDON).

Table 3.11. *The observed and predicted unit-cell dimensions for bicyclo(de,ij)-naphthalene (CSD refcode DIHDON); the molecular structure is shown in Fig. 3.25 (Gavezzotti 1991a)*

Space group	a (Å)	b (Å)	c (Å)	α (°)	β (°)	γ (°)	Lattice energy (kcal/mol)
$P2_1/c$							
Experiment	8.245	10.516	15.668	90.0	90.85	90.0	−22.5
Experiment/ optimized	7.91	10.80	13.58	90.0	95.7	90.0	—
Calculated	7.93	10.58	13.93	90.0	93.9	90.0	−22.8
$P2_1$	10.03	9.17	9.17	90.0	68.7	112.3	−21.6
$P2_1$	10.62	10.33	10.33	90.0	70.20	90.0	−21.3
$P\bar{1}$	9.90	7.00	12.25	57.0	75.5	61.6	−20.3

different predictive methods described here, the authors prefer in their publications to quote the predicted structure against the "optimized" experimental structure, but for the purposes of this review we shall just consider the raw experimental results, although the experimental structure minimized through the same lattice-energy refinement procedure is an equally fair comparison. The choice of comparison depends on whether the whole methodology, or specifically the search approach, is being examined. The unit-cell values for best predicted structure are very close to the known experimental values, but a number of other predicted structures (possibly, as yet unknown polymorphs?) have very similar lattice energies.

The results for this molecule are very impressive, but for other molecules the performance of the predictive method is not as clear cut, with significant errors being found in the size of the unit-cell angles (Gavezzotti 1991a). Further refinement of the methodology and the assumptions therein is likely to improve the results. One of the key assumptions is concerning the transition from molecular nuclei to full structure generation. That decision, based upon the cluster energy, is made before the full periodic conditions have been applied. A strong interaction within a nucleus does not necessarily mean that this motif will appear in the full structure. The current approach is being developed to consider other symmetry operations, such as centering and mirror-plane symmetry operations, and to consider situations when there is more than one molecule in the crystallographic asymmetric unit.

In a similar process, Holden et al. (1993) considered an approach in terms of molecular coordination spheres. From a detailed examination of the Cambridge Crystallographic Database (Section 3.4), coordination numbers were identified for various space groups. The search was restricted to molecules containing just C, H, N, O, and F with only one molecule in the asymmetric unit (that is, no solvates or complexes). The most common coordination number was fourteen. For example, the space group $P\bar{1}$ consists of two molecules along each axis, two along each of the three shortest face diagonals and two along the shortest body diagonals.

The construction of possible full crystal packing patterns follows in three stages. Initially a line of molecules is established by moving a second molecule towards the central molecule until some specified interaction energy criterion is achieved. A two-dimensional grid is then organized by moving a line of molecules towards the central line. The final step involves moving a two-dimensional grid parallel to the central grid. The orientation of the molecules within the two-dimensional grids and the three-dimensional packing arrangement depend on the symmetry operations within the space group being investigated. The crystal system to which the space group belongs determines the technique adopted for the approach of the grids toward each other. In an orthorhombic space group, the unit cell angles are fixed, so only the dimensions have to be determined. In the case of a monoclinic system the approach of the grids has to be shifted relative to the appropriate axis so that the unique angle can be obtained as well as the unit-cell dimensions. The MOLPAK program developed by the authors of this approach will produce a three-dimensional map of the minimum unit-cell volume as a function of the orientation of a rigid body. The conformation of the rigid body is produced using semi-empirical quantum mechanics calculations (Section 3.7.1).

Once a series of trial structures has been generated, these are refined using standard lattice-energy minimization methods. To reduce computation time the authors prefer, however, not to calculate an electrostatic contribution to the lattice energy, as detailed in Section 3.3, but rather to use alternative A and B nonbonded parameters (equation (3.3)), which compensate for the lack of a separate electrostatic contribution. The performance of this method can be judged from the compound 2,4,6-trinitro-N-methyldiphenylamine (CSD refcode VIHGAU), the molecular structure of which is given in Fig. 3.26. Table 3.12 has the observed structure, quoted against predicted values for various space groups. The lowest-energy theoretical structure is in excellent agreement with the experi-

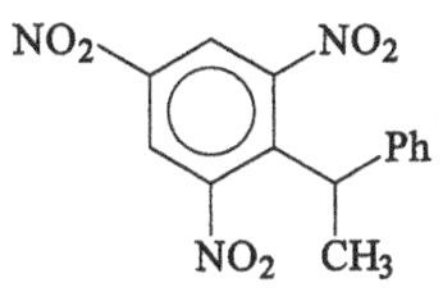

Fig. 3.26. The molecular structure of 2,4,6-trinitro-*N*-methyldiphenylamine (CSD refcode VIHGAU).

Table 3.12. *The observed and calculated unit cell data for the compound 2,4,6-trinitro-N-methyldiphenylamine (CSD refcode VIHGAU); the molecular structure is shown in Fig. 3.26 (Holden et al. 1993)*

Space group	a (Å)	b (Å)	c (Å)	α (°)	β (°)	γ (°)	Lattice energy (kcal/mol)
$P2_1/c$							
Experiment	12.654	7.371	15.083	90.0	101.76	90.0	—
Experiment/ optimization	12.772	7.344	15.021	90.0	100.39	90.0	−34.71
Calculated	12.773	7.344	15.021	90.0	100.38	90.0	−34.71
$P\bar{1}$	10.963	8.800	8.194	99.23	72.34	112.3	−33.99
$P2_1$	8.305	9.032	10.248	90.0	70.20	90.0	−32.42
$P2_12_12_1$	15.067	13.49	6.976	90.0	90.0	90.0	−32.49

mental arrangement (Holden et al. 1993). It is interesting to note that the difference between all the potential structures is only a few kcal and the prediction closest to the observed structure is only 0.7 kcal/mol more stable than the next on the list.

In recent publications a Monte Carlo cooling approach has been applied to describe the packing of one-dimensional stacking aggregates, including perylenedicarboximide pigments, as shown in Fig. 3.27 (Perlstein 1994). Starting aggregates are generated using five identical molecules randomly oriented at set separation distances along a defined axis. The central molecule has four surrounding molecules set at $\pm r$ and $\pm 2r$, where r is the separation distance between the centers of gravity of the molecules. The Monte Carlo cooling is carried out from 4000 K to 300 K. Each Monte Carlo step involves randomly rotating the central molecule and regenerating the other molecules along the axis at the set distances. The energy of this arrangement is then computed and compared with the previous energy and either accepted or rejected, based on the acceptance criteria of the Monte Carlo algorithm. This process is repeated twenty times for both rotational angles and the separation distance r. The temperature is

Table 3.13. *Comparison between the calculated and experimental values for the interplanar spacing d, the longitudinal displacement l and the transverse displacement t for some perylenedicarboximides (Perlstein 1994)*

Structure	R substituent	d	l	t	Source
DICNUY	$-CH_2CH_2CH_2CH_3$	3.379	3.13	1.19	Monte Carlo
		3.402	3.10	1.11	Experimental
		3.396	2.98	1.04	Monte Carlo
DICPIO	$-CH_2$-phenyl	3.425	3.08	1.10	Experimental
SAGWEC	-azobenzene	3.416	0.79	3.84	Monte Carlo
		3.480	0.81	3.96	Experimental

Notes: Structure identifier is the Cambridge Structural Database reference code. See Fig. 3.27 for basic structure and axis definition for d, l, and t. Data in Ångstroms.

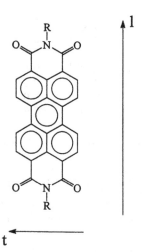

Fig. 3.27. The molecular structure of perylenedicarboximides and the reference translational and longitudinal diplacements.

then cooled by 10 percent and the process repeated. At a temperature of 300 K the lowest-energy structure found is accepted as a local minimum. This whole procedure is repeated until 350 local minima (which should include the global minimum) have been found.

A summary of the results for the stacking is given in Table 3.13. The results include both predicted and experimental values for the interplanar separation distance d, the longitudinal displacement l, and the transverse displacement t. The reference axes are shown in Fig. 3.27. The results show good agreement between experimental and calculated values. Although this is not a full three-dimensional structure prediction, this technique is one that has potential to contribute to the problem of predicting three-dimensional structures. It is of particular interest that these results have been obtained for molecules with flexible end groups. The nonbonded interaction and torsional terms were taken from a well known molecular mechanics package and partial atomic charges derived by the Gasteiger method (Gasteiger and Marsili 1980).

Recently an approach to the crystal-packing problem, based around a modified Monte Carlo simulated-annealing process, was described (Karfunkel and Gdanitz 1992). The packing problem is addressed in three stages. The first step is the key to the method, the Monte Carlo procedure producing a large number of trial structures. The second step reduces the number of these structures by clustering together structures that are "similar." Based on this reduced set of trial structures, standard lattice-energy minimizations produce fully refined three-dimensional structures.

The key step is the modified Monte Carlo simulation, which involves the separation of the variables that define the spatial extensions of the crystal from those concerned with the angular degrees of freedom (Gdanitz 1992). The translational vectors of the other molecules in the cell are expressed in polar coordinates so that the distances between the centers of the central molecule and those of the surrounding molecules can be considered as part of the angular variables. This removal of the variables responsible for spatial extension of the crystal from the Monte Carlo search space reduces the tendency for the crystal to remain as a gas. The angular variables that define the Monte Carlo search space include the unit-cell angles, the orientation angles of the central and surrounding molecules and the angular component of the polar coordinate definition of the surrounding molecules. Essentially each Monte Carlo step chooses a set of independent angular variables. For each set of angular variables the parameters affecting the spatial extension of the crystal are optimized. This is initially done by relieving any bad contacts, as defined by the van der Waals radii (Section 3.2), followed by optimization with respect to the crystal energy.

The results of this methodology are extremely hopeful, with predictions showing excellent agreement with experiment. This is reflected in Table 3.14 where the calculations for 4,8-dimethoxy-3,7-diazatricyclo(4.2.2.2-2,5-) dodeca-3,7,9,11-tetraene (Fig. 3.28), CSD refcode BAWHOW, are presented. The postulated structures show excellent agreement with experiment; the more sophisticated the charge description, the better the description of the unit-cell angles. The MNDO electrostatic potential-fitted charges give a much better description of the unit-cell angles than the same simulation using the cruder Gasteiger charges. In more recent work, these authors have gone on to examine other materials and have preferred to use potentially-fitted charges from high-level *ab initio* quantum-chemistry calculations (Leusen 1996).

At present, there are a number of methods that permit tentative predictions of solid-state structures from a molecular structure. The assumptions made in these approaches tend to restrict the routine application of the methods to industrially

Table 3.14. *The observed and calculated structures for the molecular structure of 4,8-dimethoxy-3,7-diazatricyclo(4.2.2.2-2,5-) dodeca-3,7,9,11-tetraene (CDS refcode BAWHOW); the molecular structure is shown in Fig. 3.28*

	a (Å)	b (Å)	c (Å)	α (°)	β (°)	γ (°)	Total energy (kcal/mol)
Experiment	6.37	7.16	6.35	100.0	106.0	73.5	—
Minimization	6.56	7.05	6.50	102.0	108.0	72.0	86.49
Minimization/ reduced	6.50	6.56	7.05	72.0	78.1	72.4	—
1	6.50	6.56	7.05	72.0	78.0	72.4	86.52
2	6.21	6.31	7.96	89.5	74.2	64.0	86.81
3	6.21	6.31	8.66	72.2	62.1	64.0	86.87
4	5.89	7.09	8.10	105.0	120.0	101.0	86.92
1*	6.36	6.61	7.03	66.7	77.0	74.1	—

Note: The calculations were made using MNDO ESP charges except for 1*, which was calculated using Gasteiger charges (Karfunkel and Gdanitz 1992).

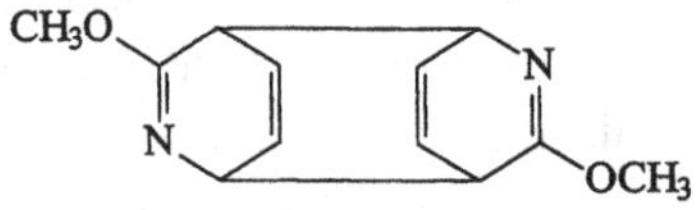

Fig. 3.28. The molecular structure of 4,8-dimethoxy-3,7-diazatricyclo(4.2.2.2-2,5-) dodeca-3,7,9,11-tetraene (CSD refcode BAWHOW).

important materials. A number of issues still have to be addressed. These include:

- The use of rigid probes as input structures hinders the application to conformationally flexible molecules.
- The description of the electrostatics varies from simple point-charge methods (Gasteiger, for example) to potential-derived charges from higher-level *ab initio* methods. The dependence of the quality of the results on these charge derivation methods has still to be fully explored and explained.
- The differences between the calculated structure that is closest to the known structure and other predicted structures (which are not experimentally observed) are in some cases only fractions of a kcal/mol. The magnitude of the differences might be a consequence of difficulties in the search methods or in the accuracy of

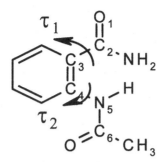

Fig. 3.29. The molecular structure of *o*-acetoamidobenzamide.

force fields, or the shallowness of the potential surface being examined. It may of course be a result of the existence of as yet unknown and/or unstable polymorphs.

• The need for an unified force field that describes the molecular structure (in particular, torsional freedom) to the same degree of acccuracy as the intermolecular interactions.

All the aforementioned difficulties have been discussed in a recent theoretical investigation of conformational polymorphism (Buttar et al. 1998). In this study, conformational maps were derived from theoretical methods and compared against molecular structures from different polymorphs. One example (*o*-acetoamidobenzamide) (Fig. 3.29) exists in two polymorphic forms, α and β. Both forms possess strong N—H:::O=C intermolecular hydrogen bonds. The α form differs from the β in that it possesses an additional intramolecular hydrogen bond, which results in different conformations. Fig. 3.30 shows the conformational map for the two torsions that define the conformational differences between α and β. Clearly, the α conformation is considerably more stable than the β. The map shows that the α and β occur close to, but not actually at, minima on the potential energy surface. Interestingly, there are other minima (for example $\tau_1 = -135$, $\tau_2 = -60$) on the map with energies lower than the β conformer, which are not observed experimentally. This raises the intriguing possibility of as yet undiscovered conformational polymorphs. However, these may not be observed, as this particular conformation may not pack in an effective manner.

All the theoretical methods applied suggest that the α conformation is more stable than the β conformation by approximately 10 kcal/mol. Since the β structure is observed, it might be assumed that the lattice must stabilize this energetically unfavorable conformation. It might therefore be expected that the β form would have a lattice energy around 10 kcal/mol more stable than α. However lattice energy calculations show that the β polymorph is only 2.5 kcal/mol more stable than the α form. The difference between the total energies of the two polymorphs (conformational and intermolecular) is around 7.5 kcal/mol. This is considerably larger than the differences normally seen in polymorph prediction approaches (see Table 3.14).

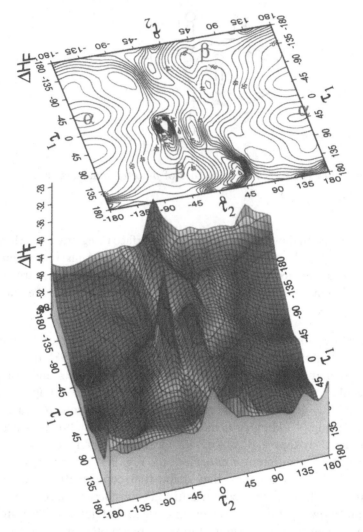

Fig. 3.30. The conformational map for the two torsions that define the conformational differences between the α and β polymorphs of o-acetoamidobenzamide.

3.7.3. *Structure Determination from Powder Diffraction Patterns – Rietveld Refinement in Combination with Theoretical Methods*

Although it is sometimes difficult to get a single crystal of sufficient size and quality in the desired polymorph for traditional single-crystal studies, it is usually possible to get a powder of the material of interest. From a known crystal structure (unit-cell dimensions, atomic coordinates) an x-ray diffraction trace for that material can be simulated. The positions of the peaks are dependent on the repeat spacings in the crystal and the intensities are a consequence of the atom positions. Since it is possible to simulate the x-ray diffraction trace for known structures, it is clearly possible to simulate the patterns for postulated trial structures and alter these trial structures to "fit" the experimental pattern. This procedure of simulation/fit pattern/alter trial structure is the basis of the Rietveld refinement method (Rietveld 1969). A number of other parameters related to machine function and sample characteristics also have to be considered when matching the patterns.

Traditionally structure solution using the Rietveld method has been restricted to high-symmetry inorganic structures where peak overlap is minimal (Attfield et al. 1988). Recently, there has been an increase in the number of studies on molecular materials (Lightfoot et al. 1992; Delaplane et al. 1993; Fitch and Jobic 1993). Starting structures for refinement have conventionally been generated using traditional single-crystal techniques, which involve the separation and integration of individual peaks. These approaches tend to struggle with low-symmetry systems, which are the most popular with organic molecules (Section 3.2) where peak overlap is significant. Methods currently being developed allow better trial structure generation using a combination of molecular-orbital calculations (Section 3.7.1) and crystal-packing calculations (Fagan, Roberts, and Docherty 1994; Fagan et al. 1995), as well as Monte Carlo algorithms (Harris et al. 1994; Newsam, Deem, and Freeman 1992). Maximum entropy methods (Tremayne et al. 1992) have also been developed in order that greater amounts of information can be extracted from these patterns.

Although structure solution from powders requires some experimental data (the x-ray diffraction trace), it has some benefits over the *ab initio* approach to structure solution. The use of the diffraction trace as an experimental observable reduces concerns over the location of the global minimum and force-field accuracy, which have hindered the routine application of *ab initio* methods.

3.7.3.1. *Crystal-Packing Calculations in Combination with Rietveld Methods*

The solution of a structure by the Rietveld method has been considered as a multistage problem (Fagan et al. 1994; Fagan et al. 1995). The methodology is summarized in Fig. 3.31 and below:

- The first of these involves the collection of the data, the indexing of the resulting pattern and the determination of the unit-cell dimensions and the space group.
- This is followed by an initial model construction, which involves molecular-orbital calculations to generate a reasonable molecular structure.
- The third stage involves the introduction of the molecular structure into the unit cell and examination of potential packing motifs consistent with the space-group symmetry.
- The final stage involves the actual Rietveld refinement, where slight changes to the packing arrangement/molecular structure are carried out so that the simulated pattern from the proposed structure gets closer to the observed pattern.

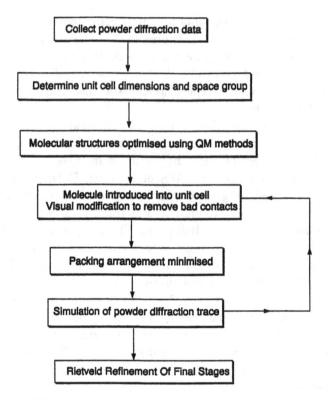

Fig. 3.31. The basic methodology for structure solution from powders.

During the second stage, the molecular structure is introduced into the unit cell in various arrangements and the lattice energy minimized with respect to the molecular orientation and translation. Unlike the refinement processes described in the previous section, the unit-cell dimensions are held fixed. The advantage over the *ab initio* prediction methods is that in this approach the simulated/ experimental powder trace match is used as a final check and thus concerns over global minimization, force-field accuracy, and the description of the elec- trostatics are reduced.

This validity of the method has been demonstrated by comparing packing arrangements based on a powder analysis with structures obtained through independent single-crystal studies. Amongst the molecules studied was the com- mercially important triphendioxazine system. 6,13-Dichlorotriphendioxazine ($C_{18}H_8N_2O_2Cl_2$), as shown in Fig. 3.32, is the basic chromophore unit of a number of commercially important dyestuffs and no structural information was available on these types of systems at the start of this study. The sample was obtained as a highly crystalline powder from nitrobenzene. High resolution powder diffraction data using a Debye-Scherrer scattering geometry was recorded on beamline 2.3 at the SERC's Synchrotron Radiation Source (SRS) at Daresbury laboratory.

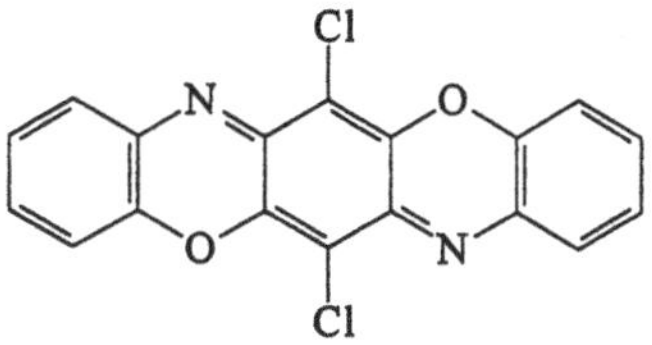

Fig. 3.32. The molecular structure of 6,13-dichlorotriphendioxazine.

Table 3.15. *Crystallographic data for 6,13-dichlorotriphendioxazine determined for powder and single-crystal x-ray diffraction*

Empirical formula	$(C_{18}H_8N_2O_2Cl_2)$	
Molecular weight	355.2	
Density (mg/cm^{-3})	1.646	
	Single crystal	Powder
Morphology	Prism	—
Size (μm)	$370 \times 150 \times 70$	ca 45
Space group	$P2_1/c$	$P2_1/c$
a (Å)	8.700	8.717
b (Å)	4.870	4.887
c (Å)	17.064	17.174
β (°)	97.50	97.86
Volume (Å^3)	716.8	723.6
Z	2	2
Absorption coefficient	0.467	—
Temperature (K)	298	298
Wavelength (Å)	0.7106	1.2023
2θ range	3–50	2–50
Standard reflections	$(-3\ 0\ 6)$, $(1\ 0\ 10)$	—
R factor	0.073	0.136

Unit-cell dimensions were obtained after the indexing of patterns using the programs of Werner, Visser and Louer (Shirley 1978). The space group was then determined by consideration of the systematic extinction conditions. The lattice parameters and packing coefficients were consistent with two molecules in the unit cell (Table 3.15). An initial model for the molecular structure was built for 6,13-dichlorotriphendioxazine using standard fragments. An initial optimization of this structure using molecular mechanics was followed by further refinement using the molecular-orbital package MOPAC (Section 3.7.1).

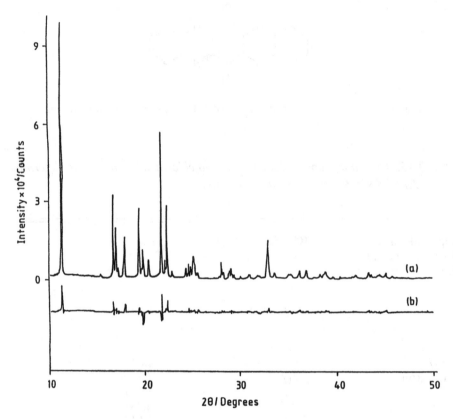

Fig. 3.33. The observed x-ray diffraction pattern (*a*) and difference between simulated and observed patterns (*b*) for 6,13-dichlorotriphendioxazine.

This molecular structure was then introduced into the unit cell. Initial bad contacts were removed through rotation of the molecule (and those related by symmetry) into a reasonable position. On the basis of standard crystal-packing calculations, the lattice energy was minimized. The x-ray diffraction pattern was then simulated and the proposed structure manipulated through further rotations and minimization cycles until the best fit between calculated and experimental patterns was found. For a comprehensive study this can be a time-consuming procedure and further work in this area has demonstrated the value of an automated, systematic search procedure for this stage (Hammond et al. 1996; Hammond et al. 1997).

All the refinement procedures were carried out using the Rietveld refinement program DBWS. The initial parameters refined included the zeropoint correction for the detector, the peak profile, the background correction, peak asymmetry, and the unit-cell dimensions. The crystal shape was also taken into account. The final *R* factor for this structure was 0.013. The final refinement plot is shown in Fig. 3.33.

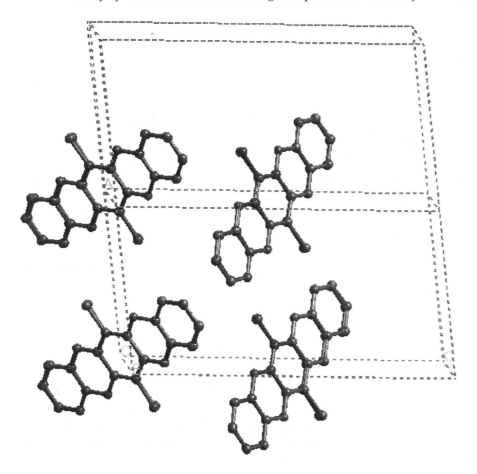

Fig. 3.34. The overlay of the structures for 6,13-dichlorotriphendioxazine determined from powder (dark gray) and by an independent single crystal structure determination (light gray).

An *independent* single-crystal study (Potts 1994) from the same sample was undertaken as a way to verify the structure determined from powder diffraction. Vacuum sublimation at temperatures around 400°C produced dark-red, prism-shaped crystals. The structure was solved by direct methods using SHELXTL/PC (version 4.1) and refined using a full-matrix least-squares fit. Hydrogen atoms not located from the Fourier difference maps were geometrically fixed. The final R factor for this structure was 0.073. The unit-cell dimensions are listed in Table 3.15.

Figure 3.34 shows the overlay of the unit cell for the structure solved by powder diffraction with the packing of the arrangement determined by single-crystal methods. The structures are very similar, with only slight deviations. The structure of 6,13-dichlorotriphendioxazine is held together through π–π interactions down the **b** axis. Non-bonded Cl--H contacts at 2.8 Å form dimers in the **ac** plane.

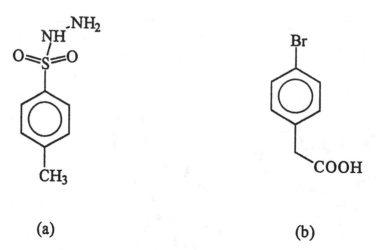

Fig. 3.35. The molecular structures of 4-toluene sulfonyl hydrazine (a) and p-bromo-phenylacetic acid (b).

3.7.3.2. Monte Carlo Algorithms in Combination with Rietveld Methods

Monte Carlo methods have recently been applied to solve previously known structures in order to check the methodology and to predict unknown structures (Harris et al. 1994; Newsam, Deem, and Freeman 1992). The Monte Carlo approach involves guessing the plausible atomic arrangements, checking their suitability against the diffraction pattern and finally refining the structure using conventional Rietveld methods. Each configuration (trial structure) is generated by a random displacement (subject to a specified constraint). The powder diffraction pattern is then simulated for this configuration and the scale factor optimized. The agreement factor for this new configuration is then compared with the previous configuration and accepted or rejected depending on a user-defined probability. The probability and maximum displacements are chosen so that around 40 percent of the trial moves are accepted. After a sufficient number of configurations the best trials (those with the best agreement factors) are considered for full Rietveld refinement.

The method has been confirmed for the known structure 4-toluene sulfonyl hydrazine (Fig. 3.35(a)) (Harris et al. 1994). In this system only the sulfur atom was considered initially. After 1,000 moves the best sulfur fractional coordinates were found to be 0.6690, 0.1610, 0.4161. These were then refined using Rietveld methods to 0.6692, 0.1486, 0.4031. The position of the sulfur was fixed and the position of the phenyl ring refined. The phenyl ring was considered as a rigid body and rotated at random around an axis constrained to pass through the known S atom. After 1,000 moves the best position for the phenyl ring was determined. This was used as a starting model for traditional refinement and the final refined structure was within the experimental error of the previous

Table 3.16. *The atomic coordinates for the sulfur atom in 4-toluene sulfonyl hydrazine after the Monte Carlo step, subsequent refinement and full refinement (Harris et al. 1994)*

Refinement stage	X	Y	Z
Monte Carlo step	0.6690	0.1610	0.4161
Single atom refinement	0.6692	0.1486	0.4031
Full Rietveld refinement	0.6835	0.1583	0.4085

single-crystal structure determination. An overlay of the unit cell determined by powder and single crystal (filling the four Ph–S units in the cell) gave an RMS fit of 0.0645. This reflects the almost perfect match of the structures determined in the two different methods. The coordinates for the sulfur atom are compared in the various stages in Table 3.16.

This approach has been applied to solve the previously unknown structure of *p*-bromophenyl acetic acid (Fig. 3.35(*b*)) (Harris et al. 1994). The unit cell was determined using indexing programs as described previously and the space group assigned on the basis of systematic absences. In a similar manner to the previous system, the Br atom was considered by itself and then as part of a larger unit consisting of the Br–Ph–C fragment of the structure. The best Monte Carlo move was then refined and the final *R* factor was 0.066. Increasingly, more complex molecules are examined by this procedure. The promise of this approach has recently been demonstrated in the crystal structure determination of Red Fluorescein (Tremayne et al. 1997).

3.8. Conclusions

Molecular modeling and computational chemistry are playing an increasingly important role in the study of molecular materials. The visualization of the complex and often elegant three-dimensional structures permits a greater understanding of molecule/solid-state structure relationships.

Crystallization is essentially a molecular recognition phenomenon on a grand scale. Polymorphism is the consequence of the recognition of a different balance of the subtle intermolecular interactions possible in different solid-state arrangements. Lattice-energy calculations allow the relative stabilities of different solid-state packing motifs to be examined and the important intermolecular interactions to be characterized.

External crystalline properties such as crystal shape can be linked to the packing arrangements through understanding the surface structure and the important intermolecular interactions between growth units. Habit modification can be interpreted as a form of "molecular trickery." The impurity fools the surface by completing the expected intermolecular bonding pattern through presenting similar functional groups. Once incorporated, the impurity disrupts growth along a direction by interfering with the normal intermolecular bonding pattern. Energy calculations can be employed to determine the ease of incorporation of the impurity and its subsequent disruption of growth.

Ultimately, the aim of molecular modeling and computational chemistry is the design (before synthesis) of novel molecules and materials. In molecular crystal chemistry this requires the *ab initio* prediction of solid-state structures. Although much progress has been made and the potential of such an ability is enormous, progress has been hindered by problems with global minimization and force-field accuracy. Predictions remain restricted, with rigid molecular structures as input to the methodologies.

The determination of structural information from x-ray powder diffraction has seen a renaissance in recent years. The strengths of computational chemistry in its rapid generation of "chemically sensible" trial structures complement the need for good starting data for Rietveld refinement. The use of the experimental diffraction trace as a check reduces some of the concerns over force-field accuracy and global minimization that have plagued *ab initio* predictions.

This methodology permits the examination of the solid-state structure of industrially relevant materials, where traditional single-crystal structure determination is not possible. It is in the study of these materials that these techniques offer considerable promise.

References

Aakeroy, C. B. and Seddon, K. R. (1993). The hydrogen bond and crystal engineering. *Chemical Society Reviews*, pp. 397–407.

Allen, F. H., Kennard, O., and Taylor, R. (1983). Systematic analysis of structural data as a research tool. *Accounts of Chemical Research*, **16**, 146–53.

Allen, F. H., Kennard, O., Watson, D. G., Brammer, L., Guy Opren, A., and Taylor, R. (1987). Tables of bond lengths determined by x-ray and neutron diffraction. Part 1. Bond lengths in organic compounds. *Journal of the Chemical Society Perkin Transactions* II, p. S2.

Allinger, N. L. and Burket, U. (1982). *Molecular Mechanics*. ACS Monograph 177. Washington, DC: American Chemical Society.

Atkinson. C., Heybourne, M. J. H., Iley, W. J., Russell, P. C., Taylor, T., and Jones D. P. (1987). Powder for use as a granular detergent composition – obtained from sodium carbonate and sodium sulphate using a crystal growth modifier. European Patent EP221776A.

Attfield, J. P., Cheetham, A. K., Cox, D. E., and Sleight, A. W. (1988). Synchrotron x-ray and neutron powder diffraction studies of the structure of α-$CrPO_4$. *Journal of Applied Crystallography*, **21**, 452–7.

Berkovitch-Yellin, Z. (1985). Toward an *ab-initio* derivation of crystal morphology. *Journal of the American Chemical Society*, **107**, 8239–53.

Black, S. N., Williams, L. J., Davey, R. J., Moffat, F., McEwan, D. M., Sadler, D. E., Docherty, R., and Williams, D. J. (1990). Crystal chemistry of 1-(4-chlorophenyl)-4,4-dimethyl-2-(1H-1,2,4-triazol-1-yl)pentan-3-one, a paclobutrazole intermediate. *Journal of Physical Chemistry*, **94**, 3223–6.

Bravais, A. (1866). *Etudes Crystallographiques*. Paris: Gauthiers, Villars.

Brock, C. P. and Dunitz, J. D. (1994). Towards a grammar of crystal packing. *Chemical Materials*, **6**, 1118–27.

Buttar, D., Charlton, M. H., Docherty, R., and Starbuck, J. (1998). Theoretical investigations of conformational aspects of polymorphism. Part 1: *o*-Acetamodobenzamide. *Journal of the Chemical Society Perkin Transactions*, **2**, 763–71.

Chang, S. and Hamilton, A. D. (1988). Molecular recognition of biologically interesting substrates: Synthesis of an artificial receptor employing six hydrogen bonds. *Journal of the American Chemical Society*, **110**, 1318–19.

Charlton, M. H., Docherty, R., and Hutchings, M. G. (1995). Quantitative structure–sublimation enthalpy relationship studied by neural networks, theoretical crystal packing calculations and multilinear regression analysis. *Journal of the Chemical Society Perkin Transactions*, **2**, 2023–9.

Clark, M., Cramer, R. D., and Van Opdenbosch, N. (1989). Validation of the general purpose tripos 5.2 force-field. *Journal of Computational Chemistry*, **10**, 982–1012.

Clydesdale, G., Roberts, K. J., and Docherty, R. (1994*a*). Modelling the morphology of molecular crystals in the presence of disruptive tailor-made additives. *Journal of Crystal Growth*, **135**, 331–40.

Clydesdale, G., Roberts, K. J., Lewtas, K., and Docherty, R. (1994*b*). Modelling the morphology of molecular crystals in the presence of blocking tailor made additives. *Journal of Crystal Growth*, **141**, 443–50.

Cohen, M. L. (1989). Materials from theory, *Nature*, **338**, 291.

Delaplane, R. G., David, W. I. F., Ibberson, R. M., and Wilson, C. C. (1993). The *ab initio* crystal structure determination of α-malonic acid from neutron powder diffraction data. *Chemical Physics Letters*, **201**, 75–8.

Desiraju, G. R. (1989). *Crystal Engineering – The Design Of Organic Solids*. Amsterdam: Elsevier.

Desiraju, G. R. (1995). Supramolecular synthons in crystal engineering – a new organic synthesis. *Angewandte Chemie International Edition English*, **34**, 2311–27.

Dewar, M. J. S. (1993). The semi-*ab initio* (SA) approach to chemistry. *Organic Mass Spectrometry*, **28**, 305–10.

Docherty, R., Clydesdale, G., Roberts, K. J., and Bennema, P. (1991). Application of Bravais–Friedel–Donnay–Harker, attachment energy and Ising models to understanding the morphology of molecular crystals. *Journal of Physics D Applied Physics*, **24**, 89–99.

Docherty, R. and Jones W. (1997). Theoretical methods for structure determination. *Organic Molecular Solids: Properties and Applications*, Chapter 3, pp. 113–48. London: CRC Press.

Docherty, R. and Roberts K. J. (1988). Modelling the morphology of molecular crystals: Application to anthracene, biphenyl and β-succinic acid. *Journal of Crystal Growth*, **88**, 159–68.

Docherty, R., Roberts, K. J., Saunders, V., Black, S. N., and Davey, R. J. (1993). Theoretical analysis of the polar morphology and absolute polarity of crystalline urea. *Faraday Discussions*, **95**, 11–25.

Donnay, J. D. and Harker D. (1937). A new law of crystal morphology extending the law of Bravais. *American Mineralogist*, **22**, 446–67.

Etter, M. C. (1991). Hydrogen bonds as design elements in organic chemistry. *Journal of Physical Chemistry*, **95**, 4601–10.

Fagan, P. G., Roberts K. J., and Docherty R. (1994). Investigating the crystal structure of triphendioxazine using a combination of high resolution powder diffraction and computational chemistry. *Institution of Chemical Engineers Research Event*, **2**, 719–21.

Fagan, P. G., Roberts, K. J., Docherty R., Chorlton, A. P., Jones, W., and Potts, G. P. (1995). An *ab-initio* approach to crystal structure determination using high resolution powder diffraction and computational chemistry techniques: application to 6,13-dichlorotriphendioxazine. *Chemistry of Materials*, **7**(12), 2322–6.

Fan, E., Vicent, C., Geib, S. J., and Hamilton, A. D. (1994). Molecular recognition in the solid state: Hydrogen-bonding control of molecular aggregation. *Chemistry of Materials*, **6**, 1113–17.

Fanfani, L., Tomassini, M., Zanazzi, P. F., and Zanari, R. (1975). Theoretical structures for pyridazino [4,5-d] pyridazine and 1,4,5,8 tetramethoxypyridazino [4,5-d] pyridazine. *Acta Crystallographica*, **B31**, 1740–5.

Fitch, A. N and Jobic, H. (1993). The crystal structure of nonborane. *Journal of the Chemical Society Chemical Communications*, pp. 1516–17.

Friedel, M. G. (1907). Etudes sur la loi de Bravais. *Bulletin de la Société Française de Mineralogie*, **30**, 326–455.

Gasteiger, J. and Marsili, M. (1980). Iterative partial equalization of orbital electronegativity – a rapid access to atomic charges, *Tetrahedron*, **36**, 3219–22.

Gavezzotti, A. (1991*a*). Generation of possible crystal structures from the molecular structure for low polarity organic compounds. *Journal of the American Chemical Society*, **113**, 4622–9.

Gavezzotti, A. (1991*b*). Molecular packing and other structural properties of crystalline oxohydrocarbons. *Journal of Physical Chemistry*, **95**, 8948–55.

Gavezzotti, A. and Filippini, G. (1992). Molecular packing of crystalline azahydrocarbons. *Acta Crystallographica*, **B48**, 537–45.

Gavezzotti, A. and Filippini, G. (1994). Geometry of intermolecular $X - H \cdots Y(X, Y = N, O)$ hydrogen bond and the calibration of empirical hydrogen bond potentials. *Journal of Physical Chemistry*, **98**, 4831–7.

Gdanitz, R. J. (1992). Prediction of molecular crystal structures by Monte Carlo simulated annealing without reference to diffraction data. *Chemical Physics Letters*, **190**, 391–6.

Gibbs, J. W. (1906). *Collected Works*. New York: Longmans Green.

Grant, G. H. and Richards, G. W. (1995). *Computational Chemistry*. Oxford: Oxford Science Publications.

Hammond, R. B., Roberts, K. J., Docherty, R., Edmondson, M., and Gairns, R. (1996). X-form metal free phthalocyanine: Crystal structure determination using a combination of high resolution x-ray powder diffraction and molecular modelling techniques. *Journal of the Chemical Society Perkin Transactions*, **2**, 1527–8.

Hammond, R. B., Roberts, K. J., Docherty, R., and Edmondson, M. (1998). Computationally assisted structure determination for molecular materials from x-ray powder diffraction data. *Journal of Physical Chemistry B*, **101**, 6532–6.

Harris, K. D. M., Tremayne, M., Lightfoot, P., and Bruce, P. G. C. (1994). Crystal structure determination from powder diffraction methods by Monte Carlo methods. *Journal of the American Chemical Society*, **116**, 3543–7.

Hartman, P. and Bennema, P. (1980). The attachment energy as a habit controlling factor. *Journal of Crystal Growth*, **49**, 145–56.

Hartman, P. and Perdok, W. G. (1955). On the relations between structure and morphology of crystals I. *Acta Crystallographica*, **8**, 49–52.

Holden, J. R., Du, Z., and Ammon, L. (1993). Prediction of possible crystal structures for C, H, N, O and F containing compounds. *Journal of Computational Chemistry*, **14**, 422–37.

Hunter, C. A. and Sanders, J. K. M. (1990). The nature of π–π interactions. *Journal of the American Chemical Society*, **112**, 5525–34.

Jones, J. E. (1923). On the determination of molecular fields: 1. From the variation of the viscosity of a gas with temperature. *Proceedings of the Royal Society A.*, **106**, 441–7.

Karfunkel, R. and Gdanitz, R. J. (1992). *Ab-initio* prediction of possible crystal structures for general organic molecules. *Journal of Computational Chemistry*, **13**, 1171–83.

Kitaigorodskii, A. I. (1973). *Molecular Crystals and Molecules*. New York: Academic Press.

Kitaigordskii, A. I., Mirskaya, K. V., and Tovbis, A. B. (1968). Lattice energy of crystalline benzene in the atom–atom approximation. *Soviet Physics Crystallography*, **13**, 176–80.

Lacy, M. E. (1990). Neural network technology and its application in chemical research. *Tetrahedron Computer Methodology*, **3**, 119–28.

Lahav, M., Berkovitch-Yellin, Z., Van Mil, J., Addadi, L., Idelson, M., and Leiserowitz, L. (1985). Stereochemical discrimination at organic crystal surfaces. 1. The systems serine/threonine and serine/allothreonine. *Israel Journal of Chemistry*, **25**, 353–61.

Leusen, F. J. J. (1996). *Ab-initio* prediction of polymorphs. *Journal of Crystal Growth*, **166**, 900–3.

Lifson, S., Hagler, A. T., and Dauber, P. (1979). Consistent force field studies of intermolecular forces in hydrogen bonded crystals. 1. Carboxylic acids, amides and the C=O:::H hydrogen bonds. *Journal of the American Chemical Society*, **101**, 5111–21.

Lightfoot, K., Tremayne, M., Harris, K. D. M., and Bruce, P. G. (1992). Determination of a molecular crystal structure by x-ray powder diffraction on a conventional laboratory instrument. *Journal of the Chemical Society Chemical Communications*, **14**, 1012–13.

Litvinov, I. A. (1990). Molecular and crystal structure of triethylammonium 6,7-benzo-3-*o*-hydroxyphenyl-1,4-diphenyl-2,8,9-trioxa-4-phospha-1-boratobicyclo [3.3.1] nonane. *Zhurnal Strukturnui Khimii*, **31**, 351–62.

Maddox, J. (1988). Crystals from first principles. *Nature*, **335**, 201.

Maginn, S. J., Compton, R. G., Harding, M., Brennan, C.M., and Docherty, R. (1993). The structure of 2,4,6-trichloro-1,3,5-triazine. Evidence of anisotropy in chlorine/nitrogen interactions. *Tetrahedron Letters*, **34**, 4349–52.

Mayo, S. L., Olafson, B. D., and Goddard, W. A. III (1990). DREIDING: A generic force field for molecular simulations. *Journal of Physical Chemistry*, **94**, 8897–8909.

Momany, F. A., Carruthers, L. M., McGuire, R. F., and Scheraga, H. A. (1974). Intermolecular potentials from crystal data. III – Determination of empirical potentials and application to the packing configurations and lattice energies in crystals of hydrocarbons, carboxylic acids, amines and amides. *Journal of Physical Chemistry*, **78**, 1595–1620.

Newsam, J. M., Deem, M. W., and Freeman, C. M. (1992). Direct space methods of structure solution from powder diffraction data. *Accuracy in Powder Diffraction II*, special Publication 846, 80–90, NIST, May.

Perlstein, J. (1994). Molecular self assemblies. 3. Quantitative predictions for the packing geometry of perylenedicarboximide translation aggregates and the effects of flexible end groups. Implications for monolayers and three dimensional crystal structure predictions. *Chemistry of Materials*, **6**, 319–26.

Phoenix, L. (1966). How trace additives inhibit the caking of inorganic salts. *British Chemical Engineer*, **11**, 34–8.

Potts, G. D. (1994). The Crystal Chemistry of Organic Pigments. PhD Thesis, University of Cambridge.

Price, S. L. (1996). Application of realistic electrostatic modelling to molecules in complexes, solids and proteins. *Journal of the Chemical Society Faraday Transactions*, **92**, 2997–3008.

Price, S. L. and Stone A. J. (1984). A six-site intermolecular potential scheme for the azabenzene molecules derived by crystal structure analysis. *Molecular Physics*, **51**, 569–83.

Rietveld, H. M. (1969). A profile refinement method for nuclear and magnetic structures. *Journal of Applied Crystallography*, **2**, 65–71.

Roberts, K. J., Sherwood, J. N., Yoon, C. S., and Docherty, R. (1994). Understanding the solvent-induced habit modification of benzophenone in terms of molecular recognition at the crystal/solution interface. *Chemistry of Materials*, **6**, 1099–1102.

Royer, J., Decoret, C., Tinland, B., Perrin, M., and Perrin, R. (1989). Theoretical determination of the relative stabilities of organic substances. Application on para-substituted phenols. *Journal of Physical Chemistry*, **93**, 3393–6.

Shirley, R. (1978). *Computing in Crystallography*, eds. H. Schenk, O. Olthof-Hazekamp, H. van Koningsveld, and G. C. Bassi, pp. 221–34. Delft: Delft University Press.

Simons, J. (1991). An experimental chemist's guide to ab-initio quantum chemistry. *Journal of Physical Chemistry*, **95**, 1017–29.

Stewart, J. J. P. (1990). MOPAC, a semi empirical molecular orbital program. *Journal of Computer Aided Molecular Design*, **4**, 1–105.

Taylor, R. and Kennard, O. (1982). Crystallographic evidence for the existence of C—H:::O, C—H:::N and C—H:::Cl hydrogen bonds. *Journal of the Chemical Society*, **104**, 5063–70.

Taylor, R. and Kennard, O. (1984). Hydrogen bond geometry in organic crystals. *Accounts of Chemistry Research*, **17**, 320–6.

Treymane, M., Kariuki, B. M., and Harris, K. D. M. (1997). Structure determination of a complex organic solid from x-ray powder diffraction data by a generalized Monte Carlo method: The crystal structure of red fluorescein. *Angewandte Chemie International Edition English*, **36**, 770–2.

Treymane, M., Lightfoot, P., Glidewell, C., Harris, K. D. M., Shankland, K., Gilmore, C. J., Bricogne, G., and Bruce P. G. (1992). The combined maximum entropy and likelihood method to *ab-initio* determination of an organic crystal structure from x-ray data. *Materials Chemistry*, **2**, 1301–2.

Westwell, M. S., Searle, M. S., Wales, D. J., and Dudley, H. W. (1995). Empirical correlations between thermodynamic properties and intermolecular forces. *Journal of the American Chemical Society*, **117**, 5013–15.

Whitesides, G. M., Simanek, E. E., Mathias, J. P., Seto, C. M., Chin, D. N., Mammen, M., and Gordon, D. M. (1995). Noncovalent synthesis: Using physical organic chemistry to make aggregates. *Acc. Chem. Res.*, **28**, 37–44.

Williams, D. E. (1966). Nonbonded potential parameters derived from crystalline aromatic hydrocarbons. *Journal of Chemical Physics*, **45**, 3770–8.

Williams, D. E. and Cox, S. R. (1984). Nonbonded potentials for azahydrocarbons: The importance of the coulombic interaction. *Acta Crystallographica*, **B40**, 404–17.

Wright, J. G. (1987). *Molecular Crystals*. Cambridge: Cambridge University Press.

Wulff, G. (1901). Zur frage der Geschwindigkeit des Wachsthums und der Aufloesung der Krystallflaechen. *Zeitschrift für Kristallographie und Mineralogie*, **34**, 499–530.

Zupan, J. and Gasteiger, J. (1993). *Neural Networks for Chemists*. Weinheim: Verlag Chemie.

4

Towards an Understanding and Control of Nucleation, Growth, Habit, Dissolution, and Structure of Crystals Using "Tailor-Made" Auxiliaries

ISABELLE WEISSBUCH, MEIR LAHAV, AND
LESLIE LEISEROWITZ

4.1. Introduction

Although crystals and crystallization as a process play an important role in pure and applied science, our ability to design crystals with desired properties and our control over crystallization based on structural understanding are still limited. Indeed, our ignorance of the overall process of crystal growth is brought to the fore by comparison with the controlled crystal growth provided in biomineralization. Organisms can mold crystals of specific morphologies, sizes, and orientations' in the form of composites of minerals and organic materials with characteristics vastly different from those of their inorganic counterparts (Löwenstam and Weiner 1989; Addadi and Weiner 1992; Mann 1993). Nevertheless, the technological advances in controlling the microstructure of inorganic materials on the subnanometer scale have led to tremendous success in the preparation of ceramics, synthetic layered microstructures, and semiconducting materials. Paradoxically, advances in the field of organic materials have not been as striking, although the techniques for the preparation of films such as Langmuir–Blodgett and self-assembled multilayers on solid support hold promise as pyroelectric, piezoelectric, and frequency doubling devices, etc. The thread that binds these diverse topics involves molecular interactions at interfaces.

The surfaces of a growing crystal can be thought of as composed of "active sites," which interact stereospecifically with molecules in solution, in a manner similar to enzyme–substrate or antibody–antigen interactions. The repetitive arrangements at crystal surfaces, and the knowledge we have of their structures, offer simpler means to pinpoint molecular interactions. It is possible to use the concept of molecular recognition at interfaces to address open questions in the areas of crystal nucleation and growth, crystal polymorphism, and interactions with the growth environment.

Molecular Modeling Applications in Crystallization, edited by Allan S. Myerson. Printed in the United States of America. Copyright © 1999 Cambridge University Press. All Rights reserved. ISBN 0 521 55297 4.

Landmark papers for the prediction of crystal morphology were provided by Hartman and Perdok (1955*a*, *b*; Hartman 1973) and Hartman and Bennema (1980), who laid the ground rules for a quantitative determination of crystal morphology based on the internal crystal structure. Regarding the effect of additives and solvent on crystal growth and morphology, a variety of important experimental studies have been reported in the latter half of this century. For example, Wells (1946) performed elegant experiments, particularly on polar crystals, to demonstrate the effect of solvent on crystal growth and shape. Changes in habit induced in NaCl crystals by the presence of α-amino acids were reported by Fenimore and Thraikill (1949). Buckley (1951) described a variety of crystal habit modifications, in particular the effect of dyes, that led to change in morphology and to colored inclusions known as hourglass or *Maltese cross* shapes. Whetstone (1953) explained the nature of such dye adsorption phenomena in inorganic crystals. He invoked a structural match involving Coulomb interactions between the additive and the structure of the crystal surface onto which adsorption occurred. In the case of molecular crystals, the sugar technologists (Smythe 1967, 1971; Mantovani et al. 1967) studied the effect of raffinose on the growth of sucrose during extraction of the latter from molasses. By carrying out kinetic studies, they established that raffinose hinders growth along specific directions of sucrose crystals and proved the presence of its selective occlusion. The control, however, on nucleation, growth, dissolution, and morphology of molecular crystals remained primarily a matter of "mix and try." There was no general approach for the design of additives targeted for binding to particular crystalline phases or preselected faces of any crystal. It was not, and still is not, possible to wield the exquisite control Mother Nature exerts in the field of biological mineralization.

In recent years a stereochemical approach has been adopted involving growth and dissolution of molecular crystals in the presence of "tailor-made" auxiliaries, which interact with specific crystal faces and may be crystal growth inhibitors or nucleation promotors (Addadi et al. 1985; Weissbuch et al. 1991; Weissbuch et al. 1995). With the assistance of such molecular auxiliaries it is possible to examine a variety of processes. We list first those that have involved crystal growth inhibitors. Crystals may be engineered with desired morphologies (Berkovitch-Yellin et al. 1982*a*). A correlation may be made between the chirality of tailor-made additives and the macroscopic properties of affected host crystals. By such means, the absolute configuration of chiral molecules or polar crystals may be assigned (Addadi et al. 1986). With occlusion of the additive into the crystal, the symmetry of the crystal may be reduced and so make a "centrosymmetric" crystal emit a second harmonic signal on irradiation with a laser beam (Weissbuch et al. 1989). Molecular crystals can be etched at desired faces on dissolution (Shimon, Lahav, and Leiserowitz 1985, 1986*a*) and the effect of solvent on crystal growth may be elucidated (Wireko et al. 1987; Shimon et al. 1990, 1991). Crystallization of only one enantiomorph from a racemic mixture may be induced (Addadi et al. 1982*b*, *c*) and the precipitation of a particular crystal polymorph brought about (Staab et al. 1990; Weissbuch, Leiserowitz, and

Lahav 1994*b*; Weinbach et al. 1995). The surface structure of crystals growing in the presence of additives may be monitored by atomic force microscopy and by grazing incidence x-ray diffraction (Cai et al. 1994; Gidalevitz et al. 1997*a*, *b*).

Here a bird's-eye view of all these different, yet connected, topics is presented. We begin with a brief description of how atom–atom potential-energy calculations may be applied for a quantitative analysis of molecular interactions at crystal interfaces.

4.1.1. *Quantitative Estimates of Molecular Interactions*

Energy calculations involving determination of crystal morphology and the effect of solvent and tailor-made additives on crystal shape and form have been made using atom–atom potential-energy functions generally composed of van der Waals and electrostatic terms. The van der Waals parameters are empirically derived from the packing arrangements of three-dimensional crystal structures and their sublimation energies (see, for example, Hagler and Lifson 1974). The electrostatic energy contribution, which is generally a major component and of longer range than the van der Waals terms, predominates at intermediate and large distances. This energy is sensitive to the relative orientations of the molecules and hence plays a decisive role in determining the structure and shape of crystals.

The electrostatic energy is described by an interatomic Coulombic interaction with parameters that depend upon charge distribution and so may be derived by quantum-chemical methods, self-consistent force-field methods (Hagler and Lifson 1974), or from experimental deformation electron-density distributions as determined from low-temperature x-ray diffraction data (Hirshfeld 1977*a*, *b*). Over the past twenty years accurate, low-temperature x-ray data, sometimes combined with neutron diffraction data, from single crystals of small organic molecules has increasingly provided an excellent measure of the details of molecular electron density distributions (Hirshfeld 1977*a*, *b*, 1991; Coppens 1982). These distributions are generally depicted in the form of an electron-density difference map:

$$\delta\rho(r) = \rho(r) - \rho^m(r),$$

where $\rho(r)$ denotes the actual electron density at point r in the crystal and $\rho^m(r)$ denotes the model density of the promolecule, generally composed of spherically symmetric, neutral, and non-bonded atoms.

Although the deformation density maps by themselves yield interesting information, for other applications the results they contain must be expressed in a different form. For instance, the electrostatic properties of the molecule can be extracted from such $\delta\rho(r)$ distributions for the calculation of molecular or group dipole moments, the electrostatic potential about the molecule, or the electrostatic interaction energy between molecules. In order to make such energy calculations from the deformation density, it is convenient to partition the

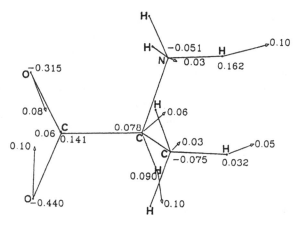

Fig 4.1. (S)-Alanine. Net atomic charges (e) and dipole moments (eÅ).

molecular charge density among the several atoms to serve as a basis for a calculation of molecular electrostatic properties. Such a partitioning scheme introduced by Hirshfeld (1977*b*) – the Stockholder recipe – yields atomic fragments, the net charge and dipole and quadrupole moments of which are transferable from molecule to molecule with reasonable accuracy. Such moments, derived either from experimental or theoretical charge distributions, have been used for electrostatic calculations of hydrogen-bonding molecules. An example is provided by the low-temperatuare (23 K) x-ray diffraction study of (S)-alanine by Destro, Marsh and Bianchi (1988). They determined the deformation density distribution by least squares, based on the pseudoatom model of Stewart (1976). From the x-ray diffraction data of Destro et al., the multiple atomic moments of alanine were derived (Eisenstein, M., private communication) according to the Stockholder recipe, as shown in Fig. 4.1. The differences in the polarity of the NH and CH bonds are evident, as well as the net charge on the zwitterionic CO_2^- and NH_3^+ groups (−0.61 and 0.44 electron units respectively). The charges compare convincingly with the values (−0.59, 0.42) derived by Destro et al. from their pseudoatom refinement. The atomic moments of alanine have been used to help understand the difference in freezing temperature of water drops deposited on crystals of (R, S)-alanine and (S)-alanine (Gavish et al. 1992; Wilen 1993; Lahav et al. 1993). Electrostatic parameters of other functional groups such as CO_2H, $CONH_2$, C—F, and C—H, derived from low-temperature x-ray diffraction data, have been applied for the analysis of molecular packing characteristics (Berkovitch-Yellin and Leiserowitz 1980, 1982*c*; Jacquemain et al. 1992).

4.1.2. *Calculation of the Theoretical Crystal Form and of the Effect of Additives on Crystal Morphology*

As already alluded to, the basic rules for a quantitative determination of crystal morphology were laid down by Hartman and Perdok (1955*a*, *b*; Hartman, 1973) and Hartman and Bennema (1980). According to their analyses, the crucial relation is between the layer energy E_L, which is the energy released when a layer is formed, and the attachment energy E_{att}, which is defined as the energy per molecule released when a new layer is attached to the crystal face. E_{att} controls the growth rate perpendicular to the layer, while E_L measures the stability of the layer. The working hypothesis is that the morphological importance of a crystal face is inversely proportional to its attachment energy. The energies, E_L and E_{att}, may be computed using atom–atom potential energy functions. By applying such a procedure, it became possible to calculate the energy values of E_L and E_{att} for various low-index faces of a crystal, from which the "theoretical crystal form" may be derived. This method has been applied for a variety of crystals, organic (Berkovitch-Yellin 1985; Berkovitch-Yellin et al. 1985; Boek et al. 1991; Docherty et al. 1993; Roberts and Walker 1996) or inorganic (Woensdregt 1993).

The effect of tailor-made additives on crystal growth has been modeled by energy calculations, according to the following procedure. First, the crystallographically different surface sites of the various faces are determined on which the additive replaces a substrate molecule with a minimum loss in binding energy. The binding energy at a surface site for a substrate molecule is defined as $E_b = E_L + E_{att}$. The binding energy for the adsorbed tailor-made additive is $E_b' = E_L' + E_{att}'$, where E_L' and E_{att}' are its layer and attachment energies respectively. Such calculations may be reliably used for a precise determination of the preferred surface adsorption sites of the additive and its molecular conformation in case of ambiguity. This procedure, however, is reliable for a quantitative determination of the change in morphology only if the additive can be eventually occluded into the crystal without inducing pronounced distortions in the arrangement of the surrounding substrate molecules. No direct experimental information is yet available demonstrating that an additive that would induce lattice distortions would be eventually occluded into a regular molecular site. This statement holds even if experiments have shown that additives have been occluded into the crystal bulk and oriented in the same way as a substrate molecule. Such an additive may reside only on the surface of the crystal mosaic blocks. Nevertheless, modeling the adsorption of tailor-made additives on crystal surfaces has yielded fruitful information, as will be described here for various systems.

4.2. Morphological Crystal Engineering

The habit of a crystal is defined by the relative rates of its growth in different directions; the faster the growth in a given direction, the smaller the face developed perpendicular thereto. Consequently, when growth is inhibited in a direction perpendicular to a given face, the area of this face is expected to increase relative to the areas of other faces of the same crystal. Differences in the relative surface areas of the various faces can therefore be directly correlated to the inhibition in different growth directions. Dramatic morphological changes observed during growth of organic crystals in the presence of additives have revealed a high degree of specificity in the interaction of the foreign material

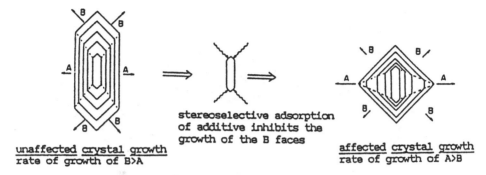

stereoselective adsorption
of additive inhibits the
growth of the B faces

unaffected crystal growth
rate of growth of B>A

affected crystal growth
rate of growth of A>B

Scheme 4.1. Scheme showing changes in crystal morphology as a function of adsorption of additives onto specific faces.

disturbed growth

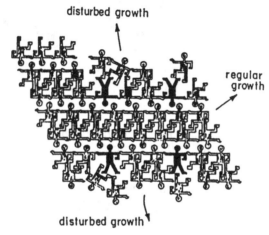

regular growth

disturbed growth

Scheme 4.2. Selective adsorption of the tailor-made additive at crystal faces and the concomitant inhibition of growth.

with the different structures of surfaces of the crystalline matrix. Such a change in crystal morphology is depicted in Scheme 4.1.

As mentioned in the previous section, the role played by impurities on the habit modification of crystals (Kern, 1953) has long attracted the attention of crystal growers, but such studies did not, in the main, involve morphological change by design. This became possible with the use of "tailor-made" additives. From a systematic study of a variety of organic compounds crystallized in the presence of additives of molecular structure similar to the structures of the corresponding substrate molecules (Addadi et al. 1985) and so labeled "tailor-made" additives, it was possible to deduce a stereochemical correlation between the structures of the affected crystal surfaces and the molecular structure of the additive. We could infer that the "tailor-made" additive is adsorbed on the growing crystals, but only at certain surfaces and then with the part of the adsorbate that differs from that of the substrate emerging from the crystal. This selective adsorption of the tailor-made additive at crystal faces and the concomitant inhibition of their growth is depicted in Scheme 4.2. Once this

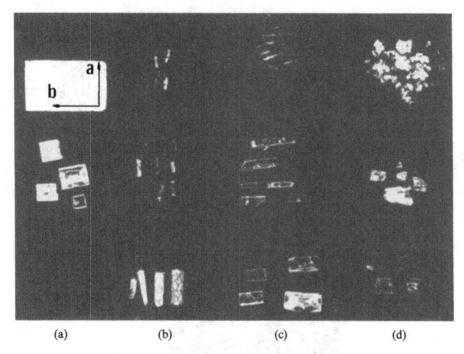

Fig. 4.2. Crystals of benzamide: (*a*) Pure. (*b*)–(*d*) Grown in the presence of additives: (*b*) benzoic acid; (*c*) *o*-toluamide; (*d*) *p*-toluamide.

mechanism was established, it became possible to exploit it to modify system-atically the morphology of crystals by tailoring additives that bind at preselected faces and thus inhibit growth in a predictable manner.

Benzamide, $C_6H_5CONH_2$, provides an example where the host–additive inter-actions at the growing solid–liquid interface can be pinpointed and correlated with the change in crystal morphology (Berkovitch-Yellin et al. 1982*a*). This compound crystallizes from ethanol as plates (Fig. 4.2(*a*)), in the monoclinic space group $P2_1/c$. The molecules in the crystal form hydrogen-bonded cyclic dimers, further interlinked by hydrogen bonds to yield ribbons parallel to the **b** axis (Fig. 4.3(*a*)). The ribbons are stacked along the 5.6 Å **a** axis to form stable (001) layers. These tightly packed layers juxtapose along the **c** direction, inter-acting via weak van der Waals interactions between phenyl groups, thus account-ing for the {001} platelike shape of the crystals. Benzoic acid C_6H_5COOH additive, adopting the stable O=C—OH *syn* planar conformation, can replace a molecule of benzamide at the end of a ribbon (Fig. 4.3(*a*)); however, at the site of the additive, an attractive N—H··O bond (−6 kcal/mole) is replaced by repulsion (1–2 kcal/mole) between adjacent oxygen lone-pair electrons of the bound additive molecule and of the oncoming benzamide molecule, leading to an overall loss in energy of 7–8 kcal/mole (Berkovitch-Yellin et al. 1985). As predicted, the presence of benzoic acid in solution inhibits growth of the benza-mide crystals along **b**, transforming the pure platelike crystals into needles elon-gated along the **a** axis (Fig. 4.2(*b*)). Inhibition of growth along the **a** direction

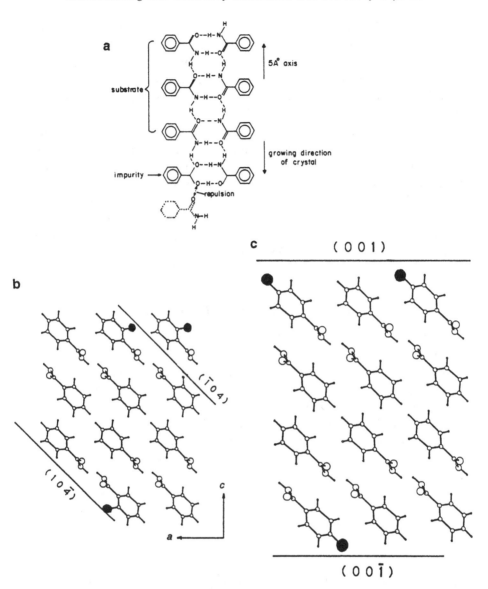

Fig. 4.3. (*a*) Schematic representation of the ribbon motif of hydrogen-bonded dimers of benzamide molecules interlinked along the 5 Å **b** axis, showing the effect of benzoic acid inhibiting growth along **b**. (*b*) Packing arrangement of benzamide, viewed along the **b** axis, showing the effect of *o*-toluamide inhibiting growth along the **a** direction. (*c*) Packing arrangement of benzamide, viewed along the **b** axis, showing the effect of *p*-toluamide inhibiting growth along the **c** direction.

was accomplished by adding *o*-toluamide, o-$H_3CC_6H_4CONH_2$, to the crystallization solution. The crystals consistently grew as bars elongated in **b** (Fig. 4.2(*c*)). The additive *o*-toluamide can easily be adsorbed in the hydrogen-bonding chain without disturbing growth in the **b** direction. The *o*-methyl group emerges, from the (10$\bar{4}$) side face (Fig. 4.3(*b*)), and thus interferes with growth in the **a** direction, along which the dimers are stacked. The relative orientation of benzamide at the {001} faces is shown in Fig. 4.3(c). Thinner and thinner plates (Fig. 4.2(d)) are obtained by adding increasing amounts of p-toluamide, p-$H_3CC_6H_4CONH_2$, th p-methyl substituent of which perturbs the already weak van der Waals interactions between the phenyl layers in the c direction (Fig. 4.3(c)).

The present approach, which is based upon specific molecular interactions at the solid–liquid interface, was applied to a variety of molecular crystals including α-amino acids, dipeptides, primary and secondary amides, carboxylic acids, sugars, and steroids. It therefore appears to be general and has since been successfully applied to a large class of materials, be they organic (Lewtas et al., 1991; Skoulika, Michaelidis, and Aubry, 1991; Antinozzi, Brown, and Purich 1992), inorganic (Black et al. 1991; Cody and Cody 1991; Ristic, Sherwood, and Wojeshowski 1993; Rifani et al. 1995), or biological (Markman et al. 1992).

4.3. Tailor-Made Etchants for Organic Crystals

Growth and dissolution are not reciprocal processes. Growth forms are bounded by the slowest-growing faces, dissolution forms by the fastest-dissolving orientations (de Boer 1993). Nevertheless, we found that a "tailor-made" inhibitor of growth of a given crystal face affects the rate of dissolution of that same crystal face (Shimon et al. 1985, 1986a).

It has been long recognized that partial dissolution of crystals in the presence of additives may induce formation of well shaped etch pits on particular faces of the crystal. For example, when crystals of calcite were partially dissolved in the presence of (D) or (L) tartaric, malic or lactic acids, etch pits with enantiomorphous shapes were observed (Honess 1927). Owing to the complexity of the etching mechanism, no simple model involving a stereochemical correlation between the structure of the etchant, that of the to-be-etched face, and the symmetry and the geometry of the etch pit, had been evolved, so that selection of etchants was done generally by trial and error.

Crystal dissolution invariably begins at sites of emerging dislocations (Horn 1951; Gevers, Amelinckx, and Dekeyser 1952). At such sites more than one type of facial surface is exposed to the solvent. When an additive that binds selectively to a given face of the crystal is present in the solution, the crystal dissolves in different directions at rates that differ from those in the pure solution. Subsequently, etch pits are formed at the dislocation centers on those faces at which the additives are bound. It is also correct to say that there is a difference between etching and dissolution. If the latter is inhibited, the former is made more prominent. The possibility of designing molecules that can interact stereospecifically with a given face of a crystal should, in principle, make those molecules efficient etchants of the face. Therefore, it was anticipated that the same molecules, which act as "tailor-made" growth inhibitors of a given face, operate as "tailor-made" etchants of the same face when the crystal is partially dissolved.

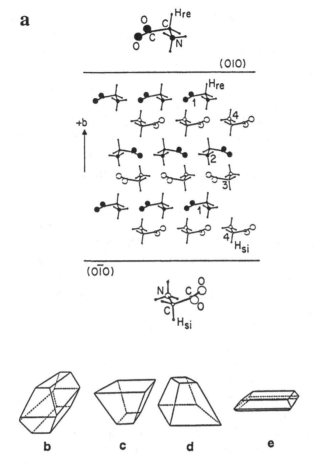

Fig. 4.4. (*a*) Packing arrangement of α-glycine viewed along the **a** axis, showing the four symmetry-related molecules; the C—H_{re} bonds of the molecules, almost parallel to **b**, point towards the (010) face and the corresponding C—H_{si} bonds towards the (0$\bar{1}$0) face. (*b*)–(*e*) Crystal morphologies of α-glycine when grown in the presence of: (*b*) no additive; (*c*) R-α-amino acids; (*d*) S-α-amino acids; (*e*) Racemic α-amino acid additives.

The generality of this approach was tested on several molecular crystals (Shimon et al. 1985, 1986*a*), which we illustrate with one example.

We consider the packing characteristics of the α form of glycine, $+H_3NCH_2CO_2^-$, a prochiral molecule containing two hydrogen atoms at the central carbon atom that are enantiotopic; namely, replacing one of these hydrogens by a different group yields a chiral molecule. α-Glycine packs in an arrangement (Fig. 4.4(*a*)) of space-group symmetry $P2_1/n$. Of the four symmetry-related molecules (1, 2, 3, 4, Fig. 4.4(*a*)), 1 and 2, related by twofold screw symmetry, have their C—H_{re} bonds pointing in the $+$**b** direction and emerging from the (010) face. By symmetry, molecules 3 and 4, related to 1 and 2 by a center of inversion, have their C—H_{si}^* bonds pointing in the $-$**b** direction and emerging

* We have used the Prelog and Helmchen (1982) nomenclature to specify the diastereotopic C—H ligands.

from the (0$\bar{1}$0) face. Only the (R)-amino acid additives can substitute a glycine molecule at the 1 and 2 sites, and then only on face (010), while only (S)-amino acids can be adsorbed at sites 3 and 4 on face (0$\bar{1}$0). This constraint arises from the steric requirement that the additive molecule be recognized on the {010} surfaces as a "host" molecule so that the α-amino acid side chain can replace the H atom of glycine only at the C—H bond emerging from the crystal surface. Otherwise, repulsion would occur between the additive and the surrounding molecules on the crystal surface. α-Glycine crystallizes from water as bipyramids (Fig. 4.4(b)). Crystallization experiments in the presence of (R)-α-amino acid additives led to the formation of pyramids with an (010) basal plane, because growth in the +**b** direction is inhibited (Fig. 4.4(c)). (S)-α-Amino acid additives induce the enantiomorphous morphology (Fig. 4.4(d)). Racemic α-amino acid additives cause the formation of {010}* plates (Fig. 4.4(e)) because growth at both the +**b** and −**b** sides of the crystal is inhibited (Weissbuch et al. 1983).

The faces relevant to the etching process are of the type {010}, being those inhibited on growth. Platelike crystals of glycine with such well developed faces were submitted to partial dissolution in an undersaturated solution of glycine containing variable amounts of other α-amino acids. In the presence of racemic additives, the rate of dissolution was much slower than in their absence. Moreover, the inhibition was more pronounced, the bulkier the additive side chain (Shimon et al. 1986a, 1988). When resolved R-alanine, $^+$H$_3$NCH(CH$_3$)CO$_2^-$, was present in the solution, well developed etch pits were formed only on the (010) face (Fig. 4.5(a)). These pits exhibit twofold morphological symmetry with surface edges parallel to the **a** and **c** axes of the crystal. The enantiotopic (0$\bar{1}$0) face dissolved smoothly (Fig. 4.5(b)), exactly as it does when the crystal is dissolved in an undersaturated solution of pure glycine. As expected, S-alanine induced etch pits on the (0$\bar{1}$0) face. Racemic alanine etched both opposite {010} faces.* The well established role played by dislocations for inducing etch pit formation was also demonstrated. A pure crystal of glycine was cleaved through its basal plane to expose the (010) and (0$\bar{1}$0) faces. These two parts of the crystal, on being etched with racemic alanine additive, displayed etch pits in equivalent regions of the two cleaved faces (Fig. 4.5(c), (d)).

The same type of chiral selectivity holds for an enantiomorphous pair of crystals dissolved in the presence of a chiral resolved molecule, when only one of the enantiomorphs is etched. Such enantioselective etching experiments provide a simple method to differentiate between enantiomorphous single crystals that do not express hemihedral faces. The method has been demonstrated for a class of racemic α-amino acids that crystallize as conglomerate mixtures (that is, each enantiomer crystallizes in its own enantiomorphous crystal), such as glutamic acid.HCl and asparagine.H$_2$O (Shimon et al. 1986a). Moreover, it also proved possible to bring about a pronounced separation of enantiomorphs by the enantioselective dissolution of a conglomerate mixture in the presence of "tailor-made" optically pure polymeric additives (Shimon et al. 1988).

* The symbol {hkl} refers to all the symmetry-related faces, whereas (hkl) refers to only the face specified by h, k, and l.

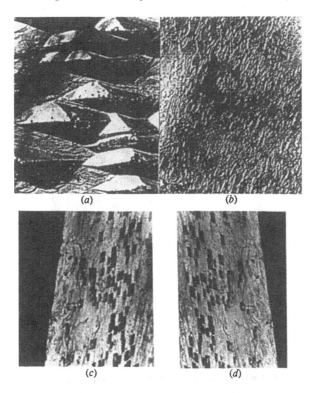

Fig. 4.5. Optical microscope photographs of the {010} faces of an α-glycine crystal after etching: (*a*) and (*b*) The (010) and (01̄0) faces as etched by R-alanine. (*c*) and (*d*) The (010) and (01̄0) faces of a cleaved crystal after etching with racemic alanine.

4.4. Assignment of Absolute Configuration of Crystals and Molecules

4.4.1. *Polar Crystals*

The standard method for the direct assignment of absolute configuration of polar crystals, composed of chiral, racemic, or nonchiral molecules, was introduced by Bijvoet in 1948 with the use of anomalous x-ray scattering. Over the past decade, several independent methods for the assignment of absolute configuration of polar crystals have been applied, including the effect of tailor-made molecular additives on crystal morphology and structure, as well as gas–solid reactions (Paul and Curtin 1987) and ferroelectric measurements in liquid crystals (Walba et al. 1988). The induced change in crystal morphology with tailor-made additives, the anisotropic distribution of the additive inside the host crystal, and the role played by additives in the formation of etch pits can be taken advantage of for the assignment of absolute configuration of chiral molecules and polar crystals. In this way, it has been possible to assign the absolute molecular structure of a variety of molecular systems, such as α-amino acids, sugars, and steroids (Addadi et al. 1986). Although one might argue that these studies have only confirmed what was already known, they demonstrate the interplay

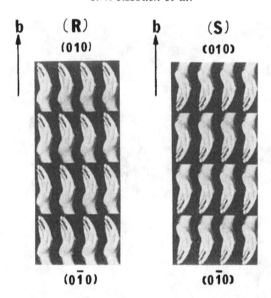

Scheme 4.3. Two enantiomorphous sets of hands arranged in a lattice.

between molecular chirality and various macroscopic phenomena, and also satisfy the curiosity of students of stereochemistry with an inadequate knowledge of x-ray scattering. Furthermore, the method has been applied to assign the absolute structures of polar crystals, the constituent molecules of which are either a racemic mixture or nonchiral, and may be the only viable method if the constituent molecules are hydrocarbons.

To demonstrate the method, we first focus on polar crystals composed of chiral-resolved molecules. The method involves the principle of fixing the orientation of the constituent molecules *vis-à-vis* the polar axis, or axes, of the crystal and subsequently determining the absolute configuration of the chiral molecules. This principle is depicted in Scheme 4.3, which shows two enantiomorphous sets of hands arranged in a lattice; the right hands forming an (R)* crystal and the left hands an (S)* crystal. The fingers of the hands are exposed at the (010) face of the right-handed "molecules" and, by symmetry, at the (0$\bar{1}$0) face of the left-handed "molecules." Thus, by determining at which face of the crystal specimen the fingers or wrists are exposed, the handedness of the constituent molecules may be assigned. This may be performed by applying the two-step adsorption–inhibition mechanism described above with appropriate tailor-made additives.

We illustrate this approach with the example of *N*-(E-cinnamoyl) alanine (Berkovitch-Yellin et al. 1982*b*). The molecules of *N*-(E-cinnamoyl)-S-alanine 1*a* crystallize in space group $P2_1$. The packing arrangement delineated by the crystal faces as grown from methanol solution is shown in Fig. 4.6. The molecules are arranged such that the carboxyl groups emerge at the two {1$\bar{1}$1} faces

---

* For convenience the parentheses enclosing the chirality of the molecule specifies a crystal; absence thereof, such as R, S, specifies the molecule only.

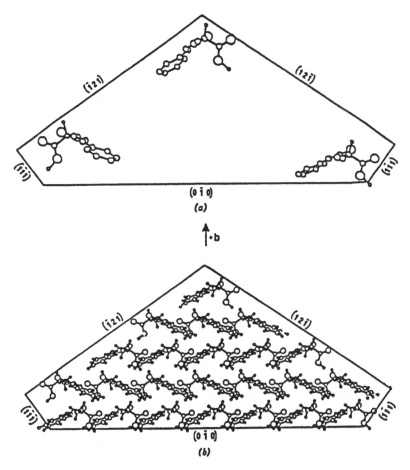

Fig. 4.6. Packing arrangement of *N*-(E-cinnamoyl)-(S)-alanine delineated by the faces observed in the pure crystal. To demonstrate the effect of the tailor-made additives, the upper figure shows the orientation of only three host molecules *vis-à-vis* the crystal faces.

and the C(chiral)—H bonds are directed along the $+\mathbf{b}$ axis. In a study on the effect of tailor-made additives on the morphology of pure N-(*E*-cinnamoyl)-(S)-alanine 1*a* (Fig. 4.7(*a*)), it was found, as expected, that the methyl ester (1*b*), of the same absolute configuration as host molecule (1*a*), induced large $\{1\bar{1}1\}$ faces (Fig. 4.7(*b*)) since the O—CH$_3$ group replaces an O—H group emerging from such faces.

Tailor-made additive *N*-(E-cinnamoyl)-(R)-alanine 1*c*, of configuration opposite to that of the host induced, as expected, formation of an (010) face (Fig. 4.7(*c*))

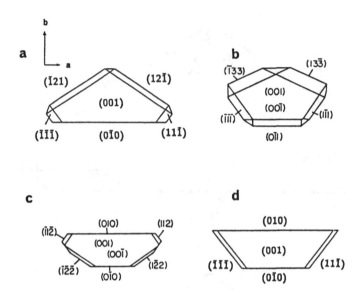

Fig. 4.7. Morphologies of measured crystals of *N*-(E-cinnamoyl)-(S)-alanine; (*a*)–(*c*) refer to crystals grown from methanol: (*a*) pure crystal; (*b*) grown in the presence of the methyl ester 1*b*, (*c*) grown in the presence of E-cinnamoyl-(R)-alanine 1*c*; (*d*) pure crystal grown from acetic acid.

because of an interchange of CH_3 and H at the guest asymmetric carbon, so that the C—CH_3 guest moiety replaces the emerging host C—H bond.

It was found that solvent may act in a manner similar to the tailor-made additives, for example, by growing (1*a*) from acetic acid (Shimon *et al.* 1991) instead of methanol. Acetic acid is a solvent that can selectively bind at the exposed carboxylic acid groups of the two {1Ī1} faces, forming a hydrogen-bonded dimer (2*a*). Acetic acid can also bind to the $CHCO_2H$ moiety of cinna-moyl-alanine via a cyclic dimer (2*c*) on the (010) face. The dimer (2*b*) is the motif adopted by acetic acid in its own crystal structure (Jonsson 1971) forming a C—H$\cdots$O (carbonyl) interaction (Taylor and Kennard 1982; Berkovitch-Yellin and Leiserowitz, 1984; Steiner 1996). Crystallization of *N*-(E-cinnamoyl)-alanine (1*a*) from glacial acetic acid yields crystals with the morphology shown in Fig. 4.7(*d*), in keeping with expectation.

A nice example of how anomers, interconvertible in solution, may be used for the assignment of absolute configuration is provided by the influence on the

a   **b**

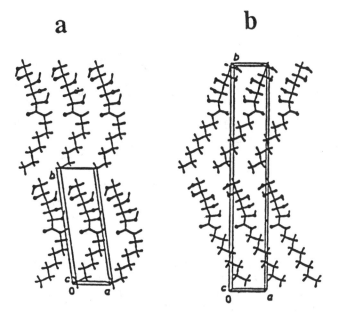

Fig. 4.8. Head-to-tail packing arrangement of: (*a*) *N*-(*n*-heptyl)-(D)-gluconamide in space group *P*1; (*b*) *N*-(*n*-octyl)-(D)-gluconamide in space group *P*2₁.

morphology of α-lactose hydrate crystals by the anomeric β-lactose molecules present in solution (Michaelis and van Kreveld 1966; Wisser and Bennema 1983).

Direct assignment of absolute configuration of chiral-resolved molecules by the difference in wetting properties of hydrophobic and hydrophilic crystal faces was made possible (Wang et al. 1991) using crystals of the class belonging to chiral-resolved alkyl gluconamides $C_nH_{2n+1}NHCO(CHOH)_4CHOH$, $n = 7$–10. The molecules crystallize as plates in an arrangement akin to that depicted in Scheme 4.3, namely head-to-tail, as shown in Fig. 4.8 (Zabel et al. 1986). Thus one face of the plate crystal is hydrophobic and the opposite hemihedral face is hydrophilic. Measurements of contact angle with a variety of polar solvents deposited on the opposite platelike faces of specimen crystals could establish which face is hydrophobic and which hydrophilic. This information fixes the orientation of the constituent molecules along the polar **b** axis of the crystal specimen and consequently their chirality. These wettability experiments also led to crystal growth experiments on these systems to help determine the role of the solvent on crystal growth, as discussed in Section 4.7.

The last example involves etching experiments, which have the advantage that they may be performed on pure specimen crystals that have not been grown in the presence of a tailor-made additive. Here we describe the assignment of the absolute structure of racemic (R,S)-alanine by enantioselective etching (Shimon et al. 1986*b*). This compound crystallizes in the polar space group *Pna*2₁. In the crystal, which is needlelike along the **c** axis, the R- and S-molecules are oriented with respect to the polar needle **c** axis so that the carboxylate CO_2^- groups are

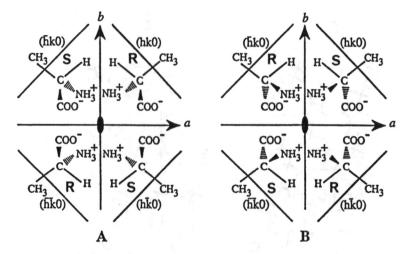

Fig. 4.9. Schematic (*mm*2 point group) representation of the orientation of the four molecules of (R,S)-alanine *vis-à-vis* the symmetry-related {210} faces, when viewed along the polar **c** axis. These four molecules are oriented such that only one has its C—CH$_3$ group emerging from a given {210} face; (A) and (B) represent the two alternative orientations of the molecules for the corresponding orientations of the polar crystal structure for a given direction of the **c** axis.

exposed at one end of the polar axis and the amino NH$_3^+$ groups at the opposite end. The crystals exhibit different (hemihedral) faces at opposite ends of the **c** axis in the sense that one end face is flat and the opposite capped. Conventional x-ray crystallography does not allow one to assign the absolute molecular orientation with respect to the polar axis and thus at which ends the CO$_2^-$ and the NH$_3^+$ groups are exposed. Etching has been successfully applied for such an assignment by making use of the well developed {210} side faces of the needle. The orientations of the (R) and (S)-molecules *vis-à-vis* the four symmetry-related {210} side faces for the two possible orientations of the crystal structure with respect to the polar axis are shown in Fig. 4.9. Naturally all four symmetry-related molecules are exposed at each of the four symmetry-related {210} faces, but only one of the four molecules is oriented such that its C—CH$_3$ group emerges from a particular {210} crystal face as shown in Fig. 4.9. Once these orientations in a specimen crystal are determined, the absolute molecular arrangement *vis-à-vis* the polar axis is assigned. Partial dissolution in the presence of S-α-amino acids such as serine, $^+$H$_3$NCH(CH$_2$OH)CO$_2^-$, or phenylalanine, $^+$H$_3$NCH(CH$_2$C$_6$H$_5$)CO$_2^-$, should replace an (S)-alanine substrate molecule at only the ($\bar{2}$10) and (2$\bar{1}$0) faces for the orientation shown in Fig. 4.9(*a*), but would affect the (210) and ($\bar{2}\bar{1}$0) pair for the opposite orientation of Fig. 4.9(*b*). This expectation was demonstrated experimentally, etch pits being formed on only one pair of {210} side faces (Fig. 4.10), so fixing the absolute structure of the specimen crystal, and thus of all crystals grown from aqueous solution because the crystals exhibit hemihedral end faces.

Fig. 4.10. Scanning electron micrograph of one of the {210} side faces of (R,S)-alanine etched by (S)-threonine; only two adjoining faces are shown, one etched and the other smooth.

The absolute orientation of specimen polar crystals of (R,S)-alanine could also be assigned from growth experiments in the presence of additives designed to affect either of the two polar ends of the crystal. Crystallization experiments in the presence of alanine ethyl ester that would bind at the fast-growing CO_2^- end of the crystal yielded much shorter needles. *N*-methyl alanine, which was expected to affect the slow-growing capped end of the crystal, yielded needles as long as from the pure solution, fixing thus the orientation of the molecules in the crystal (Shimon et al. 1986*b*).

Growth experiments in the presence of additives, complemented by surface-energy calculations, have also been used for assignment of absolute structure. For example, the absolute orientation of specimen polar crystals of *m*-chloronitro-benzene (*m*-CNB) was determined by crystal-growth experiments in the presence of *p*-chloronitrobenzene (*p*-CNB) and *p*-dinitrobenzene (*p*-DNB) additives (Chen et al. 1994). The pure *m*-CNB crystals grown from melt and from acetone solutions do not exhibit hemihedral faces, although the space group is *Pna2₁*. Growth in the presence of *p*-CNB and *p*-DNB yielded asymmetrically shaped crystals affected along the polar **c** axis. Qualitative calculations of the binding energy of the additives to the crystal were in agreement with the experiment, insofar that the additive could be bound at only one specific end of the polar axis. Docherty et al. (1993) calculated a polar morphology of urea (space group $P\bar{4}2_1m$) that was in good agreement with the crystal shape obtained from sublimation. A classical type of energy calculation, in which one assumes that the charge distribution of the molecules in the bulk and of those being attached at the surface are the same, would not yield a polar morphology. By attaching different atomic charges to the bulk molecules and to those being attached at the surface, through the use of *ab initio* quantum-chemical calculations, the polar morphology was computed. However, the assignment of the absolute crystal structure of urea has not yet

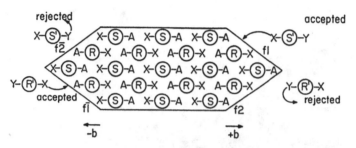

Scheme 4.4. Scheme showing the adsorption of chiral-resolved additives onto enantiotopic faces of a centrosymmetric crystal.

been confirmed by experimental methods. The crystals of urea grown from benzene or water, discussed in Section 4.7, do not exhibit polar morphologies.

4.4.2. *Use of Centrosymmetric Crystals for Assignment of Molecular Chirality*

In contrast to the situation for chiral crystals, it had not been generally appreciated that in centrosymmetric crystals the orientations of the constituent molecules are unambiguously assigned with respect to the crystal axes, from a conventional structure determination. Thus the known orientation of the two enantiomeric molecules in a centrosymmetric crystal can be exploited for the direct assignment of absolute configuration of chiral-resolved molecules, provided the structural information embedded in the racemic crystal is transferred to a chiral additive molecule. The direct assignment of the absolute configuration of such chiral-resolved additives may thus be determined through the morphological changes they induce selectively on one set of enantiotopic faces of centrosymmetric crystals with appropriate packing features.

A prerequisite for application of this method is that within the centrosymmetric racemic crystal a specific functional group attached to an R-molecule points toward the face $f1$ but not toward the opposite face $f\bar{1}$ (Scheme 4.4).

By symmetry, the same functional group attached to an S molecule will emerge at the enantiotopic face $f\bar{1}$, but not $f1$. Crystallization of a centrosymmetric crystal in the presence of a chiral additive R′, designed so that it will fit in the site of an R molecule on the growing crystal faces $f\bar{1}$ or $f\bar{2}$ but not on the enantiotopic faces $f1$ or $f2$, will hinder growth in the −**b** direction but not in the +**b** direction (Scheme 4.4). By virtue of symmetry, the enantiomeric additive S′ will inhibit growth perpendicular to faces $f1$ and $f2$, while the racemic additive R′, S′ will inhibit growth in both directions, +**b** and −**b**. This approach is not limited to racemic crystals, but the constituent molecules should at least be prochiral for ease of interpretation. For example, the absolute configuration of all naturally occurring α-amino acids, but for proline, may be assigned by their effect on the morphology of the α form of glycine, a prochiral molecule (Weissbuch et al. 1983). We shall illustrate the above principle by making use

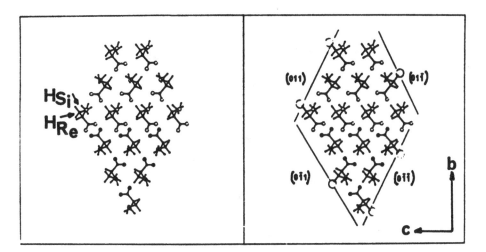

Fig. 4.11. Stereoscopic view of the packing of (R,S)-serine crystal along the **a** axis. Each *bc* layer is composed of R or S molecules only. For clarity only half of each R (open circles) and S (filled circles) layer is shown. The positions of the four {011} faces are shown with respect to the crystal structure. Threonine additive molecules have been stereospecifically inserted, with the β-methyl groups indicated by large circles.

of centrosymmetric crystals of (R, S)-serine to assign the absolute configuration of threonine as additive (Weissbuch et al. 1985*b*).

Racemic serine crystallizes in space group $P2_1/a$. Within the structure (Fig. 4.11) the $C_b \rightarrow H_{Si}$ bond vector of the rigid methylene group of the R-serine molecule has a major component along $+\mathbf{b}$ and, by symmetry, the $C_b \rightarrow H_{Re}$ bond vector of the S-serine has a major component along $-\mathbf{b}$. Thus, their replacement by a methyl, as in threonine, will inhibit growth in the **b** direction. That is, an R-threonine molecule with a side-chain *b*-carbon of chirality S will inhibit growth along $+\mathbf{b}$, while the —CH$_3$ group of S-threonine will replace the H_{Re} hydrogen of S-serine and hence inhibit growth along $-\mathbf{b}$. (R,S)-serine forms tabular crystals with point symmetry $2/m$ (Fig. 4.12(*a*)); the crystals affected by either R- or S-threonine exhibit reduced morphological point symmetry 2 (the mirror plane is lost) and are enantiomorphous (Fig. 4.12(*b*), (*c*)). When R,S-threonine is used as the additive, the morphological point symmetry $2/m$ is left unchanged because the effects induced by each additive separately combine. The crystals turn into rhombs, with a clear increase in the areas of the {011} side faces relative to those of the pure form (Fig. 4.12(*d*)). The changes in morphology seen with the additives in Fig. 4.12(*b*)–(*d*) with respect to that of the pure crystals (Fig. 4.12(*a*)) can be interpreted only in terms of selective adsorption of the R enantiomer at faces (011) and (01$\bar{1}$), and of S at (0$\bar{1}$1) and (0$\bar{1}\bar{1}$). According to calculated surface binding energies (Weissbuch et al. 1985*a*), R-threonine may easily be adsorbed only on faces (011) and (01$\bar{1}$) and, by symmetry, S-threonine only on (0$\bar{1}$1) and (0$\bar{1}\bar{1}$). These results fix the absolute chirality of the chiral-resolved threonine additive.

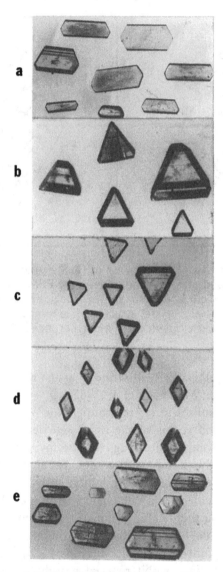

Fig. 4.12. Photographs of (R,S)-serine crystals that all have a common orientation of the lattice axes and a vertical **b** axis: (*a*) Pure; (*b*)–(*d*) grown in the presence of: (*b*) R-threonine; (*c*) S-threonine; (*d*) R,S-threonine; (*e*) R,S-*allo*-threonine.

4.5. Symmetry Lowering in Crystals Following Selective Occlusion of Tailor-Made Additives

Crystallization is a commonly used process for purification of inorganic and organic materials, and so the mode of occlusion of impurities or additives inside growing crystals is of paramount technological and theoretical importance. The results presented in the previous section, on the interaction of "tailor-made additives" with the different faces of crystals, has bearing on the general question

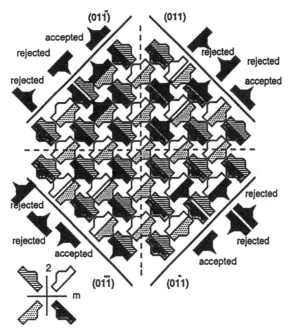

Fig. 4.13. Schematic representation of adsorption and occlusion of additive molecules (in black) through the four slanted {011} faces in a centrosymmetric crystal space group $P2_1/c$, containing four symmetry-related molecules. Their relative orientations are also depicted in the insert (point symmetry $2/m$). In each of the four crystal sectors only one of the symmetry-related molecules can be replaced by the additive on adsorption, which means loss of all symmetry elements but translation. The crystal symmetry in each sector is thus reduced from $P2_1/c$ to $P1$.

regarding the distribution and arrangement of occluded "tailor-made additives" within crystals, and, indeed, on the symmetry and structure of crystalline solid solutions.

A basic concept pertaining to this question is that in general the two-dimensional (2-D) symmetry relating molecules at crystal surfaces is generally lower than the 3-D symmetry of the bulk crystal and that the surface structures of nonsymmetry-related faces have different arrangements. For simplicity, we illustrate this aspect with a schematic arrangement of space symmetry $P2_1/c$ shown in Fig. 4.13. Note the corresponding symmetry operations of the $2/m$ point group given in the insert. The crystal is delineated by four symmetry-related, diagonal faces, labeled {011}. At these four surfaces, the three symmetry relations belonging to space group $P2_1/c$, that is, the center of inversion, the twofold screw, and the glide, are broken; thus, their two-dimensional symmetry is limited to translation, plane symmetry $P1$. We now examine the effect of tailor-made additives on these faces. One can easily see that an additive molecule, in black, bearing an appropriately modified group, will be able to substitute for a substrate molecule at any diagonal face, say (011), but at only one of the four different surface sites, because only at that site does the modified group not disturb the regular pattern of interactions at the crystal surface. Occlusion of such a molecule through the

(011) face will lead to a crystal sector of symmetry $P1$, since the occluded molecules therein are related by translation symmetry only. Occlusion of additive through all four diagonal faces {011} will lead to a crystal composed of four sectors, where these four sectors are related to each other by the $2/m$ point symmetry of the original host crystal.

Consequently, in general, an additive molecule will be anisotropically distributed within the grown crystal, preferentially occluded through different subsets at surface sites on the various faces, leading to a mixed crystal composed of sectors coherently intertwined. The selective occlusion of the additive into a sector will thus lead to a reduction of the crystal symmetry to the symmetry of the face through which it was adsorbed. This principle holds for each crystal sector, although the occluded additive may occupy only a small fraction of the bulk sites. It also holds for the whole crystal, but we note that the different crystal sectors are related to each other by the symmetry elements of the point group of the pure crystal structure. The reduced symmetry of each sector may thus be masked, depending upon the history of crystal growth, whether the sectors are examined separately, and the type of examination. For example, in a crystal containing all sectors, the loss of a glide plane or twofold screw axis should be immediately evident from the lack of the corresponding systematic absences in the x-ray or neutron diffraction pattern (naturally assuming that the crystal is of mosaic texture, which is the general case, and so the different mosaic blocks diffract independently), but not the loss of a rotation axis, mirror plane or center of inversion. Detection of the latter would require at least measurement of the complete set of diffraction data from the individual sector to test for a reduction in Laue diffraction symmetry, generally to be followed by crystal-structure refinement.

Reduction in crystal symmetry following the above principles was predicted in several host–additive systems and experimentally demonstrated by a variety of methods including changes in crystal morphology, high-performance liquid chromatography (HPLC), frequency doubling of laser light (second harmonic generation – SHG), solid-state asymmetric photodimerization, and optical birefringence. Symmetry lowering has also been directly shown by x-ray and neutron diffraction. We now review examples to illustrate the arguments and principles described.

4.5.1. *Detection of Reduced Crystal Symmetry by Enantiomeric Distribution of Additives*

The mode of occlusion of additive leading to reduction in symmetry, depicted schematically in Fig. 4.13, may be used as an aid to understand the selective occlusion of R,S-threonine into the crystals of (R,S)-serine (Weissbuch et al. 1985b). The morphological changes induced in (R,S) serine, as discussed in the previous section (see Fig. 4.11), imply that R-threonine, if occluded, would be occluded through the (011) and (01$\bar{1}$) faces. By symmetry, S-threonine would be occluded through the (0$\bar{1}$1) and (0$\bar{1}$$\bar{1}$) faces. Thus the threonine additives, race-

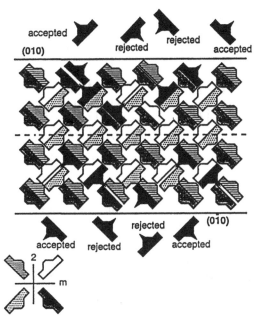

Fig. 4.14. Schematic representation of adsorption and occlusion of additive molecules (in black) through the top and bottom {010} faces in a crystal of space group $P2_1/c$, resulting in the loss of the glide and inversion symmetry. The crystal symmetry in each half is thus reduced from $P2_1/c$ to $P2_1$.

mic in solution, must on occlusion into (R,S) serine crystal, be enantiomerically segregated into domains of opposite chirality along the **b** axis during crystal growth, as was indeed experimentally shown by high-performance liquid chromatography (HPLC) methods (Weissbuch et al. 1985*b*).

Symmetry lowering of a somewhat similar nature involved the α form of glycine grown in the presence of α-amino acid additives (Weissbuch et al. 1983). As an aid to understanding, we present Fig. 4.14, which is a model representation of reduction in symmetry from $P2_1/c$ to $P2_1$ when molecules are occluded through the (010) and (0$\bar{1}$0) faces. The mode of adsorption of α-amino acid additives onto the {010} faces of α-glycine has already been described in Section 4.3. In terms of that analysis, only the R-α-amino acid additive may be occluded through the (010) face, replacing a glycine molecule at sites 1 and 2 only (Fig. 4.4(*a*)), yielding a pyramidal morphology (Fig. 4.4(*c*)). By symmetry, only the S-α-amino acid additives may be occluded into sites 3 and 4 through the (0$\bar{1}$0) face of α-glycine (Fig. 4.4(*a*)), yielding an enantiomorphous morphology (Fig. 4.4(*d*)). Thus racemic additives cause the formation of {010} plates (Fig. 4.4(*e*)), because growth at the $+$**b** and $-$**b** sides of the crystal is inhibited. These plates were found to contain 0.02 to 0.2 percent racemic α-amino acid additive within the crystal bulk with the two enantiomers totally segregated in the two crystal sectors at the $+$**b** and $-$**b** halves. As expected, the R-enantiomers populate the $+$**b**, and the S-enantiomers the $-$**b** half of the crystal according to HPLC

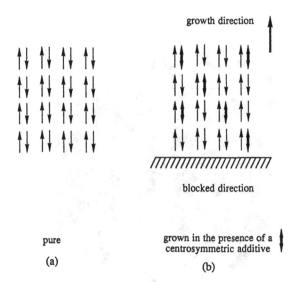

Scheme 4.5. The selective occlusion of centrosymmetric guest molecules within a centrosymmetric crystal composed of antiparallel pairs of molecules.

measurements (Weissbuch et al. 1983). In terms of the argument given above, the crystal symmetry of each half of the crystal must be reduced from $P2_1/n$ to $P2_1$, and the two sectors are enantiomorphous. We note that the absolute configuration of the α-amino acid additive can be directly assigned, not only from the induced changes in crystal morphology, but also from its anisotropic distribution on occlusion within the crystal. Symmetry lowering of a similar nature was also demonstrated in another system, which involved enantiomeric segregation of the dipeptide glycyl-leucine inside growing crystals of glycyl-glycine $^+H_3NCH_2CONHCH_2CO_2^-$ (Weissbuch et al. 1985a).

4.5.2. Detection of Reduced Crystal Symmetry by Nonlinear Optics

Symmetry lowering in host–additive systems, such as a loss of inversion, glide, or twofold screw symmetry, which would introduce (additional) polarity into the crystal, may be probed by nonlinear optical effects, such as second harmonic generation (SHG), which is active in an acentric material. One requirement is that either the host or additive molecule has a large molecular hyperpolarizability tensor β, leading to large optical nonlinearity. Consequently, a centrosymmetric pure crystal structure, which consists of antiparallel pairs of host molecules with high β coefficients (Scheme 4.5(a)), would yield a SHG active crystal upon site-selective occlusion by a centrosymmetric guest molecule, as shown in Scheme 4.5(b). Because the growth of the crystal takes place at the top exposed face, the additive is adsorbed and occluded through only one of the pair of surface sites in the growth directions. We demonstrated the potential of this approach with centrosymmetric host crystals of *p*-(*N*-dimethyl-

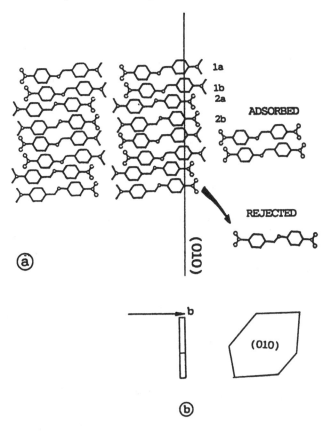

Fig. 4.15. (*a*) Packing arrangement of a polymorph of *p*-(*N*-dimethylamino)-benzyl-idene-*p*'-nitroaniline crystal, showing how the dinitro-additive may be adsorbed on the (010) face. The additive may be adsorbed only at sites of the type 1, but not at the type 2 sites. (*b*) Two views of the morphology of the platelike crystals.

amino)-benzylidene-*p*'-nitroaniline (*3a*) (Fig. 4.15), which became acentric and SHG active upon site-selective occlusion of the guest molecule *p,p*'-dinitro-benzylideneaniline (*3b*) which is pseudocentrosymmetric (Weissbuch et al. 1989). In this system the host and guest molecules respectively have large and negligible β coefficients.

$$3a \qquad\qquad 3b$$

The reverse situation, where the guest has large hyperpolarizability, is illustrated by an α-glycine host crystal containing, as additive, α-amino acids that have a high β coefficient, such as the *p*-nitrophenyl derivatives of α,δ-diamino butyric acid (*4a*), ornithine (*4b*), and lysine (*4c*).

4a   4b   4c

The amount of occluded additive was less than 0.2 percent (wt/wt of glycine) but sufficient to yield colored crystals. In terms of the arguments presented above, {010} platelike crystals of glycine, grown in the presence of R,S-additives, should have symmetry reduced from $P2_1/n$ to $P2_1$. In fact, the symmetry was reduced even further to $P1$, according to the SHG measurements (Weissbuch et al. 1989), providing additional information on the nature of the crystal growth process.

Other systems of interest are the noncentrosymmetric orthorhombic crystals of space group symmetry $P2_12_12_1$. On symmetry grounds, such crystals are nonpolar along the three principal crystallographic axes and therefore frequency doubling is forbidden in these directions. We anticipated that such crystals, when grown in the presence of appropriate additives so as to reduce the crystal symmetry to monoclinic $P2_1$, should become SHG active along the remaining 2_1 axis but not along those two axes that have lost their twofold screw symmetry. In fact, single crystals of (S)-glutamic acid.HCl grown in the presence of appropriate additives were SHG active along all three axes (Weissbuch et al. 1989), indicating symmetry lowering to $P1$, once again providing information on the nature of the crystal growth process.

4.5.3. *Detection of Reduced Crystal Symmetry by Optical Birefringence*

The properties of optical birefringence in crystals have been used to demonstrate a reduction in crystal class. McBride and Bertman (1989) took advantage of the fact that crystals that belong to a high symmetry class, such as tetragonal, trigonal, or hexagonal, are optically uniaxial, whereas crystals that belong to a lower crystal class, such as triclinic, monoclinic, or orthorhombic, are optically biaxial. They studied the tetragonal crystals of di(11-bromoundecanoyl) peroxide (5a), denoted as Br...Br, in the presence of guest (5b) where a Br atom is replaced by CH_3 denoted as Br...CH_3.

5a   5b

In the host crystal, Br...Br molecules assemble into layers that have Br atoms on both the upper and lower surfaces. Successive layers stack Br to Br atoms with 90° rotation about a fourfold screw axis in the stacking direction. These crystals, which grow as square {001} plates delineated by four {110} side faces, are not birefringent for light traveling along the fourfold screw axis perpendicular to the plate, appearing dark between crossed polarized filters. Crystals of Br...Br containing 15 percent of Br...CH_3 are birefringent, the plate revealing four sectors under crossed polarizers. Birefringence in each sector results from unsymmetrical incorporation of the Br...CH_3 additives during crystal growth. As McBride states in his paper, this method of optical microscopy is convenient for surveying large numbers of crystals, to catalog their growth patterns, and to identify reduced-symmetry domains that may be characterized by diffraction. Such an approach has been followed up by Gopalan et al. (1993), who have observed reduction in symmetry both by optical birefringence and by x-ray crystallographic studies of different sectors in the mixed crystals of $NaBrO_3$/$NaClO_3$ and $NaNO_3$/$Pb(NO_3)_2$. Garnets where Al^{3+} ions were replaced by Fe^{3+} ions also show a reduction in lattice symmetry from cubic down to triclinic (Allen and Buseck 1988).

4.5.4. *Probing Intermolecular Forces through Reduced Crystal Symmetry*

There is interest in investigating the minimal modification of the tailor-made additive, with respect to the host molecule, that would still be recognized and discriminated by the growing crystal surface, and lead to reduction in symmetry. Indeed, solid solutions composed of host and additive molecules of similar structure and shape have been generally believed to exhibit the same symmetries as those of the host crystals. A reduction in crystal symmetry involving a structure determination by x-ray or neutron diffraction should allow one to probe fine interactions between host and additive at the crystal surfaces.

We first discuss solid solutions of carboxylic acids (XCO_2H) in primary amides ($XCONH_2$), where an NH_2 group is substituted by an OH moiety (Scheme 4.6). A strong inhibition of growth develops along the direction of the $O{=}C{-}N{-}H_a{\cdots}O{=}C$ hydrogen bond, H_a being the amide hydrogen atom in antiperiplanar conformation to the $C{=}O$ system. Inhibition arises when a carboxylic acid additive attempts to force the antiperiplanar lone-pair electron lobes of its OH group against a carbonyl oxygen of the underlying molecule, an atom that would normally form a hydrogen bond with the amide's antiperiplanar hydrogen. Incorporation of a carboxylic acid in this orientation would substitute a 2 kcal/mole repulsion for a 6 kcal/mole attraction. The carboxylic acid additive molecule will thus avoid surface sites that require the OH groups to be oriented toward the surface and be preferentially adsorbed at sites where the OH group emerges from the surface (Berkovitch-Yellin et al. 1985).

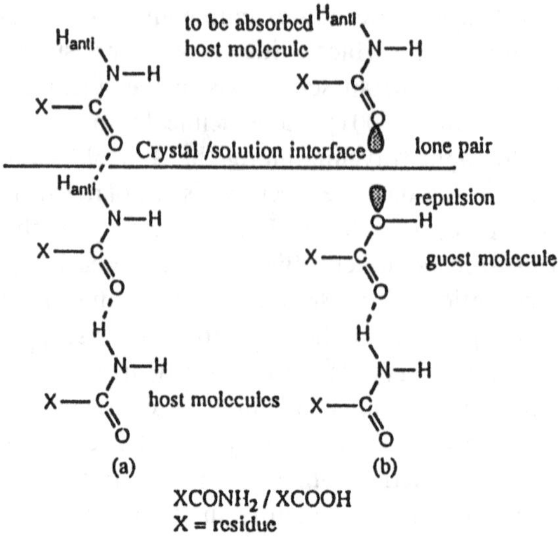

Scheme 4.6. The effect of selective adsorption of a carboxylic acid additive on growth of a primary amide crystal.

Reduction in crystal symmetry has been demonstrated in two amide–carboxylic acid systems by neutron diffraction and solid-state photodimerization. The former technique was applied to asparagine–aspartic acid (Weissinger-Lewin et al. 1989), where as much as 15 percent additive was occluded into the crystal by virtue of the numerous hydrogen bonds between guest and host; the dimerization method was applied to the host–additive system E-cinnamamide–E-cinnamic acid (Vaida et al. 1989) where small amounts (<2 percent) of cinnamic acid guest were occluded into the host crystal.

(S)-Asparagine $H_2NOCCH_2CH(NH_3^+)CO_2^-$ crystallizes from water as a monohydrate with a tight three-dimensional net of hydrogen bonds in a $P2_12_12_1$ structure (Fig. 4.16(a)). The morphology is prismatic, with 18 developed faces (Fig. 4.16(b), left). Crystallization of (S)-asparagine in the presence of S-aspartic acid $HO_2CCH_2CH(NH_3^+)CO_2^-$ yields {010} plates (Fig. 4.16(b), right). In terms of the arguments already given for amide–acid systems, the guest aspartic acid molecule should be more easily adsorbed at sites 1 and 3, on the growing (010) surface, than at sites 2 and 4 (Fig. 4.16(c)); the reverse situation holds for the opposite (01̄0) face. If, on growth of the mixed crystal, the (01̄0) face is blocked so that the amide and acid molecules would be occluded only through the (010) face, the symmetry of the mixed crystal should be reduced to $P12_11$. A low-temperature (18 K) neutron diffraction study with deuterated aspartic acid in protonated asparagine showed the expected reduction in symmetry (Weissinger-Lewin et al. 1989).

Analogously, additive E-cinnamic acid C_6H_5—CH=CH—CO_2H induces a loss of the center of inversion in the crystal of E-cinnamamide C_6H_5—CH=CH—$CONH_2$, which, in pure form, appears in a centrosymmetric monoclinic

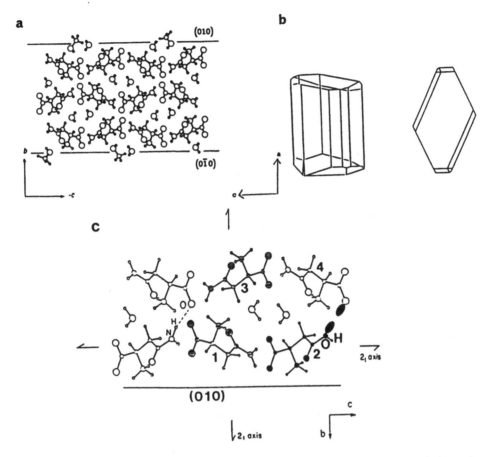

Fig. 4.16. (*a*) Packing arrangement of (S)-asparagine monohydrate viewed along the **a** axis. Four adsorbed methanol molecules are inserted on the (010) and (0$\bar{1}$0) faces, referring to crystals grown from water–methanol solution. (*b*) Morphology of (S)-asparagine monohydrate: (left) Pure crystal grown from aqueous solution; (right) grown in the presence of (S)-aspartic acid, or in methanol–water solution. (*c*) Preferential adsorption of aspartic acid on the (010) surface of asparagine (open circles) at sites of type 1 and 3 rather than 2 and 4, as the crystal is growing at the (010) face. Molecules at sites 1 and 3 are shaded, indicating that they may be occupied by either asparagine or aspartic acid. The probability for aspartic acid to be adsorbed at site 2 (full circles) is less than at site 1 because of O(hydroxyl)$\cdots$O(carboxylate) lone-pair electron repulsion between aspartic acid and asparagine at sites 2 and 4

arrangement, space group $P2_1/c$. The crystal structure (Fig. 4.17(*a*)) is composed of hydrogen-bonded dimers interlinked by N—H$\cdots$O bonds to form a ribbon-like motif that runs parallel to the **b** axis. E-cinnamic acid is preferentially occluded through half of the {011} surface sites of the crystal at the opposite ends of these ribbons, resulting in a crystal (Fig. 4.17(*b*)) with two enantiomorphous halves of at most $P2_1$ symmetry. This reduced symmetry was proven photochemically (Vaida et al. 1989). Ultraviolet irradiation of E-cinnamamide

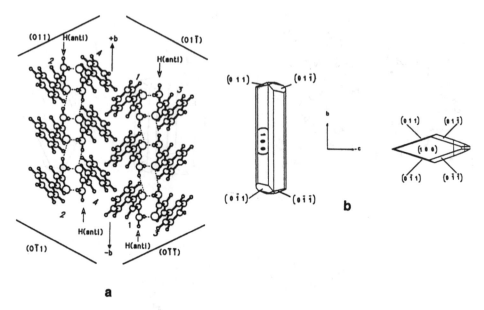

Fig. 4.17. (*a*) Packing arrangement of E-cinnamamide viewed along the **a** axis delineated by the {011} faces. (*b*) Morphology of E-cinnamamide crystals (left) pure, (right) grown in the presence of cinnamic acid.

yields centrosymmetric photodimers, by virtue of a cyclobutane ring formation involving pairs of close packed >C=C< bonds across centers of inversion (Fig. 4.18). Replacement of one of such a pair by E-cinnamic acid results in the formation of asymmetric cinnamamide–cinnamic acid photodimers of opposite chirality at the two enantiomorphous halves of the mixed crystal, with an enantiomeric ratio of 60:40 at each opposite half.

Reduction in crystal symmetry was demonstrated by photodimerization, x-ray, and neutron diffraction for the host–guest crystal structure of E-cinnamamide-2-E-thienylacrylamide (Vaida et al. 1988; Shimon et al. 1993). The host structure (Fig. 4.17(*a*)) incorporates herringbone contacts between aromatic C—H groups and π-electron clouds of neighboring phenyl rings. In order to induce site-selective adsorption and occlusion, these contacts are replaced in the host–guest system by unfavorable contacts between sulfur lone-pair electrons and the π-electron system (Fig. 4.19(*a*)). Thienylacrylamide is occluded through the enantiotopic {011} faces of the crystal (Fig. 4.19(*b*)). These faces exhibit *p*1 plane symmetry, since the four different surface sites are not related by symmetry. The guest can easily be adsorbed at only one of these four sites at which the sulfur emerges from the face, leading to *P*1 crystal symmetry upon occlusion of the guest. Thienylacrylamide is also   cluded through the {001} faces that exhibit *pm* plane symmetry. Occlusion   rough faces of this type should lead to crystal symmetry *Pc*. The mixed crystal should be thus divided into six sectors of reduced symmetry (Fig. 4.19(*c*)). The different sectors of the same

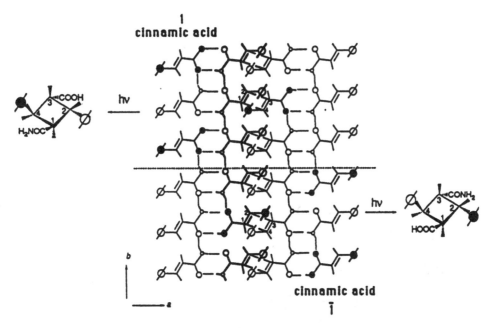

Fig. 4.18. Ribbons along the **b** axis of hydrogen-bonded molecules composed of E-cinnamamide (open circles) and occluded guest E-cinnamic acid molecules (full circles). The latter were adsorbed through site 1 at the +**b** end and site $\bar{1}$ at the −**b** end, leading to two enantiomorphous domains joined at the central dotted line. Shown also are the ribbons related by translation along the **a** axis, so depicting the close 4 Å contact between C=C double bonds related by a center of inversion. Also shown are two enantiomorphous cyclobutane photodimers between the host and occluded guest from the top and bottom halves of the crystal.

type, such as A and $\bar{A}$ are related to one another by the $2/m$ point symmetry of the host crystal. Low-temperature x-ray and neutron diffraction studies on A- and $\bar{A}$-type sectors, cut from a crystal specimen, showed symmetry $P1$ in keeping with the photodimerization studies. The proposed reduction in symmetry of the B-type sector (Fig. 4.19(c)) was proven to be correct, but only by an x-ray diffraction analysis. We may infer from these results a repulsive intermolecular interaction between the sulfur lone-pair electrons and the π-electron system of neighboring molecules, but we do not have an estimate of the energy.

We have elucidated by changes in crystal morphology, reduction in crystal symmetry, and energy calculations, the molecular conformation of a flexible additive and the effect of diastereomeric chiral additives on crystal morphology and symmetry. For example, for the centrosymmetric crystal of dipeptide glycyl-glycine when grown in the presence of glycyl-leucine additive, the latter adopts a specific conformation that minimizes unfavorable additive–host molecular interactions at the site of adsorption (Weissbuch et al. 1985a; Berkovitch-Yellin, Weissbuch, and Leiserowitz 1989).

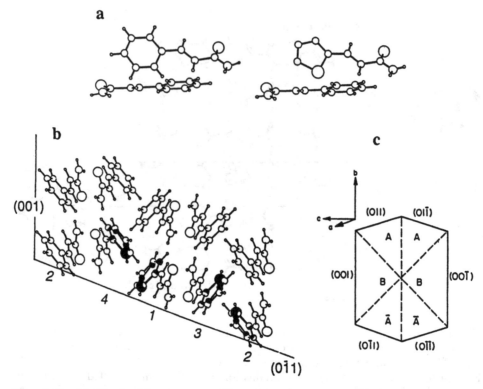

Fig. 4.19. (a) Herringbone contacts involving: (left) two cinnamamide molecules; (right) host cinnamamide and guest thienylacrylamide molecules. (b) The four different surfaces sites 1, 2, 3, 4 of E-cinnamamide at the (0$\bar{1}$1) face. Shown shaded are the S and C atoms of the superimposed 2-thienyl rings in the positions they would assume, were they to replace the phenyl rings of cinnamamide molecules at the (0$\bar{1}$1) face. Site 1 is preferred for guest adsorption on the (0$\bar{1}$1) face since the S atom at that site emerges from the crystal face. By the same argument, 1 and 2 are the preferred sites for guest adsorption on the (001) face. (c) The morphological representation of cinnamamide–thienylacrylamide crystal with six sectors of reduced symmetry.

The pronounced effect that threonine exerts on the morphology and structure of racemic serine crystals has been discussed in Sections 4.4 and 4.5. Oddly enough, additive *allo*-threonine, the diastereoisomer of threonine, does not alter very much the morphology of racemic serine crystals (Fig. 4.12(e)), nor is there any enantioselective segregation of *allo*-threonine along the **b** axis of the crystal. These observations were accounted for in a quantitative way by calculated surface binding energies, according to which S-*allo*-threonine, say, is most strongly bound on the {100} faces and somewhat less so on the (011) and (01$\bar{1}$) faces of a racemic serine crystal. We note that R- and S-*allo*-threonine additives are equally well occluded through the {100} faces because of the 2/m crystal point symmetry of (R,S)-serine, thus masking any enantioselective segregation of the additive through the {011} faces (Weissbuch et al. 1985b).

4.5.5. *Some General Rules on Symmetry Lowering through Additive Occlusion*

As a conclusion of the section on crystal symmetry lowering, we may draw some general rules. The local symmetry of a mixed crystal is determined by that of the faces developed. If the symmetry of the crystal face through which an additive is adsorbed is not lower than that of the bulk, there cannot be reduction in symmetry for the crystal sector bound by that face. In other words, the symmetry of a crystal sector cannot be lower than that of the growth sites on the bounding crystal face. This principle is so basic that, if a crystal is bound by faces of different plane symmetries, one can, at least in principle, imagine it as formed of different sectors, each with bulk symmetry as low as that of its bounding face, the sectors being related to each other by the point symmetry elements of the crystal. Thus, the "overall" symmetry of the crystal is not reduced. This is a consequence of Curie's (1894) principle as presented in his paper *Symétrie dans les phenomènes physiques*: "a physical event cannot have a symmetry lower than that of the event which caused it."

4.5.6. *Detection of Tailor-Made Additives on a Crystal Surface*

Although many experiments have been performed in which the crystal morphology can be understood in terms of strong binding of either solvent or additive molecules to specific faces, little work has been reported on the direct determination of the local structure at liquid–crystal interfaces. It is perhaps a consequence of the fact that the structure and dynamics at the solid–liquid interface are not easily amenable to direct study at the subnanometer level, although scanning-tunneling and atomic-force microscopy hold great promise. For example, atomic-force microscopy was recently used to characterize the structure of an organic diacid bound to a layered clay surface (Cai et al. 1994). X-ray diffraction has been shown to be a powerful tool to examine, at the atomic and molecular level, surfaces and interfaces of metals and semiconductors (Robinson and Tweet, 1992) and ultrathin crystalline films of organic molecules on liquid surfaces (Als-Nielsen et al. 1994). Synchrotron x-ray scattering studies have been carried out at the gold–electrolyte (Ocko et al. 1990; Wang et al. 1992a) and calcite–water (Chiarello, Wogelins, and Sturchio 1993; Chiarello and Sturchio 1994) interfaces. One important restriction for x-ray diffraction techniques to be applicable for the study of adsorbed foreign molecules on the surface of a molecular crystal is that the surface roughness should be low enough not to mask the surface diffraction signal.

Grazing-incidence x-ray diffraction (GID) experiments have been performed to determine the structure of additive (S)-methionine bound to the {010} faces of α-glycine (Gidalevitz et al. 1997a). A large ($8 \times 10 \times 16$ mm^3) crystal of α-glycine cleaved into two halves along the *ac* plane was further grown with each half exposing its cleaved face, (010) or (0$\bar{1}$0), to a saturated glycine solution containing 8 percent (S)-methionine. The measured GID pattern from the (0$\bar{1}$0) face

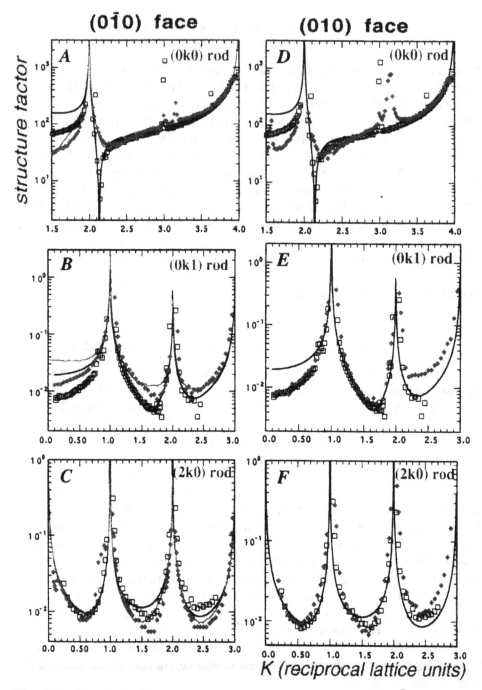

Fig. 4.20. Grazing-incidence x-ray diffraction patterns from the (0$\bar{1}$0) (left) and (010) (right) faces of α-glycine crystal after growth in aqueous solution containing (S)-methionine additive. Open squares are measured data on a pure cleaved {010} face; the black solid line is the calculated curve for the pure cleaved surface. Solid rhombs are measured data on the surface regrown in the presence of 9 wt percent of (S)-methionine; the gray solid line is the calculated curve for the surface regrown in the presence of 9 wt percent of (S)-methionine.

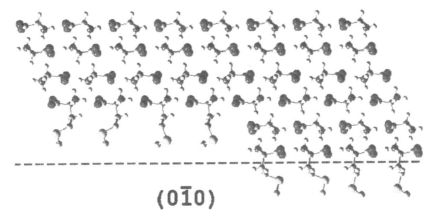

(0$\bar{1}$0)

Fig. 4.21. The proposed packing arrangement of the (0$\bar{1}$0) α-glycine face regrown in the presence of 9 wt percent of (S)-methionine. The ordered (S)-methionine molecules are bound via a pseudocenter of inversion to glycine.

(Fig. 4.20, left) was modeled in terms of a surface covered with 70 percent (S)-methionine, of which about 50 percent were ordered chains $CH_2CH_2SCH_3$ (Fig. 4.21). Surprisingly the (010) face of α-glycine was also covered with about 70 percent (S)-methionine according to the GID pattern (Fig. 4.20, right). This result was explained as arising from a pseudocentrosymmetric (S)-methionine glycine bilayer fitted onto the pure (010) face by pseudotwofold screw symmetry, yielding an imperfect lattice match to the pure (010) surface and thus an incommensurate interbilayer arrangement (Fig. 4.22A,B).

The model structure of an (S)-methionine-glycine heterobilayer bound to the (010) face of glycine implies that the (S)-methionine layer should, in principle, inhibit interlayer growth. This hypothesis is in agreement with experiments performed, in which cleaved crystals of glycine were grown under identical conditions in pure supersaturated aqueous solution and in supersaturated solution of glycine with 9 wt percent of (S)-methionine additive. Figure 4.22C shows that fastest growth along the **b** axis occurs for the cleaved crystals in pure glycine solution, distinctly slower growth along $+$**b** for a cleaved plate exposing its (010) face to glycine plus (S)-methionine solution, and almost complete inhibition of growth along $-$**b** for the crystal that exposed its (0$\bar{1}$0) face to glycine plus (S)-methionine solution.

Pseudo-twofold symmetry interlinking hydrogen-bonded bilayers in an incommensurate arrangement had also been invoked to explain packing disorder in racemic valine (Grayer Wolf et al. 1990) and the formation of twinned crystals of α-glycine grown in the presence of (S)-leucine (Li et al. 1994), when its concentration was greater than 8 mg/ml. In the latter system the two bilayers proposed to account for the crystal twinning lie in the central plane of the crystal and are each composed of heterodimers of glycine and (S)-leucine in an arrangement similar to that of the glycine and (S)-methionine hetero bilayer (Fig 4.22). These observations may be understood if the following assumption is made: the

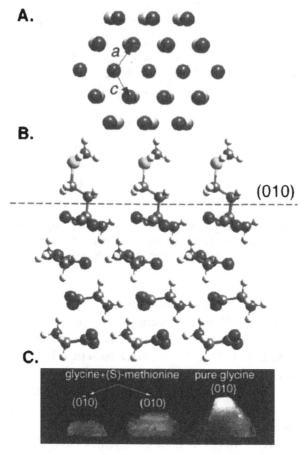

Fig. 4.22. (A) Superposition of the *ac* lattice of α-glycine upon itself after rotation by 180° about the **a** + **c** axis. (B) Proposed packing arrangement of the glycine–methionine bilayer at the (010) face. This bilayer is related to the underlying glycine bilayer by a pseudo-twofold rotation about the *a* + *c* axis. (C) Photograph of cleaved crystals of α-glycine whose {010} cleavage faces were exposed to different solutions. Before growth all three glycine plates were approximately the same size. Two plates exposing (010) (left) and (0̄1̄0̄) (center) faces were regrown in a glycine supersaturated solution in the presence of 9 wt percent (S)-methionine. The third glycine plate (right) was regrown in a pure glycine solution with the same degree of supersaturation.

α-amino acid solute molecules have a tendency to aggregate into cyclic hydrogen-bonded dimers in solution, and thus dock onto the growing crystal surface as dimers. Indeed, evidence is presented in Section 4.7 in favor of the cyclic hydrogen-bonded dimer as the growth unit on the {010} faces of α-glycine.

4.6. The Effect of Tailor-Made Additives on Crystal Texture

Surprisingly, little work has been done, to date, on the effect of tailor-made additives on mosaicity and coherence length (that is, the extent of perfect crystal order) in molecular crystals. Indirect evidence was first brought to light by low-temperature (18 K) neutron diffraction measurements on the single crystals of pure asparagine monohydrate and the solid solution of asparagine–aspartic acid monohydrate (Weissinger-Lewin et al. 1989), undertaken to demonstrate symmetry lowering in the latter system (Section 4.5). The neutron reflections of the specimen crystal of pure asparagine suffered from pronounced extinction effects, whereas the specimen crystal of asparagine–aspartic acid was almost free from extinction. These results were interpreted in terms of much higher mosaicity and smaller mosaic domains in the asparagine–aspartic acid crystals, brought about by the presence of aspartic acid. More direct evidence came from studies on biomineralization (Berman et al. 1990, 1993*a,b*), which addressed the fundamental question of whether proteins in calcite mineralized tissue alter the crystal texture in an anisotropic manner. High-resolution synchrotron x-ray diffraction was used to monitor changes in coherence length and angular mosaic spread. It was found that protein extracted from sea urchins and mollusks reduced the coherence lengths of crystals of calcium fumarate and calcium malonate in the directions perpendicular to the planes onto which the proteins were preferentially adsorbed. In contrast, the calcite crystals grown in vitro in the presence of the proteins exhibited an increase in angular spread, but no anisotropic effect in coherence length was detected. A biologically produced calcite crystal, on the other hand, showed a preferential reduction in coherence length in the direction of the **c** axis.

4.7. The Effect of Solvent on Crystal Growth and Morphology

Solvent has a strong influence on the habit of crystalline materials; however, the role played by solvent–surface interactions in enhancing or inhibiting crystal growth is still not completely resolved. To date, there have been two different approaches to clarify this point. In one theory (Bennema and Gilmer 1973; Bennema, 1992; Bourne and Davey, 1976; Elwenspoek et al. 1987), favorable interactions between solute and solvent on specific faces leads to reduced interfacial tension, causing a transition from a smooth to a rough interface and a concomitant faster surface growth. Alternatively, it has been proposed that preferential adsorption at specific faces will inhibit their growth as removal of bound solvent poses an additional energy barrier for continued growth. Our studies on the role played by tailor-made additives are in keeping with the latter approach. However, solvent–surface interactions cannot be as clearly analyzed as those between the tailor-made additives and the crystal. In order to help clarify this problem, we describe several approaches that have been adopted.

4.7.1. *Characterization of Solvent–Surface Interactions*

One of the main difficulties is to disentangle the relative contributions of the internal crystal structure and solvent–surface interactions in determining the crystal morphology. The former may be evaluated by growing the crystal by

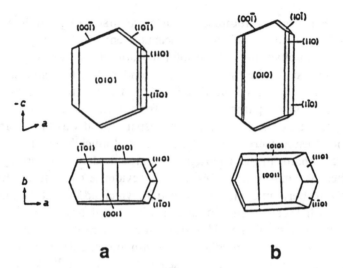

Fig. 4.23. (a) Theoretical crystal form of α-glycine. (b) Morphology of α-glycine crystals obtained by sublimation.

sublimation, if possible, or by theoretical modeling. Such information has been used to understand the effect of solvent on crystal morphology. For example the theoretical form of α-glycine (Berkovitch-Yellin 1985; Boek et al. 1991) shown in Fig. 4.23(a), is in nice agreement with the morphology of α-glycine crystals obtained by sublimation (Fig. 4.23(b)). This morphology exhibits well developed {010} faces, in keeping with the dominant hydrogen-bonding energy of the molecular bilayer in the ac plane. However, glycine crystals grown from aqueous solution display a distinctly different morphology, which is bipyramidal, with large {011} and {110} faces and less well developed {010} faces (Fig. 4.4(b)). In order to understand these differences, substrate–solvent interactions were taken into account. The large {011} and {110} faces expose CO_2^- and NH_3^+ groups well oriented to have a strong affinity for water. The {010} face exposes alternating layers of C—H groups or CO_2^- and NH_3^+ groups (Fig. 4.4(a)), and so, on average, is less hydrophilic than the {011} and {110} faces. These observations were expressed by Berkovitch-Yellin (1985) in a semiquantitative form by calculating the Coulomb potential at the various faces, and this indicated preferential adsorption of water molecules onto the polar {011} and {110} faces. Moreover, if the solute glycine molecules dock into surface sites primarily as cyclic hydrogen-bonded dimers, the {010} face will essentially expose C—H groups. This hypothesis is supported by several observations. Diffusion coefficient measurements yielded an average size of the glycine cluster of 1.8 molecules in a supersaturated solution (Chang and Myerson 1986; Myerson and Lo 1990; Ginde and Myerson 1992), but it was not established whether the clusters are in cyclic hydrogen-bonded form. Atomic-force microscopy and phase-measurement interferometric microscopy (by Cai et al. 1994) yielded a step size of ∼1 nm at the growing (010) face of glycine, which corresponds to the thickness of the

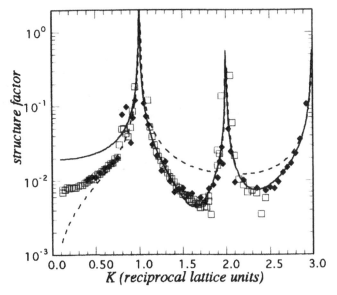

Fig. 4.24. Grazing-incidence x-ray diffraction pattern of an α-glycine $\{010\}$ surface during its growth from a saturated 0.75:0.25 ethanol–water solution. Open squares are data points measured on the pure dry $\{010\}$ cleaved surface. Solid rhombs are data points measured in situ during the growth of α-glycine. The solid line represents calculated model for the crystal terminated at the (010) face by the mixture of layers 1 and 3 and so exposing C—H groups (Fig. 4.4(a)). The dashed line represents calculated model for the crystal terminated by the mixture of layers 2 and 4 and so exposing CO_2^- and NH_3^+ groups (Fig. 4.4(a)).

hydrogen-bonded glycine bilayer. Definitive evidence that a growing $\{010\}$ face of α-glycine in an aqueous glycine solution exposes primarily C—H groups came from an in situ monitoring of the (010) face by grazing-incidence x-ray diffraction measurements (Fig 4.24) using synchrotron radiation (Gidalevitz et al. 1997b). Further support for the arguments that the $\{110\}$ and $\{011\}$ crystal faces of glycine are enhanced by their interactions with water comes from the observed crystal morphology of the α-amino acids with hydrophobic chains such as phenylalanine (Weissbuch et al. 1990), valine, and isoleucine (Torii and Iitaka 1970, 1971). Their crystal structures are composed of hydrogen-bonded bilayers, as in α-glycine, and their morphologies are platelike when grown from aqueous solution. Clearly the hydrophobic chains provide increased layer energy and a lower binding concentration of water at the sides of such bilayers than in α-glycine, providing a strong propensity for plate formation.

4.7.2. Crystalline Solvates

Another approach involved the use of crystalline solvates where the solvent of crystallization plays the dual role of solvent and crystal solute. Such crystals were grown in the added presence of tailor-made solvent, which is a slightly

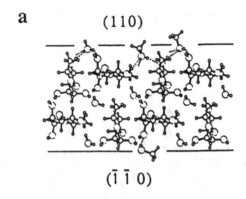

a   (110)

(1̄ 1̄ 0)

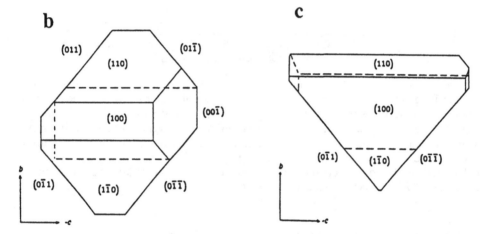

b

(011)   (01̄1̄)

(110)

(100)   (001̄)

(01̄1)   (1̄10)   (01̄1̄)

c

(110)

(100)

(01̄1)   (1̄10)   (01̄1̄)

Fig. 4.25. (*a*) Packing arrangement of α-rhamnose monohydrate crystal viewed along the **a** axis; the OH bonds of the hydrate water molecules point along the +**b**, but not the −**b**, direction; replacement of water by methanol on the {110} faces is depicted. (*b*)–(*c*) Crystals viewed along the **a** axis. Morphologies of α-rhamnose monohydrate crystals grown from: (*b*) aqueous solution; (*c*) 9:1 methanol:water solution.

modified version of the solvate solvent. It was found that, when crystalline hydrates were grown from aqueous solution in the added presence of methanol (the "tailor-made" solvent), the change in morphology was interpretable (Shimon et al. 1990, 1991) in a manner akin to the effect of a tailor-made inhibitor. This approach is illustrated with the crystal of α-rhamnose monohydrate, which contains a polar arrangement (Fig. 4.25(*a*)) in space group $P2_1$ and, when grown from pure aqueous solution, displays a bipyramidal morphology (Fig. 4.25(*b*)). The two O—H bonds of the hydrate water molecules are oriented towards the +**b** direction, but not towards the −**b** direction, of the crystal. Thus, addition of methanol, as a cosolvent, changes the morphology of the crystal, completely inhibiting growth along the +**b** direction and yielding a pyramidal crystal (Fig. 4.25(*c*)).

A second example is (S)-asparagine.H$_2$O. Crystals grown from pure water are delineated by eighteen faces (Fig. 4.16(b)). The packing arrangement (Fig. 4.16(a)) shows that one of the two O—H bonds of each water molecule emerges from the {010} faces, which are poorly expressed. Therefore, methanol molecules can be easily attached at these surface sites (Fig. 4.16(a)). The methyl group of each adsorbed methanol protrudes from the two {010} faces, impedes growth in the **b** direction and thus induces formation of large {010} crystal faces (Fig. 4.16(b)). On the {011} faces, only two of the four symmetry-related molecules can be substituted by methanol and so inhibition of {011} should be less dramatic. This reasoning proved to be in keeping with experiment (Shimon et al. 1991); growth in the presence of methanol as additive yielded {010} plates similar in shape to when asparagine is grown in the presence of aspartic acid (Fig. 4.16(b)). That methanol can indeed be selectively adsorbed on both the {010} and {011} faces was further demonstrated by experiments involving partial dissolution of asparagine.H$_2$O crystals in methanol solution, revealing etch pits, although poorly developed, on only these faces (Shimon et al. 1986a).

4.7.3. *Polar Crystals*

In another approach, use was made of the principle that in a polar crystal the attachment energy E_{att} at opposite and hemihedral faces (hkl) and $(\overline{hkl})$ are the same (neglecting effects such as possible differences in molecular charge density in the bulk and molecules to be attached at the opposite faces). Although a difference in these E_{att} values has been invoked to explain the polar morphology of urea obtained by sublimation (Docherty et al. 1993), its effect on morphology must be far less than that of solvent on crystal growth and morphology. Thus, in the absence of a difference in the effect of binding of solvent at the hemihedral faces (hkl) and $(\overline{hkl})$, these two faces should grow at the same rate (Turner and Lonsdale 1950). A pronounced difference in growth rate must then be associated with differences in solvent–surface interactions.

This principle was exploited in the family of platelike crystals composed of amphiphilic molecules that pack in layers stacked head-to-tail in a polar arrangement (chainlike amphiphilic molecules normally pack head-to-head and tail-to-tail in the crystalline state). Thus the polar platelike crystals are hydrophobic at one plate face and hydrophilic at the opposite face. The idea was to ascertain the relative binding properties of polar solvents to the opposite hemihedral faces from the macroscopic wetting properties, in terms of contact-angle measurements, and to correlate this effect with the relative rates of growth of the two crystal faces in the same solvents. In Section 4.4 we showed how the absolute configuration of the constituent molecules of chiral-resolved alkylgluconamides, which pack in the desired polar arrangement (Fig. 4.8), may be assigned from the difference in wetting properties of the hydrophilic and hydrophobic faces. In CH$_3$OH solution the polar platelike crystals of the octyl derivative ($n = 8$) were found to grow almost four times faster at the hydrophobic side than at the hydrophilic side, as shown in Fig. 4.26 (Wang, Leiserowitz, and Lahav 1992b).

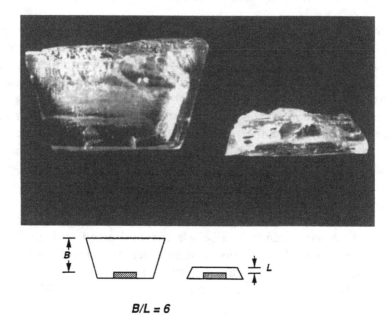

Fig 4.26. Morphology of a *N*-octylgluconamide crystal grown from methanol. The seeds can be observed as an opaque shadow at the bottom of the crystals. The thickness of added material on the hydrophobic surface is denoted as *B* and that added on the hydrophilic surface is denoted as *L*. The *B/L* ratio has an average value of about five.

This result is completely in keeping with the observation that water and other polar solvents wet the hydrophilic face more strongly than the opposite hydrophobic face (see Section 4.4). It is noteworthy that at the hydrophilic face, solvent methanol can form O—H···O hydrogen bonds with the terminal hydroxyl groups of two neighboring chains, since the translation separation distance between them is about 5 Å.

The theory that strong solvent–surface interactions inhibit crystal growth is also borne out by a comparison of the effect of tailor-made additives and acetic acid solvent on the morphology of the polar crystals of *N*-(cinnamoyl)-alanine already discussed in Section 4.4. A benchmark study on the effect of solvent on the growth of polar crystals was carried out by Wells (1949). He found that in aqueous solution, the α form of resorcinol (1,3-dihydroxybenzene), which crystallizes in space group $Pna2_1$, grows unidirectionally along the polar **c** axis in aqueous solution. The {011} faces at one end of the polar axis have O—H and C—H "proton donor" properties and the {0$\bar{1}\bar{1}$} faces at the opposite end have hydroxy-rich "proton acceptor" properties (Fig. 4.27). However, Wells could not fix the absolute direction of growth along the polar axis. This assignment was made with the use of tailor-made additives (Wireko et al. 1987). The unidirectional growth in aqueous solution takes place at the hydroxy-rich "proton acceptor" {0$\bar{1}\bar{1}$} faces. This observation was explained in terms of stronger binding of water to the C—H and O—H "proton donor" {011} faces (Wireko et al. 1987;

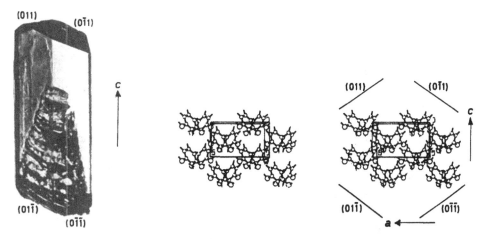

Fig. 4.27. (Left) Typical crystal of α-resorcinol grown from water. The four end faces {011} are marked. (Right) Packing arrangement of α-resorcinol. Stereoview along the **a** axis. The planes parallel to the "proton acceptor" and "proton donor" faces are denoted.

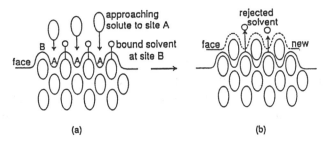

Scheme 4.7. Selective adsorption of solvent on a corrugated crystalline face.

Davey, Milisavljevic, and Bourne 1988), but these analyses were not completely irrefutable, hinting that other growth mechanisms might also be responsible. Thus, we describe next experiments on the polar crystals of γ-glycine and racemic alanine, as well as urea, where different mechanisms have been invoked to account for their growth behavior.

4.7.4. A "Relay" Mechanism of Crystal Growth as a Result of Selective Solvent Binding

An entirely different approach involved strong selective adsorption of solvents at a subset of molecular surface sites, of, say, type *B* and repulsion of solvent at the remaining set of surface sites of, say, type *A* on the crystal face. This is depicted in Scheme 4.7, where we emphasize the difference between the two types of sites by assuming a corrugated surface such that the *A*-type site is a cavity and the *B*-type site is on the outside upper surface of the cavity. Scheme 4.7(*a*) shows the *B*-type sites blocked by solvent *S* and the *A*-type sites unsolvated. Thus solute

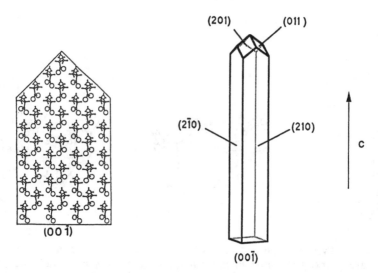

Fig. 4.28. Packing arrangement of (R,S)-alanine delineated by crystal faces, as viewed down the **b** axis. The capped faces at the $+\mathbf{c}$ end of the polar axis expose NH_3^+ and —CH_3 groups at their surfaces, whereas the opposite $(00\bar{1})$ face exposes carboxylate CO_2^- groups.

molecules can easily fit into A-type sites. But once docked into position (Scheme 4.7(b)), the roles of the A- and B-type sites are essentially reversed and the solvent molecules that originally were bound to B-type sites would be repelled since they now occupy A-type sites. This cyclic process can lead to fast growth by a kind of "relay" mechanism.

In such a situation as described here, where desolvation is rate limiting, we have implicitly made use of the idea that the free energy of incorporation of a solute molecule helps to displace bound solvent. This relay mechanism may be demonstrated by a crystal with a polar axis and the experiment involves a comparison of the relative rates of growth at the opposite poles for the different solvents, as exemplified by the growth and dissolution of the crystals of (R,S)-alanine and the γ form of glycine in different solvents (Shimon et al. 1990, 1991). These two crystals have similar packing features and so the discussion is confined to (R,S)-alanine. As already described in Section 4.4, the zwitterionic molecules are aligned so as to expose CO_2^- groups at one end of the polar **c** axis and NH_3^+ groups at the opposite end. The crystal has a polar morphology, with the CO_2^- groups emerging at the flat $-\mathbf{c}$ end, as shown in Fig. 4.28. The $(00\bar{1})$ face at the carboxylate end contains cavities. This face grows and dissolves faster in aqueous solution than the "smoother" amino face (Fig. 4.29(a)). We propose that the faster growth at the $(00\bar{1})$ side of the crystal in aqueous solution is due to the relay mechanism of growth, according to the following arguments. The water molecules may be strongly bound by hydrogen bonds to the outermost layer $(00\bar{1})$ of CO_2^- groups. In contrast, the pockets act as proton acceptors for the NH_3^+ groups of the solute molecules. Replacement of the solute NH_3^+ moiety by

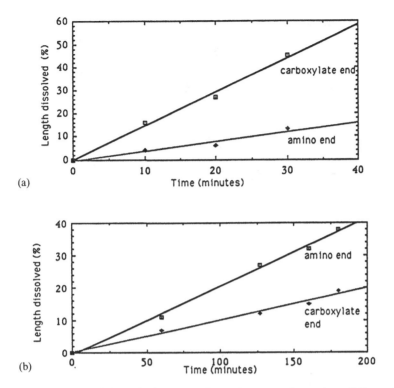

Fig. 4.29. Graph of the relative growth at the opposite poles of the polar axis of (R,S)-alanine crystals: (*a*) in water; (*b*) in 4:1 methanol:water mixture.

solvent water within the pockets yields repulsive or, at best, weakly attractive interactions. The pockets will therefore be weakly hydrated and so relatively easily accessible to approaching solute molecules (Fig. 4.30). There is, however, an alternative mechanism that may account for the faster growth at the −c end: a higher crystal attachment energy at the (00$\bar{1}$) end of the crystal than at the opposite end. We could demonstrate that the relay mechanism is the more likely on the following basis: We predicted that methanol molecules could bind into the pockets at the (00$\bar{1}$) face. The methyl group of this cosolvent molecule can form attractive, albeit weak, C—H$\cdots$O interactions within the pocket and the OH group can form a hydrogen bond to the CO_2^- group at the surface of the pocket. In keeping with prediction, crystals of (R,S)-alanine in an 80 percent methanol: water mixture grow faster at the +c amino end of the crystal than at the −c carboxylate end (Fig. 4.29(*b*)), indicating that methanol blocks solute access to the pockets more effectively than water at the (00$\bar{1}$) face.

This work implies that, for those crystals that expose sets of different surface sites at each face, the average binding energy of solvent to a face may not be a sufficient criterion for establishing the relative morphological importance of the different faces. It may be necessary to assess the ease of replacement of solvent by solute at each of the different surface sites, as well as the different modes of docking the solute molecules at the surface. For example, Boek et al. (1991)

212   *I. Weissbuch et al.*

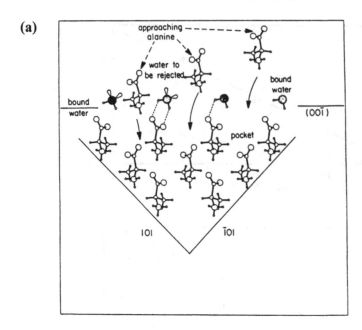

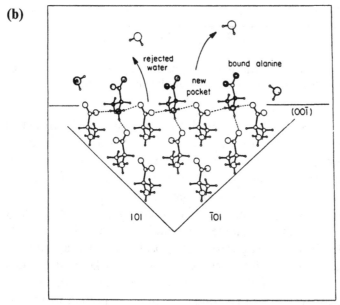

Fig. 4.30. Schematic representation of the (00$\bar{1}$) face of R,S-alanine during crystal growth. (*a*) In this view approaching solute alanine molecules are depicted about to be bound within the pockets of the (00$\bar{1}$) face. Also shown are water molecules bound to the outermost CO_2^- groups of this face. The pockets remain poorly solvated because the electron lone-pair–lone-pair oxygen–oxygen repulsions inhibit the binding of water within them (similar to what has been described earlier for the asparagine–aspartic acid system, in Section 4.5). (*b*) In this view, the newly adsorbed alanine molecules are each bound via three NH···O hydrogen bonds. The previously bound water molecules are shown being rejected by O(water)···O(carboxylate) lone-pair–lone-pair repulsions. Note the formation of new unsolvated pockets.

calculated the theoretical growth form of tetragonal urea and found that it is almost cubelike and compares well with those of crystals from ethanol or benzene. However, the theoretical form is very different in shape from the [001] needles obtained from aqueous solution. Boek and Briels (1992) performed molecular dynamics simulations of interfaces between water and crystalline urea. They argued that the observed needle morphology cannot be explained by using a simple layer model because of strong adsorption of water to the relevant faces. By examining the interface between a saturated aqueous urea solution and crystalline urea, Boek et al. (1994) explained the [001] needle morphology in terms of wrongly and randomly absorbed urea molecules to the {110} side faces, so providing an increased interfacial entropy. Further theoretical studies yielded a needle-like morphology of urea crystal, in excellent agreement with experiment (Liu et al. 1995*a*, *b*). It is noteworthy that oxamide crystals grown from aqueous solution contain about 2 percent of wrongly oriented oxamide molecules in the lattice (Swaminathan and Craven 1982).

4.8. Control of Crystalline Phase Formation

Any crystallization process must cross the nucleation barrier. It is generally assumed that, as the molecules start to associate in supersaturated solutions, they form embryos with structures resembling those of the to-be-grown crystals. For polymorphic systems embryos of all the observed polymorphs are probably formed. Each type of embryo would resemble structurally the crystal into which it will eventually develop. If this hypothesis is correct, one may use the structural information of the mature crystals to design inhibitors of a particular crystalline phase. The overall consequence of this stereospecific inhibition process can be that the unaffected phase, if less stable, will grow by kinetic control. We shall consider several systems where the logic of the above arguments applies. We first discuss the synthesis of chiral enantiomerically pure material from nonchiral reagents. This will be followed by a method for the kinetic resolution of enantiomers and the induced crystallization of a particular polymorph. Finally, the induced nucleation of a particular phase that acts as a template for the nucleation and further growth of the more stable phase into a mature crystal will be discussed.

4.8.1. *Amplification of Chirality by Crystallization with Chiral Additives*

The solid state may be used for the synthesis of chiral enantiomerically pure material from nonchiral reagents by precipitation of nonchiral molecules in appropriately packed chiral crystals, followed by a lattice-controlled reaction (Green, Lahav, and Rabinovich 1979; Addadi et al. 1980; Scheffer and Pokkuluri 1991; Vaida et al. 1991; Toda, 1996; Ohashi, 1996). However crystallization from achiral solvents generally yields {R} and {S} crystals in equal amounts. In order to obtain enantiomerically pure material, it is necessary to

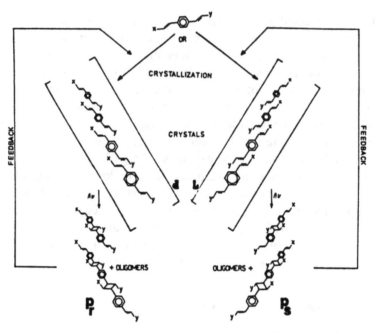

Scheme 4.8. Enantiomorphous crystals composed of nonchiral molecules and the feedback mechanism by the photodimerization products.

inhibit the nucleation of one of the two chiral crystal phases. This was achieved in the following way. Molecular dienes were "engineered" to pack in chiral crystals (Addadi and Lahav, 1979; Addadi et al. 1982*b*; van Mil et al. 1981, 1982), in which the molecules are stacked by translation such that the neighboring C=C double bonds are in appropriate proximity for photodimerization (Scheme 4.8). Ultraviolet irradiation of a single crystal yields dimers, trimers, and oligomeric products of a single chirality P_r or P_s, dependent upon the chirality of the starting crystal. Thus the handedness of the molecular stacking arrangement is retained in the structure of the oligomer. A net asymmetric induction of one of the crystalline phases was achieved when crystals were grown in the presence of oligomers of a single chirality (that is, P_r or P_s) obtained from a large single crystal. In all experiments it was observed that a large excess was formed of the phase that was enantiomorphic to the one in which the additive itself had been generated; namely, {R} crystals were obtained from a solution containing additives P_s, and vice versa.

A natural extension of this work led to the suggestion of a new method for kinetic resolution of a racemic mixture crystallizing in the form a conglomerate of {R} and {S} crystals (Addadi et al. 1981, 1982*c*), as described below.

4.8.2. *Resolution of Enantiomers*

Racemic mixtures of molecules that crystallize as a conglomerate yield two enantiomorphous equienergetic crystalline phases. Kinetic resolution of the conglomerate may be achieved with the assistance of tailor-made chiral-resolved inhibitors, which would, in solution, bind enantioselectively and stereoselectively to only one of the enantiomeric crystalline phases and so prevent its eventual growth. This principle has been demonstrated for many systems (Addadi et al. 1982a, c; Zbaida et al. 1987). There are no particular requirements on molecular packing, although we may state that, the higher the point symmetry of the crystal, the more different orientations in the lattice the constituent molecules may occupy, making the inhibition process more pronounced since the additive would be adsorbed on more faces. The kinetic resolution may encompass preferred precipitation from a racemic mixture in solution of one enantiomorph at the expense of the more stable racemic crystal phase. Naturally, the energy gap between the racemic crystal and the mixture of resolved enantiomorphs must be relatively small. For example, racemic histidine.HCl precipitates, at 25°C, as a stable racemic dihydrate crystal and a metastable conglomerate of resolved monohydrate crystals. Addition of as little as one percent of the chiral-resolved poly-(p-acrylamido-S-phenyalanine) additive induced resolution with quantitative yield of the desired enantiomer (Weissbuch et al. 1987).

4.8.3. *Induced Crystallization of a Particular Polymorph*

We shall consider two different scenarios involving control of polymorphism; in one type of system, advantage is taken of differences in crystal polarity, whilst, in the other, use is made of differences in crystal layer energy.

We examine preferred precipitation of a crystal containing a polar axis, at the expense of the dimorph that is centrosymmetric and so nonpolar. The concept is that, in crystals with a polar axis, all molecules are aligned in the same direction relative to the polar axis. In centrosymmetric crystals, neighboring molecules are arranged in an antiparallel manner. These two arrangements are depicted in Scheme 4.9. Thus an appropriate tailor-made additive will inhibit growth of the centrosymmetric form at the opposite ends of the crystal, and so prevent its appearance, but will inhibit growth of the polar form only at one end of the crystal.

This concept has been demonstrated in two different systems. In that of *N*-(2-acetamido-4-nitrophenyl) pyrrolidine (PAN) (*6a*), the metastable polar form (Fig. 4.31(*b*)) displays optical second harmonic generation.

6a   *6b*

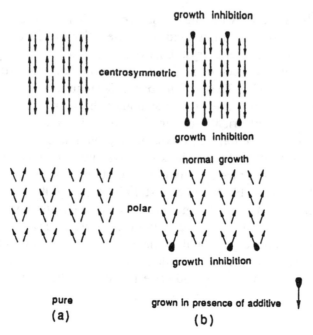

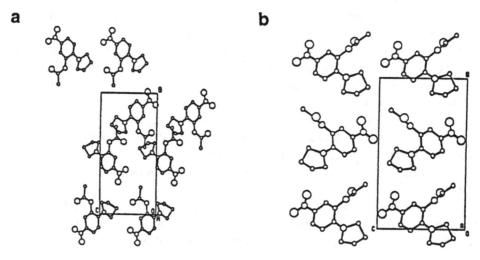

Scheme 4.9. Selective adsorption of a tailor-made additive on centrosymmetric and polar crystals.

Fig. 4.31. Crystalline packing arrangements of PAN (6*a*): (*a*) stable form which is pseudocentrosymmetric; (*b*) metastable form where the polar axis is along the vertical direction.

As little as 0.03 percent of the polymer (6*b*) inhibited precipitation of the pseudo-centrosymmetric stable form of PAN (Fig. 4.31(*a*)), allowing crystallization of the metastable form only (Staab et al. 1990).

    The other system involved preferential crystallization from aqueous solution of the γ form of glycine, when grown in the presence of the additive racemic

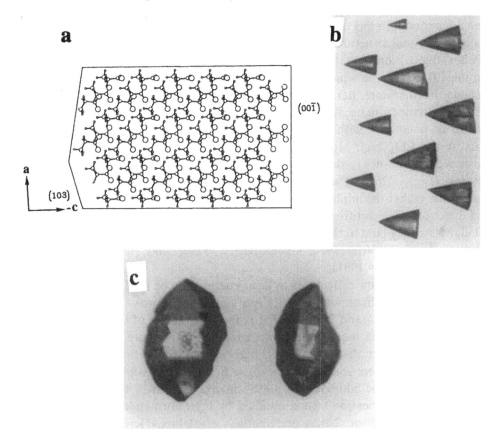

Fig. 4.32. (*a*) Packing arrangement of the γ form of glycine, as delineated by the faces grown from aqueous solution containing acetic acid. (*b*) Photographs of γ-glycine crystals grown from aqueous solutions in the presence of 3 wt percent of hexafluorovaline additive. (*c*) Photographs of α-glycine crystals grown from aqueous solution in the presence of low concentration (1 wt percent) of hexafluorovaline additive, showing inhibition of the four {011} faces.

hexafluorovaline $(CF_3)_2CHCH(NH_3^+)CO_2$ (Weissbuch et al. 1994b). As already mentioned, glycine usually crystallizes in the centrosymmetric α form from aqueous solutions, but Iitaka (1961) reported that glycine can precipitate in a polar γ form from acetic acid or ammonia solutions. The γ-form appears in space group $P3_1$ (or $P3_2$), arranged in a way similar to the (R,S)-alanine crystal (Fig. 4.28), so that the molecules expose CO_2^- groups at one end of the polar trigonal axis and NH_3^+ groups at the opposite end (Fig. 4.32(a)). It was found that racemic hexafluorovaline is an effective inhibitor of growth of α-glycine since it binds to the four {011} faces and so blocks growth along both **b** and the fast growing **c** directions (Fig. 4.32(c)). This behavior is in contrast to that of the common α-amino acid additives that bind primarily to the {010} faces and so retard growth of α-glycine only along the **b** direction, inducing formation of the {010} plates, as described earlier in Section 4.5. The abnormal behavior of hexafluorovaline may be understood in terms of its molecular structure and conformation as found in its native crystal structure (Weissbuch et al. 1990); steric repulsion imposed by the hexafluoro-isopropyl moiety prevents adsorption onto the {010} face, but not on the {011} faces, to which it is strongly bound through charge-transfer inter-actions. Furthermore, hexafluorovaline does not inhibit nucleation and growth of the γ-glycine form because the additive is bound thereto primarily at the NH_3^+ end of its polar axis, with the crystal growing unidirectionally along its polar **c** axis at the CO_2^- end. However, the hexafluorovaline molecules can also be adsorbed on the newly developed {$\bar{1}51$} side faces of the trigonal pyramidal crystals (Fig. 4.32(b)). The salient lesson to be learnt from these experiments on glycine is that the inhibitor, to be effective, should bind along the fast-grow-ing directions. The antifreeze proteins in fish that inhabit the polar seas act in a similar way; they bind to the fast-growing directions of the ice nuclei to lower the freezing point (DeVries 1984; Chakrabartty, Yang, and Hew 1984; Knight, Chang, and DeVries 1991).

Now we probe how differences in layer energy may be used for the preferred precipitation of a thermodynamically less stable polymorph. Assume the exis-tence of two forms 1 and 2, of which 1 is the more stable of the two; therefore $E_1 < E_2$, where E is the crystal lattice energy. Furthermore, we assume that both structures form distinct layer arrangements and that the layer energy E' of form 2 is the more stable, namely $E_2' < E_1'$. If it is possible, with the assis-tance of tailor-made additives, to inhibit interlayer growth to such an extent that the crystal nucleation step is initiated by layer formation, then, in princi-ple, the less stable form 2 may preferentially crystallize. Here we present such an example.

The long-chain hydrocarbons C_nH_{2n+2}, $23 \leq n \leq 36$, form ultrathin crystalline films when spread on water at 5°C, with a concentration for monolayer coverage (Weinbach et al. 1995). The crystalline properties of these films on water were characterized by grazing-incidence x-ray diffraction (GID) using synchrotron radiation. The number of layers formed proved to be dependent on chain length; for $n > 31$, monolayer material was obtained, and as the chain length was decreased from $C_{29}H_{61}$ to $C_{23}H_{48}$, the number of layers increased from two to about twenty. Schematic views of the mono- and multilayer packing arrange-ments of the alkane series are shown in Fig. 4.33 (top, middle). With the excep-

Fig. 4.33. Schematic representation of the layer-by-layer building of alkane C_nH_{2n+2} crystalline films on a water subphase, as determined by GID. Top: Side view of the molecules illustrating the decreasing number of layers with increasing n. Below each chemical formula, the film thickness, unit cell parameters a, b, c (Å), α, β, γ (°), and space group are given. For $n = 28$, 29 and for the alkanes $n = 23$, 24 with additive (see below), the values for the materials with an even number of carbons in the chain are given in parentheses. Middle: The packing arrangements as viewed along the molecular axes. For the orthorhombic structures the two layers of the unit cells along the **c** direction are given. Open circles indicate molecules in one layer and shaded circles molecules in the other layer. Bottom: Bilayers formed on addition of small amounts of alcohol additives to alkanes $n = 23$, 24.

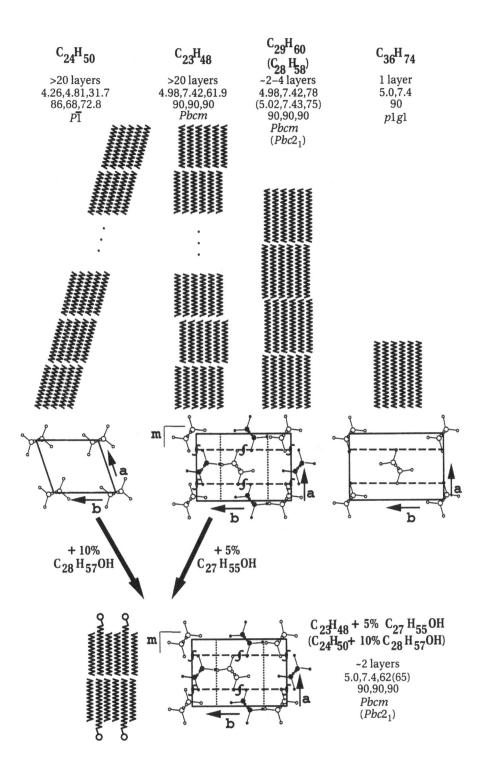

tion of the multilayer of $C_{24}H_{50}$, in which all the hydrocarbon chains are highly tilted, parallel, and pack in the triclinic space group $P\bar{1}$, the molecular chains in the other crystal structures are aligned normal to the layer plane, assuming a herringbone layer packing in orthorhombic unit cells.

Tailor-made additives were found to inhibit the multilayer growth of the alkanes; addition of 5–10 percent of the alcohols $C_{27}H_{55}OH$ and $C_{28}H_{57}OH$ to $C_{23}H_{48}$ and $C_{24}H_{50}$ yielded crystalline bilayers (Fig. 4.33, bottom). Not only was the film thickness affected, but the alcohol $C_{28}H_{57}OH$ also induced a change in the packing arrangement of $C_{24}H_{50}$ from triclinic $P\bar{1}$, in which the chains are tilted by 22° from the layer normal, to an orthorhombic cell, space group $Pbc2_1$, in which the chains are aligned normal to the layer. The appearance of the orthorhombic form may be understood in terms of crystal and layer energies, as alluded to above. According to atom–atom potential-energy calculations, the crystal lattice energy of the $P\bar{1}$ tilted chain structure (−48.0 kcal/mol) is more stable than that of the $Pbc2_1$ untilted chain structure (−47.5 kcal/mol) by 0.5 kcal/mol. However, if only a single layer is considered, their stability is reversed, with an energy difference of 0.8 kcal/mol in favor of the $Pbc2_1$ form (−44.5 compared with −45.3 kcal/mol). We may understand the preferred precipitation of the orthorhombic phase if the additive induces layer-by-layer growth from inception, but if there is an interplay of inter- and intralayer growth at the nucleation stage, then crystallization of the triclinic form should be preferred.

4.8.4. Insight on the Early Stages of Nucleation via Induced Twinning of Racemic Alanine Crystals

We examine a system in which a chiral-resolved additive induces nucleation of a chiral phase from a racemic mixture in solution, but this step is followed by the eventual precipitation of the thermodynamically stable racemic crystal through an epitaxial structural fit between the two crystalline phases. The constituent molecules are alanine. The racemic (R,S) crystal, already described in Sections 4.4 and 4.7, crystallizes in the polar space group $Pna2_1$. The molecules form parallel hydrogen-bonded chains along the polar **c** axis, exposing CO_2^- groups at one end of the polar axis and NH_3^+ groups at the opposite end (Fig. 4.28).

The enantiomorphous crystals of (R)- or (S)-alanine have very similar cell dimensions, but the hydrogen-bonded chains are arranged antiparallel in space group $P2_12_12_1$, so exposing both CO_2^- and NH_3^+ groups at the ends of the **c** axis. The energy gap between these two phases must be small and their solubilities are very similar. The crystals exhibit a morphology of point symmetry $mm2$ in keeping with the space group. Precipitation of alanine from a racemic solution, when grown in the presence of chiral-resolved α-amino acid additive, such as R-threonine, yields (R,S)-alanine crystals with an affected morphology, propeller in shape and exhibiting 222 point symmetry (Fig. 4.34(a)). Growth in the presence of the opposite enantiomeric additive S-threonine yields propellers of the mirror image (Weissbuch et al. 1994a). The crystals expose CO_2^- groups at each end of

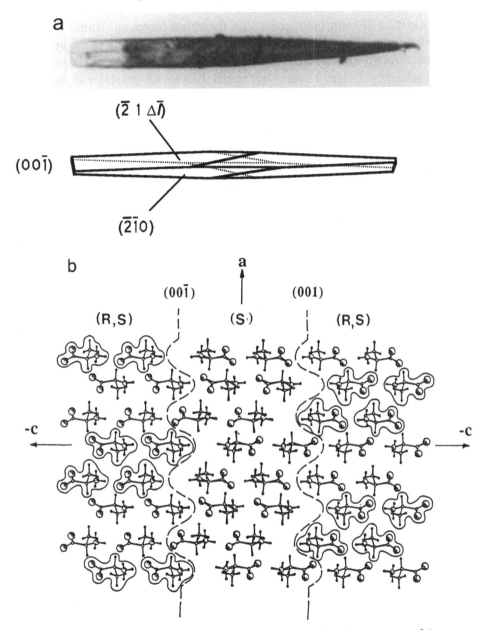

Fig. 4.34. (*a*) Twinned crystals of (R,S)-alanine grown in the presence of 1 percent R-threonine additive. (*b*) Proposed structural arrangement showing epitaxial growth of (R,S)-alanine at the (001) and (00$\bar{1}$) faces of a nucleus of (S)-alanine.

the propeller and so would appear to be twinned about the central *ab* plane, in keeping with observation that the two crystal halves appear to be stitched across this central *ab* plane. However, the two surfaces in contact across such a twin-ning plane would each expose NH$_3^+$ groups, which must be energetically unfa-vorable. One reasonable explanation for the observed effect is the inhibition of crystal nucleation of both the racemic and the chiral forms of alanine with the same handedness as the additive in solution, leading to nucleation of the opposite enantiomer through absence of steric interaction with the additive. Indeed this step was created by design. These chiral nuclei then act as templates at their two opposite {001} faces, that is, as nucleation sites for epitaxial growth of (R,S) alanine, yielding the "twinned" crystals. The proposed structural arrangement at the epitaxial interface is shown in Fig. 4.34(*b*). This arrangement also brings about the point that each of the two (R,S) to (S) interfaces has half favorable N—H$\cdots$O hydrogen bonds and half unfavorable, yet well-separated, $NH_3^+\cdots^+$ H_3N electrostatic interactions, while twinning of the two (R,S) domains without the interleaving (S) layer would generate only unfavorable contacts. The tapered arms of the propeller are brought about by the enantioselective adsorption of the chiral additive at two opposite faces of the four {210} side faces. The proposed mechanism of nucleation followed by epitaxial growth was substantiated by observed growth of (R,S)-alanine from seeds of chiral-resolved alanine and by growth experiments involving various nonracemic (R:S) mixtures. Furthermore, x-ray diffraction measurements of a twinned crystal revealed the following: a misalignment between the two halves, and reflections, but only from the crystal center, that are symmetry forbidden for space group $Pna2_1$, and so could arise from a $P2_12_12_1$ crystallite.

4.9. Conclusion

A great variety of crystalline properties can be controlled and modified with the use of tailor-made auxiliaries, including growth and dissolution rates, morphology, crystal texture, etch-pit formation, structure, and polymorphism. Surprisingly, such studies also allow one to help elucidate basic scientific questions, such as pinpointing fine molecular interactions at interfaces, the correlation between molecular chirality and macroscopic phenomena in crystals, and the role played by solvent in crystal growth and morphology. Control of crystal formation in the very early stages is still primitive and the monitoring of such a process by diffraction methods has still to be done. Very little work has been done with tailor-made additives on liquid crystals or on crystals composed of macromolecules, such as proteins. Indeed, one has only to look at the fine control achieved in biomineralization to realize that we still have a long road ahead.

Acknowledgments

The work discussed in this chapter has been supported over the years by the Minerva Foundation, Münich, the Israel Science Foundation, and the US–Israel Binational Science Foundation.

References

Addadi, L., Ariel, S., Lahav, M., Leiserowitz, L., Popovitz-Biro, R., and Tang, C.P. (1980). *Chemical Physics of Solids and their Surfaces. Specialist Periodical Reports*, **8**, 202–44.

Addadi, L., Berkovitch-Yellin, Z., Domb, N., Gati, E., Lahav, M., and Leiserowitz, L. (1982*a*). *Nature*, **296**, 21–6.

Addadi, L., Berkovitch-Yellin, Z., Weissbuch, I., Lahav, M., and Leiserowitz, L. (1986). *Topics in Stereochemistry*, **16**, 1–85.

Addadi, L., Berkovitch-Yellin, Z., Weissbuch, I., van Mil, J., Shimon, L. J. W., Lahav, M., and Leiserowitz, L. (1985). *Angewandte Chemie International Edition English*, **24**, 466–85.

Addadi, L. and Lahav, M. (1979). *Pure and Applied Chemistry*, **51**, 1269–84.

Addadi, L., van Mil, J., and Lahav, M. (1981). *Journal of the American Chemical Society*, **103**, 1249–51.

Addadi, L., van Mil, J., and Lahav, M. (1982*b*). *Journal of the American Chemical Society*, **104**, 3422–9.

Addadi, L., Weinstein, S., Gati, E., Weissbuch, I., and Lahav, M. (1982*c*). *Journal of the American Chemical Society*, **104**, 4610–17.

Addadi, L. and Weiner, S. (1992). *Angewandte Chemie International Edition English*, **31**, 153–69.

Allen, F. M. and Buseck, P. R. (1988). *American Mineralogist*, **73**, 568.

Als-Nielsen, J., Jacquemain, D., Kjaer, K., Leveiller, F., Lahav, M., and Leiserowitz, L. (1994). *Physics Reports*, **246**, 251–313.

Antinozzi, P. A., Brown, C. M., and Purich, D. L. (1992). *Journal of Crystal Growth*, **125**, 215–22.

Bennema, P. (1992). *Journal of Crystal Growth*, **122**, 110–19.

Bennema, P. and Gilmer, G. (1973). In *Crystal Growth, An Introduction*, ed. P. Hartman, pp. 263–327. Amsterdam: North-Holland.

Berkovitch-Yellin, Z. (1985). *Journal of the American Chemical Society*, **107**, 8239–53.

Berkovitch-Yellin, Z., Addadi, L., Idelson, M., Lahav, M., and Leiserowitz, L. (1982*a*). *Angewandte Chemie Supplement*, 1336–45.

Berkovitch-Yellin, Z., Addadi, L., Idelson, M., Leiserowitz, L., and Lahav, M. (1982*b*). *Nature*, **296**, 27–34.

Berkovitch-Yellin, Z. and Leiserowitz, L. (1980). *Journal of the American Chemical Society*, **102**, 7677–90.

Berkovitch-Yellin, Z. and Leiserowitz, L. (1982*c*). *Journal of the American Chemical Society*, **104**, 4052–64.

Berkovitch-Yellin, Z. and Leiserowitz, L. (1984). *Acta Crystallographica*, **B40**, 159–65.

Berkovitch-Yellin, Z., van Mil, J., Addadi, L., Idelson, M., Lahav, M., and Leiserowitz, L. (1985). *Journal of the American Chemical Society*, **107**, 3111–22.

Berkovitch-Yellin, Z., Weissbuch, I., and Leiserowitz, L. (1989). In *Morphology and Growth Units of Crystals*, p. 247. Tokyo: Errapub.

Berman, A., Addadi, L., Kvick, A., Leiserowitz, L., Nelson, M., and Weiner, S. (1990). *Science*, **250**, 664–7.

Berman, A., Hanson, J., Leiserowitz, L., Koetzle, T. F., Weiner, S., and Addadi, L. (1993*a*). *Science*, **259**, 776–9.

Berman, A., Hanson, J., Leiserowitz, L., Koetzle, T. F., Weiner, S., and Addadi, L. (1993*b*). *Journal of Physical Chemistry*, **97**, 5162–70.

Black, S. N., Bromley, L. A., Cottier, D., Davey, R. J., Dobbs, B., and Rout, J. E. (1991). *Journal of the Chemical Society Faraday Transactions*, **87**, 3409–14.

Boek, E. S. (1991). Crystal–solution interfaces, PhD Thesis, University of Twente, The Netherlands.

Boek, E. S. and Briels, W. J. (1992). *Journal of Chemical Physics*, **96**, 7010–18.

Boek, E. S., Feil, D., Briels, W. J., and Bennema, P. (1991). *Journal of Crystal Growth*, **114**, 389–410.

Boek, E. S., Briels, W. J., and Feil, D. (1994). *Journal of Physical Chemistry*, **98**, 1674–81.

Boer, R. C. de (1993). Morphology Related Phenomena, PhD Thesis, Catholic University, Nijmegen, The Netherlands.

Bourne, J. R. and Davey, R. J. (1976). *Journal of Crystal Growth*, **36**, 278–86.

Buckley, H. E. (1951). *Crystal Growth*. New York: John Wiley & Sons.

Cai, H., Hillier, A. C., Franklin, K. R. Nunn, C. C., and Ward, M. D. (1994). *Science*, **266**, 1551–5.

Chakrabartty, A., Yang, D. S. C., and Hew, C. L. (1984). *Journal of Biological Chemistry*, **264**, 11313–16.

Chang, Y. C. and Myerson, A. S. (1986). *American Institute of Chemical Engineers Journal*, **32**, 1747–9.

Chen, B. D., Garside, J., Davey, R. J., Maginn, S. J., and Matsuoka, M. (1994). *Journal of Physical Chemistry*, **98**, 3215–21.

Chiarello, R. P. and Sturchio, N. C. (1994). *Geochimica Cosmochimica Acta*, **58**, 5633–8.

Chiarello, R. P., Wogelins, R. A., and Sturchio, N. C. (1993). *Geochimica Cosmochimica Acta*, **57**, 4103–10.

Cody, A. M. and Cody, R. D. (1991). *Journal of Crystal Growth*, **113**, 508–19.

Coppens, P. (1982). In *Electron Distributions and the Chemical Bond*, eds. P. Coppens and M. B. Hall, pp. 61–92. New York: Plenum.

Curie, M. P. (1894). *Journal de Physique*, p. 393.

Davey, R. J., Milisavljevic, B., and Bourne, J. R. (1988). *Journal of Physical Chemistry*, **92**, 2032–6.

Destro, R., Marsh, R., and Bianchi, R. E. (1988). *Journal of Physical Chemistry*, **92**, 966–73.

DeVries, A. L. (1984). *Philosophical Transactions of the Royal Society*, **304**, 575–88.

Docherty, R., Roberts, K., Saunders, V., Black, S, and Davey, R. J. (1993). *Faraday Discussions*, **95**, 11–26.

Elwenspoek, M., Bennema, P., and van Eerden, J. P. (1987). *Journal of Crystal Growth*, **83**, 297–305.

Fenimore, C. P. and Thraikill, A. (1949). *Journal of the American Chemical Society*, **71**, 2714–17.

Gavish, M., Wang, J. L., Eisenstein, M., Lahav, M., and Leiserowitz, L. (1992). *Science*, **256**, 815–18.

Gevers, R., Amelinckx, S., and Dekeyser, W. (1952). *Naturwissenschaften*, **39**, 448–9.

Gidalevitz, D., Feidenhans'l, R., and Leiserowitz, L. (1997a). *Angewandte Chemie International Edition*, **36**, 959–62.

Gidalevitz, D., Feidenhans'l, R., Matlis, S., Smilgies, D. M., Christensen, M. J., and Leiserowitz, L. (1997b). *Angewandte Chemie International Edition*, **36**, 955–9.

Ginde, R. E. and Myerson A. S. (1992). *Journal of Crystal Growth*, **116**, 41–7.

Gopalan, P., Peterson, M. L., Crundwell, G., and Kahr, B. (1993). *Journal of the American Chemical Society*, **115**, 3366–7.

Grayer Wolf, S., Berkovitch-Yellin, Z., Lahav M., and Leiserowitz L. (1990). *Molecular Crystals and Liquid Crystals*, **86**, 3–17.

Green, B.S., Lahav, M., and Rabinovich, D. (1979). *Accounts of Chemical Research*, **12**, 191–7.

Hagler, A. T. and Lifson, S. (1974). *Journal of the American Chemical Society*, **96**, 5327–35.

Hartman, P. (1973). *Crystal Growth An Introduction*, pp. 367–402. Amsterdam: North-Holland.

Hartman, P. and Bennema, P. (1980). *Journal of Crystal Growth*, **49**, 145–65.

Hartman, P. and Perdok, W. G. (1955a). *Acta Crystallographica*, **8**, 49–52.

Hartman, P. and Perdok, W. G. (1955b). *Acta Crystallographica*, **8**, 525–9.

Hirshfeld, F. L. (1977a). *Israel Journal of Chemistry*, **16**, 168–74.

Hirshfeld, F. L. (1977b). *Theoretica Chimica Acta (Berlin)*, **44**, 129–38.

Hirshfeld, F. L. (1991). *Crystallography Reviews*, **2**, 169–200 (and reference cited therein).

Honess, A. P. (1927). *The Nature, Origin and Interpretation of the Etch Figures on Crystals*. New York: John Wiley & Sons.

Horn, F. H. (1951). *Philosophical Magazine*, **43**, 1210–13.

Iitaka, Y. (1961). *Acta Crystallographica* **14**, 1–10.

Jacquemain, D., Grayer Wolf S., Leveiller, F., Frolow, F., Eisenstein, M., and Lahav, M., and Leiserowitz, L. (1992). *Journal of the American Chemical Society*, **114**, 9983–9.

Jonsson, P. G. (1971). *Acta Crystallographica*, **B27**, 893–8.

Kern, R. (1953). *Bulletin de la Société Française de Minéralogie et Cristallographie*, **76**, 391.

Knight, C. A., Chang, C. C., and DeVries, A. L. (1991). *Biophysical Journal*, **59**, 409–18.

Lahav, M. and Leiserowitz, L. (1992). *Journal of the American Chemical Society*, **114**, 9983–9.

Lahav, M., Eisenstein, M., and Leiserowitz, L. (1993). *Science*, **259**, 1469–70.

Lewtas, K., Tack, R. D., Beiny, D. H. M., and Mullin, J. W. (1991). In *Advances in Industrial Crystallization*, pp. 166–79. Oxford: Butterworth-Heinemann.

Li, L., Lechuga-Ballesteros, D., Szkudlarek,B. A., and Rodriguez-Hornedo, N. (1994). *Journal of Colloid and Interface Science*, **168**, 8–14.

Liu, X. Y., Boek, E. S., Briels, W. J., and Bennema, P. (1995a). *Nature*, **374**, 342–5.

Liu, X. Y., Boek, E. S., Briels, W. J., and Bennema, P. (1995b). *Journal of Chemical Physics*, **103**, 3747–54.

Löwenstam, H. A. and Weiner, S. (1989). *On Biomineralization*. New York: Oxford University Press.

Mann, S. (1993). *Nature*, **365**, 499–505.

Mantovani, G., Gilli, G., and Fagioli, F. (1967). *Proc. 13th CITS*, Falsterbo, Sweden, pp. 289–301.

Markman, O., Elias, D., Addadi, L., Cohen, I.R., and Berkovitch-Yellin, Z. (1992). *Journal of Crystal Growth*, **122**, 344–50.

McBride, J. M. and Bertman, S. B. (1989). *Angewandte Chemie International Edition English*, **28**, 330–3.

Michaelis, A.S. and van Kreveld (1966). *Netherlands Milk and Dairy Journal*, **20**, 163–81.

Myerson, A. S. and Lo, P. Y. (1990). *Journal of Crystal Growth*, **99**, 1048–52.

Ocko, B. M., Wang, J., Davenport, A., and Isaacs, H. (1990). *Physical Review Letters*, **65**, 1466–9.

Ohashi, Y. (1996). *Current Opinion in Solid State & Materials Science*, **1**, 522–32.

Paul, I. C. and Curtin, D. Y. (1987). In *Organic Solid State Chemistry*, ed. G. R. Desiraju, Chapter 9, pp. 331–66. Amsterdam: Elsevier.

Prelog, V. and Helmchen, G. (1982). *Angewandte Chemie International Edition English*, **21**, 567–83.

Rifani, M., Yin, Y. Y., Elliott, D. S., Jay, M. J., Jang, S.-H., Kelley, M. P., Bastin, L., and Kahr, B. (1995). *Journal of the American Chemical Society*, **117**, 7572–3.

Ristic, R. I., Sherwood, J. M., and Wojeshowski, K. (1993). *Journal of Physical Chemistry*, **97**, 10774–82.

Roberts, K. J. and Walker, E. M. (1996). *Current Opinion in Solid State & Materials Science*, **1**, 506–13.

Robinson, I. K. and Tweet, D. J. (1992). *Reports on Progress in Physics*, **55**, 599–652.

Scheffer, J. R. and Pokkuluri, P. R. (1991). In *Photochemistry in Organized and Constrained Media*, ed. V. Ramamurthy, Chapter 5, pp. 185–246. New York: VCH Publishers.

Shimon, L. J. W., Lahav, M., and Leiserowitz, L. (1985). *Journal of the American Chemical Society*, **107**, 3375–7.

Shimon, L. J. W., Lahav, M., and Leiserowitz, L. (1986a). *Nouveau Journal de Chemie, Special Issue*, **10**, 723–37.

Shimon, L. J. W., Vaida, M., Addadi, L., Lahav, M., and Leiserowitz, L. (1990). *Journal of the American Chemical Society*, **112**, 6215–20.

Shimon, L. J. W., Vaida, M., Addadi, L., Lahav, M., and Leiserowitz, L. (1991). In *Organic Crystal Chemistry*, IUCr Crystallographic Symposia 4, ed. J. B. Garbarczyk and D. W. Jones, pp. 74–99. Oxford: Oxford University Press.

Shimon, L. J. W., Vaida, M., Frolow F., Lahav, M., Leiserowitz, L., Weissinger-Lewin, Y., and McMullan, R. K. (1993). *Faraday Discussions*, **95**, 307–27.

Shimon, L. J. W., Wireko, F. C., Wolf, J., Weissbuch, I., Addadi, L., Berkovitch-Yellin, Z., Lahav, M., and Leiserowitz, L. (1986b). *Molecular Crystals and Liquid Crystals*, **137**, 67–86.

Shimon, L. J. W., Zbaida, D., Addadi, L., Leiserowitz, L., and Lahav, M. (1988). *Molecular Crystals and Liquid Crystals*, **161**, 199–221.

Skoulika, S., Michaelidis, A., and Aubry, A. (1991). *Journal of Crystal Growth*, **108**, 285–9.

Smythe, B. M. (1967). *Australian Journal of Chemistry*, **20**, 1115–31.

Smythe, B. M. (1971). *Sugar Technology Review*, **1**, 191–231.

Staab, E., Addadi, L., Leiserowitz, L., and Lahav, M. (1990). *Advanced Materials*, **2**, 40–3.

Steiner, T. (1996). *Crystallography Reviews*, **6**, 1–57.

Stewart, R. F. (1976). *Acta Crystallographica*, **A32**, 565–74.

Swaminathan, K. S. and Craven, B. M. (1982). *American Crystallographics Association, Series 2*, **10**, 13.

Taylor, R. and Kennard, O. (1982). *Journal of the American Chemical Society*, **104**, 5063–70.

Toda, F. (1996). In *Comprehensive Supramolecular Chemistry*, Vol. 6, eds. D. D. MacNicol, F. Toda, and R. Bishop. Chapter 15, pp. 466–516. Oxford: Pergamon Elsevier Science.

Torri, K. and Iitaka, Y. (1970). *Acta Crystallographica*, **B26**, 1317–26.

Torri, K. and Iitaka, Y. (1971). *Acta Crystallographica*, **B27**, 2237–46.

Turner, E. and Lonsdale, K. (1950). *Journal of Chemical Physics*, **18**, 156–7.

Vaida, M., Popovitz-Biro, R., Leiserowitz, L., and Lahav, M. (1991). In *Photochemistry in Organized and Constrained Media*, ed. V. Ramamurthy, Chapter 6, pp. 247–302. New York: VCH Publishers.

Vaida, M., Shimon, L. J. W., Weisinger-Lewin, Y., Frolow, F., Lahav, M., Leiserowitz, L., and McMullan, R. K. (1988). *Science*, **241**, 1475–9.

Vaida, M., Shimon, L. J. W., van Mil, J., Ernst-Cabrera, K., Addadi, L., Leiserowitz, L., and Lahav, M. (1989). *Journal of the American Chemical Society*, **111**, 1029–34.

van Mil, J., Addadi, L, Gati, E., and Lahav, M. (1982). *Journal of the American Chemical Society*, **104**, 3429–34.

van Mil, J., Gati, E., Addadi, L., and Lahav, M. (1981). *Journal of the American Chemical Society*, **103**, 1248–9.

Walba, D. M., Razuvi, H. A., Clark, N. A., and Parmer, D. S. (1988). *Journal of the American Chemical Society*, **110**, 8686–91.

Wang, J., Davenport, A., Isaacs, H., and Ocko, B. M. (1992*a*). *Science*, **255**, 1416–18.

Wang, J. L., Lahav, M., and Leiserowitz, L. (1991). *Angewandte Chemie International Edition English*, **30**, 696–8.

Wang, J. L., Leiserowitz, L., and Lahav, M. (1992*b*). *Journal of Physical Chemistry*, **96**, 15–16.

Weinbach, S. P., Weissbuch, I., Kjaer, K., Bouwman, W. G., Als-Nielsen, J., Lahav, M., and Leiserowitz, L. (1995). *Advanced Materials*, **7**, 857–62.

Weissbuch, I., Addadi, L., Berkovitch-Yellin, Z., Gati, E., Weinstein, S., Lahav, M., and Leiserowitz, L. (1983). *Journal of the American Chemical Society*, **105**, 6613–21.

Weissbuch, I., Addadi, L., Lahav, M., and Leiserowitz, L. (1991). *Science*, **253**, 637–45.

Weissbuch, I., Berkovitch-Yellin, Z., Leiserowitz, L., and Lahav, M. (1985*a*). *Israel Journal of Chemistry*, **25**, 362–72.

Weissbuch, I., Frolow, F., Addadi, L., Lahav, M., and Leiserowitz, L. (1990). *Journal of the American Chemical Society*, **112**, 7718–24.

Weissbuch, I., Kuzmenko, I., Vaida, M., Zait, S., Leiserowitz, L., and Lahav, M. (1994*a*). *Chemistry of Materials*, **6**, 1258–68.

Weissbuch, I., Lahav, M., Leiserowitz, L., Meredith, G. R., and Vanherzeele, H. (1989). *Chemistry of Materials*, **1**, 114–18.

Weissbuch, I., Leiserowitz, L., and Lahav, M. (1994*b*). *Advanced Materials*, **6**, 953–6.

Weissbuch, I., Popovitz-Biro, R., Lahav, M., and Leiserowitz, L. (1995). Lead article, *Acta Crystallographica*, **B51**, 115–48.

Weissbuch, I., Shimon, L. J. W., Addadi, L., Berkovitch-Yellin, Z., Weinstein, S., Lahav, M., and Leiserowitz, L. (1985*b*). *Israel Journal of Chemistry*, **25**, 353–61.

Weissbuch, I., Zbaida, D., Addadi, L., Lahav, M., and Leiserowitz, L. (1987). *Journal of the American Chemical Society*, **109**, 1869–71.

Weissinger-Lewin, Y., Frolow, F., McMullan, R. K., Koetzle, T. F., Lahav, M., and Leiserowitz, L. (1989). *Journal of the American Chemical Society*, **111**, 1035–40.

Wells, A. F. (1946). *Philosophical Magazine*, **37**, 180, 217, 605–29.

Wells, A. F. (1949). *Discussions of the Faraday Society*, **5**, 197–201.

Whetstone, J. (1953). *Faraday Discussions*, pp. 132–40.

Wilen, L. (1993). *Science*, **259**, 1469.

Wireko, F. C., Shimon, L. J. W., Frolow, F., Berkovitch-Yellin, Z., Lahav, M., and Leiserowitz, L. (1987). *Journal of Physical Chemistry*, **91**, 472–81.

Wisser, R. A. and Bennema, P. (1983). *Netherland Milk and Dairy Journal*, **37**, 109.

Woensdregt, C. F. (1993). *Faraday Discussions*, **95**, 97–108.

Zabel, V., Muller-Fahrnow, M., Hilgenfeld, R., Saenger, W., Pfannemuller, B., Enkelmann, V. L., and Welke, W. (1986). *Chemistry and Physics of Lipids*, **39**, 313–27.

Zbaida, D., Weissbuch, I., Shavit-Gati, E., Addadi, L., Leiserowitz, L. and Lahav, M. (1987). *Reactive Polymers*, **6**, 241–3.

5

Ionic Crystals in the Hartman–Perdok
Theory with Case Studies:

ADP (NH₄H₂PO₄)-type Structures and Gel-Grown Fractal Ammonium Chloride (NH₄Cl)

ADP (NH$_4$H$_2$PO$_4$)-type Structures and Gel-Grown
Fractal Ammonium Chloride (NH$_4$Cl)

5

Ionic Crystals in the Hartman–Perdok Theory with Case Studies:

ADP (NH$_4$H$_2$PO$_4$)-type Structures and Gel-Grown Fractal Ammonium Chloride (NH$_4$Cl)

C. S. STROM

in collaboration with

R. F. P. GRIMBERGEN, P. BENNEMA, H. MEEKES,
M. A. VERHEIJEN, L. J. P. VOGELS, AND MU WANG

To Piet Hartman, pioneer and teacher

5.1. Introduction

This chapter focuses on the role played by the periodic bond chain (PBC) theory in predicting the structural morphology, in particular of ionic structures, illustrated by theoretical and experimental research, past and present. Case studies on ammonium dihydroxy-phosphate (NH$_4$H$_2$PO$_4$) or ADP-type structures and fractal gel-grown ammonium chloride are treated in detail. Future prospects are outlined.

The ultimate goal in computational prediction of crystal morphology, when a crystal, molecular or ionic, grows according to a layer-by-layer mechanism, is twofold. First, to estimate the relative growth rates, that is, the ratios $R_{h_i k_i l_i}/R_{h_j k_j l_j}$, for each pair of faces $(h_i k_i l_i)$ and $(h_j k_j l_j)$ on the growth form. R_{hkl} is the rate with which (hkl) grows in the normal direction. Second, to describe adequately and for all types of crystal structures, the roughening transition that the growth fronts can undergo; that is, the transition from the ordered, flat-face mode to the disordered, rough-face mode.

The PBC theory (Section 5.2) offers a realistic and well founded simplified description of the complex physical processes occurring in nature, and has in many cases enjoyed gratifying confirmation from experiment. The theory considers the actually observed morphology to be a superposition of: first, a basic generic morphology caused by internal or structural factors and called the theoretical or structural morphology; and second, a habit modification effect caused by external factors, notably the solid–fluid interaction, since in general the fluid will interact differently with the different faces. Relative growth rates delivering the structural morphology can be estimated to a smaller or larger extent.

Molecular Modeling Applications in Crystallization, edited by Allan S. Myerson. Printed in the United States of America. Copyright © 1999 Cambridge University Press. All Rights reserved. ISBN 0 521 55297 4.

Methods to take account of the solid–fluid interaction, notably statistical–thermodynamical treatments, are left untouched by the traditional PBC approach.

5.1.1. *On Ionic Structures*

In general, ionic structures are characterized by the predominance of the long-range Coulomb interactions. They can be strictly ionic structures, consisting exclusively of ions, or mixed ionic and covalent structures. The short-range van der Waals interactions are usually negligible in ionic crystals, in contrast to the molecular crystals, which are adequately described by nearest-neighbour or at most next-nearest-neighbour interactions. However, molecular crystals with pronounced atomic charge distributions show ionic behaviour and, in such structures, neither van der Waals nor Coulomb interactions prevail, and therefore neither may be neglected.

The bonding pattern of any structure in general is expressed as a binary matrix, with rows and columns labeled according to particles, and entries equaling 0 or 1 to represent absence or presence of a bond. We assemble symbolically the infinitely many cations, anions and neutral molecules of the structure in the sets $\{+\} \equiv \{q_1^+, q_2^+, \ldots, q_\infty^+\}$, $\{-\} \equiv \{q_1^-, q_2^-, \ldots, q_\infty^-\}$ and $\{0\} \equiv \{M_1, M_2, \ldots, M_\infty\}$, respectively.

The so-called adjacency matrix is then infinite, symmetric, and has the form:

$$A = \begin{pmatrix} & \{+\} & \{-\} & \{0\} \\ \{+\} & A_{++} & A_{+-} & A_{+0} \\ \{-\} & A_{-+} & A_{--} & A_{-0} \\ \{0\} & A_{0+} & A_{0-} & A_{00} \end{pmatrix}, \quad A = A^{\mathrm{T}}. \tag{5.1}$$

The bonds represented by the infinite submatrices A_{ij} are self-explanatory. In the special case of molecular crystals, only A_{00} is nonzero.

The strictly ionic structures contain no neutral molecules $(A_{+0} = A_{-0} = A_{00} \equiv 0)$, and exhibit a characteristic bonding pattern that is bipartite. Since an ionic bond exists only between opposite charges, the particles of the structure can be classified in two categories, such that a particle belonging to one class may only form a link with a particle in the other class. Hence the diagonal submatrices $A_{++} = A_{--} = 0$, and the bonding information is contained in one quarter of the adjacency matrix, $A_{+-} = A_{-+}^{\mathrm{T}}$.

The bonding pattern of mixed structures where, for example, hydrogen bonds occur, deviates from the rigid bipartite adjacency with zero diagonal blocks. The submatrices A_{++} and A_{--} may be nonzero but are usually sparse as they acquire weaker covalent bonds. If neutral molecules are present, A_{+0}, A_{-0} or A_{00} may be nonzero. Accordingly, a larger portion of A is needed to reflect the bonding information of mixed covalent structures.

5.1.2. *On Growth Layers*

The growth habit is expressed by the central distances, that is, the distances between the "origin" of the crystal and the crystal faces. These are proportional to the growth rates, which are assumed to be constant.

A first step toward the goal of predicting relative growth rates consists in determining the growth layers themselves; that is to say, the face orientations $(h_i k_i l_i)$ and molecular/ionic compositions of layers or sublayers parallel to $(h_i k_i l_i)$. These compositions define the various growth fronts, and are constrained to satisfy crystallographic and chemical requirements. Various distinct growth fronts are possible associated with the same face orientation. The "optimal" growth front is expected to be preferentially realized in nature. With regard to the possible growth layers, the first stage of the PBC theory supplies a first-principles methodology, enabling a rigorous derivation of the face orientations and ionic/molecular compositions of all candidate growth layers (Section 5.4).

The concept of "network" or "connected net" of chemical bonds, to be used as a vehicle in describing the properties of growth layers, was developed in 1980–1987. A formulation treating the network bonds as binary quantities enabled a rigorous derivation of the possible growth layers from the combinatorial information of the structure (Section 5.4). A formulation based on bond energies led to a statistical-mechanical description of the roughening transition theory, and paved the way for a realistic description of the interfacial structure (Section 5.8).

5.1.3. *On Relative Growth Rates*

The second stage of the PBC theory concerns itself with the determination of the relative growth rates when the internal structural factors are taken into account. The growth rates depend on scores of physical quantities, mainly on the energy released per chemical formula when a growth layer becomes attached to the rest of the structure. But, because we only need to know the ratios of these growth rates, we could arguably afford to employ dubious calculational means, provided we ensured that the effect of the introduced discrepancies left invariant the ratios of the growth rates of the important faces.

In other words, we could consider

- Neglecting Van der Waals interactions even when they are not negligible with respect to the Coulomb, as long as they amount to a more or less constant percentage of the electrostatic interactions.
- Approximating, for example, long-range interactions by nearest-neighbour and next-nearest-neighbour expressions, even when we can hardly expect adequate convergence.
- Introducing adjustable parameters and ad hoc cutoffs in order to, if not enforce the system to converge in all cases, at least persuade it to converge in most cases.

These approximations have no ill effect as long as they lead us to a set of so-called "theoretical" growth rates R_{hkl}^{th} related to the real, as yet unknown, growth rates R_{hkl} in such a way that $R_{h_ik_il_i}^{th}/R_{h_jk_jl_j}^{th} \approx R_{h_ik_il_i}/R_{h_jk_jl_j}$ for the prominent faces. Section 5.4.4 decribes exact and efficiently convergent energy computations.

5.2. Historical Review of Structure and Morphology: The PBC Theory

Forty-two years ago Hartman (1956a, b, 1973a, 1979) and Perdok launched the periodic bond chain (PBC) theory, inspired by earlier work of Donnay and Harker (1937) and Niggli (1923), encompassing in its domain of validity, in principle, all types of structures. That theory has been applied to ionic structures with satisfaction, where other approaches to crystallization based on nearest-neighbour interactions fail to capture all the relevant physics of the infinite-range ionic forces. The PBC theory is characterized by the concept of a "structural" morphology, for which the structure is responsible, established in two steps. First, the substrate is determined; the possible growth layers are not just random selections of growth units, but have orientations and compositions strictly defined in terms of periodic bond chains. The outcome leads to the data for the second step: the habit-controlling energy quantities are determined in order to identify the prevalent growth layers in each orientation; the most important habit-controlling quantity is the attachment energy, discussed below.

5.2.1. *Structural Morphology: Periodic Bond Chains and F Slices*

Historically, Hartman and Perdok were the first to introduce the term "slice" to indicate "a stoichiometric portion of a crystal structure that is parallel to a lattice plane and that contains two or more periodic bond chains" (Hartman and Perdok 1955a, b, c; Hartman 1958a, b). In "Relations between morphology and structure of crystals" Hartman (1973b) outlines the evidence for the existence of PBCs.

... the condition that [during flat face growth] the edge [of a crystal] remains straight is that parallel to the edge there must be in the crystal structure an uninterrupted chain of strong bonds that are formed during the crystallization process.

Also by analogy, the observation that some faces grow flat shows the existence of these PBCs. The intersecting PBCs are parallel to the face in question, and carry information on the "mutual spatial relations of the bonds." The reason for taking into account strong bonds (in the first coordination sphere) is: "bonds outside the first coordination sphere have little influence on the crystal growth process, because this process can be considered as a consecutive chemical reaction."

A recent summary of the PBC theory is found in Hartman and Chan (1993):

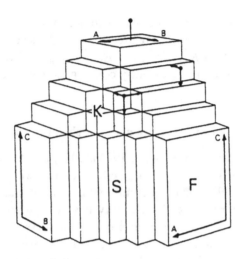

Fig. 5.1. Hypothetical crystal with three PBCs: A ∥ [100], B ∥ [010], and C ∥ [001]. *F* faces are (100), (010), and (001); *S* faces are (110), (101), and (011), while (111) is a *K* face. (Reprinted from Hartman (1973a, p. 375), with permission from the author.)

. . . crystal growth can be considered as the formation of bonds between crystallizing particles such as atoms, ions or molecules. Because these particles have to come close together in order to form the crystal, only *strong bonds* are considered, defined as *bonds in the first coordination sphere*. In molecular structures, the bonds are the hydrogen bonding, dipolar forces or Van der Waals contacts between atoms of neighboring molecules. In other words, a crystal is bound by the straight edges that are parallel to directions in which the growth units are bonded by uninterrupted chains of strong bonds . . . three categories of faces can be identified. *F faces* or flat faces: A slice (layer) d_{hkl} contains two or more PBCs. *S faces* or stepped faces: A slice d_{hkl} contains only one PBC. *K faces* or kinked faces: A slice d_{hkl} contains no PBC at all.

The *F*, *S* and *K* faces are schematically shown in Fig. 5.1 (Hartman 1973a):

The three categories have accordingly three different growth mechanisms. The *F* faces are able to grow according to a layer mechanism so that their growth rate is the slowest. Therefore they determine the theoretical habit. The growth of *K* faces can occur anywhere on the surface without nucleation, so the growth rate is the highest and the faces do not occur on the crystal. The *S* faces are intermediate between *F* and *K*. They would need one-dimensional nucleation, but because in practice at all temperatures there are sufficient kinks present, they also grow fast and normally do not occur on the crystal.

The resultant growth habit is a cooperative effect of structure and environment. The "structural" morphology represents the growth habit that would result if the crystal could grow in vacuum; for that reason it is also sometimes referred to as vacuum morphology. The environment works on the structural morphology to bring about habit modifications. Hence the growth habit obtained from vapour growth or growth out of weakly interacting solutions comes the closest to the structural habit.

The classification of the various factors affecting the growth habit of a crystal in two main categories can be found in review articles (Hartman 1969), and a recent account is in Hartman and Chan (1993):

First, the internal or structural factors include the crystal packing, defects and dislocations. Second, the external factors are all the physico-chemical parameters of the crystallization conditions such as supersaturation, temperature, impurities and solvents. Generally, slow growth is expected to favor formation of crystal habits less likely affected by external factors. Under special growth conditions, such as rapid crystallization and high supersaturations, the external factors could lead to a habit change.

The following advantage is gained by concentrating in the first instance on the structural morphology, that is the growth habit for which internal factors are responsible:

Predictive crystal habit theory . . . could enable habit modification to be carried out in a more systematic and logical way . . . A knowledge of the theoretical growth habits is essential in differentiating the effects of solvents on the observed crystal growth habits. Therefore, knowing the theoretical crystal shape, any deviation observed can be attributed more precisely to those crystal faces on which solvent effects are involved. The theoretical growth habit can be predicted from structural and energetic considerations.

5.2.2. Habit-Controlling Energy Parameters and Relative Growth Rates

The following energy equality plays a key role in this second phase of the Hartman–Perdok theory:

$$E^{cr} = E^{sl}_{hkl} + E^{att}_{hkl} + E^{dip}_{hkl}, \quad E^{dip}_{hkl} \equiv -2\pi(d^{\perp}_{hkl})^2/V, \tag{5.2}$$

$$d^{\perp}_{hkl} = (d_x h + d_y k + d_z l)\sqrt{K}/\sqrt{G},$$

with

$$K \equiv 1 - \cos^2\alpha - \cos^2\beta - \cos^2\gamma + 2\cos\alpha\cos\beta\cos\gamma,$$

$$\begin{aligned}
G \equiv {}& \left(\frac{h}{a}\right)^2 \sin^2\alpha + \left(\frac{k}{b}\right)^2 \sin^2\beta + \left(\frac{l}{c}\right)^2 \sin^2\gamma \\
& + 2\frac{hk}{ab}(\cos\alpha\cos\beta - \cos\gamma) + 2\frac{hl}{ac}(\cos\alpha\cos\gamma - \cos\beta) \\
& + 2\frac{kl}{bc}(\cos\beta\cos\gamma - \cos\alpha),
\end{aligned}$$

where V is the unit cell volume, $\mathbf{d} \equiv d_x\mathbf{a} + d_y\mathbf{b} + d_z\mathbf{c} = \Sigma_i q_i\mathbf{r}_i$ is the electric dipole moment of the unit cell generating the F slice, and $d^\perp$ is the component of $\mathbf{d}$ perpendicular to the (hkl) face.

The crystal energy E^{cr} is the total energy of sublimation, the energy per unit-cell content per mole released when the crystal is formed from the crystallizing particles (Hartman 1973a). When the temperature is left out of consideration, E^{cr} is always a constant, independent of slice boundary or face orientation (Hartman 1987).

The slice energy E^{sl}_{hkl} is the energy per unit cell content per mole when a slice d_{hkl} is formed (Hartman 1973a). Let ϕ_{ij} be the interaction energy between atoms i and j. The slice energy is defined (Hartman 1958a, b) as $E^{\mathrm{sl}} = \frac{1}{2}\Sigma_{ij}\phi_{ij}$, where $i,j = 1,\ldots,N, i$ are the atoms of the unit cell and j are atoms in the entire F slice, including the unit cell at the origin.

The attachment energy is the amount of energy released per unit-cell content per mole when the new layer attaches itself to the rest of the structure. It equals the sum of the interactions of the 0-slice unit cell with an infinite parallel stacking of slices. Let E^m_0 be the interaction energy between the unit cell of the 0th slice and the mth slice. The attachment energy is given by Hartman (1973a)

$$E^{\mathrm{att}}_{hkl} = \sum_{m=1}^{\infty} E^m_0(hkl). \tag{5.3a}$$

The surface energy (Hartman 1973a) E^{surf}_{hkl} is the amount of work necessary to split a crystal isothermally and reversibly into two half crystals, producing the surfaces (hkl) and $(\bar{h}\bar{k}\bar{l})$, at infinite separation. And the specific energy $\sigma_0(hkl)$ is the surface energy per unit area on (hkl),

$$E^{\mathrm{surf}}_{hkl} = \sum_{m=1}^{\infty} m E^m_0(hkl), \quad \sigma_0(hkl) = (d_{hkl}/V)E^{\mathrm{surf}}_{hkl}, \tag{5.3b}$$

where, as before, V is the unit cell volume.

Finally, the dipole energy E^{dip}_{hkl} arises from the existence of a component of the dipole moment normal to the face (Hartman 1982). The site potential at $\mathbf{R}$ due to the mth slice has the form $U_m(\mathbf{R})$ when the slice is dipoleless, but $V_m(\mathbf{R}) + C$ when the slice is polar, where C is a constant. The site potential due to a finite collection of slices is by definition convergent. The site potential caused by an infinite stacking of slices in the case of a dipoleless substrate is convergent, $\Sigma^\infty_m U_m(\mathbf{R}) = $ finite. However, when the substrate is polar, the site potential caused by infinitely many slices diverges, being the sum of a convergent term, $\Sigma^\infty_m V_m(\mathbf{R}) = $ finite, and a divergent term, $\Sigma^\infty_m C$. Furthermore, it is easily seen that the interaction energy of a charge-neutral collection of particles $\{q\}$, for example the particles belonging to a unit cell, with an infinite stacking of slices converges in either case because $\Sigma_{mq} qC \equiv 0$. So the slice and attachment energies of both apolar and polar substrates converge equally well. Whereas for apolar faces $d^\perp = 0$ so that the crystal energy becomes equal to the sum of the slice and attachment energies, for polar substrates it is given by equation (5.2).

Note that all of the above energy quantities depend on the particular slice composition in (hkl), but are by definition independent of the overall sign of the face indices; that is to say, no distinction is possible between $+(hkl)$ and $-(hkl)$.

Alternative definitions of slice and attachment energy exist, involving an overall scaling factor of 2 and leading in effect to the same energy-balance relation $E^{cr} = E^{layer} + 2E^{att}$, with $E^{layer} \equiv 2E^{sl}$ (Berkowits-Yellin 1985).

The habit-controlling energy is the energy considered primarily responsible for determining the morphology. The lower the magnitude of the habit-controlling energy, the lower the relative growth rate of a certain face, and the larger its surface area. The magnitude of the habit-controlling energy is an indicator of morphological importance. Thus the relation between growth rate or central distance, R_{hkl}, and magnitude of the habit-controlling energy associated with the optimal F slice in (hkl) is monotonicity.

Hartman (1973*a*) postulated that the minimum $|E^{att}_{hkl}|$ among all the F slices of (hkl) is a measure of the growth rate in general, and for F faces it is proportional to the growth rate. A closer examination of the relation between growth rate and attachment energy was carried out by Hartman (1980*a*, *b*) and Hartman and Bennema (1980). The assumed proportionality has found its way into most routine applications and it is an oversimplification offering a workable formula in practice. So the attachment energy is taken as the habit-controlling energy for the growth form, at any rate when $E^{dip}_{hkl} = 0$. Surface polarity is further discussed in Section 5.3. Especially in the case of noncentrosymmetric structures, polar growth layers are a common occurrence. The more stable F slices are associated with the lowest dipole energies E^{dip}_{hkl}, or otherwise surface reconstruction takes place to minimize the dipole moment. It is not clear whether the appropriate habit-controlling quantity should be taken as the attachment energy or the supplement of the slice energy, $E^{suppl}_{hkl} \equiv E^{cr}_{hkl} - E^{sl}_{hkl} = E^{att}_{hkl} + E^{dip}_{hkl}$. Both choices can be justified (Van der Voort and Hartman 1990*b*).

In the equilibrium form the central distances are proportional to the specific surface energy, and furthermore for very small crystals the growth form is expected to resemble the equilibrium form more closely (Hartman 1973*a*, private communication, 1995).

5.2.3. *Historical Setting*

The PBC theory arose against the background of earlier concepts and geometrical prescriptions. According to the earlier Bravais–Friedel–Donnay–Harker (BFDH) formulation, there is no restriction in principle on the number of possible flat-face orientations. In the BFDH formulation, any face could be a flat face provided d_{hkl} is large enough; and for a given orientation a single growth layer is possible, that is the planar-cut slice d_{hkl}. The growth rate is inversely proportional to the interplanar distance.

In his paper entitled "On the validity of the Donnay–Harker law", Hartman (1978*a*) placed the PBC theory in a historical setting in the context of the BFDH and Niggli approaches earlier brought forward. A brief and very rough way of recasting Hartman's refined exposition goes as follows: The BFDH law states that the growth rate $R_{hkl} \simeq 1/d_{hkl}$. If, in calculating the attachment energy, one is justified in neglecting higher-order interactions than nearest neighbours, then no distinction can be made between the attachment and surface energies, so roughly: $E^{att} \simeq E^{surf} \simeq \sigma V/d_{hkl}$. If one is dealing with fairly isotropic structures, then the specific surface energy σ is to a large extent isotropic, that is to say, independent of the face orientation, so that the geometrical prescription of BFDH arises as a special approximate case of Hartman's attachment energy formulation.

5.2.4. *Discussion of Technical Issues*

A PBC is a series of growth units and bonds forming a simple or composite (uninterrupted) chain of stoichiometric composition.

The direction [*uvw*] of a PBC is given by the difference in cell indices between the translationally related molecules. It is therefore imperative that all translationally related molecules differ by the same primitive lattice translation [*uvw*], so that the PBC may acquire a unique direction [*uvw*]. The following scheme shows an example of a valid PBC, starting with molecule $M_1[110]$ and ending with the translationally related molecule $M_1[0\bar{2}1]$ having orientation $[0\bar{2}1] - [110] = [\bar{1}\bar{3}1]$, as well as a chain lacking a unique direction, therefore not qualifying as a PBC; the latter chain may be assigned three orientations, namely $[201] - [011] = [2\bar{1}0]$, $[10\bar{1}] - [011] = [1\bar{1}\bar{2}]$, and $[10\bar{1}] - [2\bar{1}0] = [\bar{1}\bar{1}\bar{1}]$:

$$M_1[110] \rightarrow M_2[020] \rightarrow M_1[0\bar{2}1]\ldots, \text{valid pbc;}$$
$$M_1[011] \rightarrow M_2[203] \rightarrow M_1[201] \rightarrow M_3[000] \rightarrow M_1[10\bar{1}]\ldots, \text{invalid pbc.}$$

Since [*uvw*] must be a *primitive* lattice translation, *u*, *v* and *w* must be mutually prime integers with reference to a primitive unit cell. If the unit cell of the structure under study is not primitive, then one must also take account of the appropriate translations [*uvw*] with fractional indices, or perform at the outset a transformation of axes to a primitive unit cell. Hartman's original definition of PBC further stipulates that the PBC composition must be stoichiometric and that in centrosymmetric structures the dipole moment of the PBC must be confined to the PBC direction. This restriction, implying that the PBCs of centrosymmetric structures have axial symmetry, is now considered unnecessary.

An *F* slice is a combination of two or more intersecting PBCs, where the PBC directions define the slice orientation. It consists of all the atoms belonging to both PBCs and constitutes a growth front. Several growth fronts are possible in the same face orientation. Since each PBC is stoichiometric, so is the obtained *F* slice. Although not mentioned explicitly by Hartman, one tacitly exercises caution in combining PBCs so as to ensure that at all times the translation differences between translationally related molecules present in the *F* slice are parallel to the face (see Section 5.3).

A growth unit is a fundamental cluster of atoms or ions, which therefore cannot be cleaved at the surface. Individual ions in the case of ionic crystals, and individual molecules in the case of molecular crystals, fit that description. Larger growth units are in principle not excluded. In the mother phase ions or molecules often form loose associations. Some times, precrystallized units appear as stable clusters of ions and/or molecules, in addition – or instead of – individual molecules. If there is evidence that stable precrystallized units like dimers exist as such in the solution, then some or all of the growth units could be appropriately specified as clusters larger than individual molecules.

Hartman's motivation for defining PBCs on the basis of strong bonds is that these bonds are formed as the particles come together in the process of crystallization (Hartman 1973*b*), as the final outcome of intricate processes in which attachment events overcome detachment events. In Hartman's concept these bonds are based on proximity rather than some kind of approximation to nearest-neighbour interactions. To avoid confusion in this work, we will refer to "crystallization bonds" in the context of Hartman's theory.

For ionic structures the first coordination sphere is a well defined crystallographic concept, considering that a clear demarcation line usually exists in terms of distance (or energy), between the first and the second coordination spheres. For molecular crystals the first coordination sphere consists of molecular contacts, hydrogen bonds, and weak dipolar forces. When it comes to practical application, the PBC prescription need not always be free from ambiguity. If in cases of doubt the first coordination sphere were to be extended by including weaker bonds, the *additional* PBCs and *F* slices that would emerge, if any, would be bound to be less significant with respect to those obtained from the more restricted coordination sphere. The determination of PBCs and *F* slices in the framework of an adequate specification of the first coordination sphere is fairly model-independent. However, if inadvertently some crystallization bonds are left out of consideration, the theoretical results may be inadequate. The practical situation of the *F* face analysis of ADP-type structures in Section 5.4 offers an interesting illustration of both features.

5.3. Operational Concepts, Definitions and Procedures

Attempts to predict computationally crystal morphology in the framework of the PBC theory involve two steps: first, determine the structural morphology that is the result of internal factors; second, estimate the habit modifications brought about by the external factors, mainly the properties of the fluid out of which the crystal is growing. An essential prerequisite for undertaking both steps is to determine from first principles all possible growth layers. Before continuing, exact and unambiguous definitions of growth layers and related terms are in order (Strom et al. 1995*a*, Appendix A; Strom, Vogels, and Verheijen 1995*b*).

A *growth layer* in (*hkl*) is the physical layer by which a crystal grows, resulting in a flat (*hkl*) face.

5.3.1. *Slices*

A "slice" (Hartman and Perdok 1955*a*, *b*, *c*) (or interchangeably layer or lamina) is a portion of the structure, parallel to (*hkl*), infinite in two dimensions, with a finite thickness in the third, and satisfies certain general requirements. A crystallographic requirement is that the average layer thickness must be equal to d_{hkl} or fraction thereof, and must be consistent with the selection conditions of the space group. A chemical requirement is that the composition of the growth layer must be stoichiometric, meaning that the slice must be composed of an integral number of chemical formula units; that is, the same proportions of chemical species as in the bulk. A requirement of probabilistic–statistical nature, otherwise known as the "flatness condition", is that all differences in lattice translations between translationally related particles must be parallel to (*hkl*). The reason for the last requirement is evident: If some translation differences oblique to (*hkl*) were to be admitted, the slice would lose its unique sense of orientation and would represent roughened growth. Another way of looking at this, is that the flatness condition ensures the uniqueness of face orientation, up to an overall sign; see Section 5.4.

The interplanar distance d_{hkl}^{prim} is the maximum possible slice thickness when the indices are expressed in terms of a primitive cell. That statement does not hold when a nonprimitive cell is employed. If the slice thickness is a fraction of the allowed maximum, the slice may also be called a subslice.

Slices can be classified according to the way they are formed by combining PBCs.

- *F face, F slice, pseudo-F face, pseudo-F slice.* When two nonparallel PBCs in directions $[uvw]_1$ and $[uvw]_2$ are combined, the resulting composition is an infinite two-dimensional portion of the structure parallel to (*hkl*) with thickness d_{hkl}. The face indices are obtained from the PBC direction indices in the obvious way, $(hkl) = [uvw]_1 \times [uvw]_2$. Clearly the overall sign of the face orientation is indeterminate, since a distinction between $+(hkl)$ and $-(hkl)$ could only depend on the order of multiplication.

  This combination has stoichiometric composition because each PBC does. Provided (*hkl*) satisfies the space-group selection conditions, that combination is called an "*F* slice" if it satisfies the flatness condition; a "pseudo-*F* slice", if it does not. In other words, an *F* slice is a slice with the additional property of being composed of two or more intersecting PBCs. A pseudo-*F* slice is composed of two intersecting PBCs, but is strictly speaking not a slice at all, because it violates flatness, since it contains lattice translations oblique to the face. A face containing at least one *F* slice is an "*F* face", whereas a face containing only pseudo-*F* slices is a pseudo-*F* face. In complete analogy, F subslices and pseudo-*F* subslices are defined with reference to thickness Rd_{hkl}^{prim}, $R < 1$.

- *S face, S slice.* An S slice is a slice that contains one and only one PBC. A face d_{hkl} is an S face if PBCs in d_{hkl} exist in one and only one direction. In complete analogy, S subslices have thickness Rd_{hkl}^{prim}, $R < 1$.
- *K face, K slice.* A face (hkl) is a K face if there exist no PBCs parallel to (hkl). A K slice is a slice d_{hkl} containing no PBCs. Thus any slice parallel to a K face has by definition K character, but of course K slices can be specified arbitrarily in any orientation. In complete analogy, K subslices have thickness Rd_{hkl}^{prim}, $R < 1$. The concept of slice or lamina in the most general case is indeed a K slice.

Slices can be classified according to the way they are mutually related by symmetry operations.

When some symmetry operator parallel to (hkl) is located halfway between opposite slice boundaries, it causes these boundaries to become identical, $B_+(hkl) \equiv B_-(hkl)$. Slices with distinct opposite boundaries, $B_+(hkl) \neq B_-(hkl)$, emerge in the absence of a symmetry operator parallel to (hkl), or when that operator happens to be closer to one of the two boundaries. A surface dipole moment $d_{hkl}^{\perp}$ may exist for two reasons: asymmetry between opposite boundaries, $B_+(hkl) \neq B_-(hkl)$ and/or the dipole moments of the individual molecules. For slices with symmetrically related opposite boundaries $d^{\perp}$ may be nonzero as a result of molecule polarity (Strom et al. 1997). On the other hand, it is possible for slices with distinct opposite boundaries to be dipoleless (Hartman and Strom 1989). In many cases the magnitude of $d^{\perp}$ is a good indicator of symmetry relations between F slices.

Symmetry relates different slices in the same or in different orientations. The following classification of slice multiplets refers to symmetrically related slices characterized by the same face indices (hkl).

A group of F or S or K parallel (sub)slices form a *multiplet* if they are mutually related by symmetry operations. Some symmetry operations can also transform an individual slice on itself. If a given F slice is invariant under all symmetry operations, it is called a *singlet*, belonging to a group by itself. Pairwise symmetric slices form a *doublet*, a group of four is a *quartet*, etc.

The most common example of a doublet is, as mentioned above, a pair of slices, say 1 and 2, with boundaries $B_{\pm}^1$ and $B_{\pm}^2$, respectively; $\pm$ denotes the opposite parallel boundaries of a slice. The boundaries of the pair of slices are related through a symmetry operation parallel to the face by $B_+^1 \equiv B_-^2$ and $B_-^1 \equiv B_+^2$.

The higher the multiplicity of the group, the lower the symmetry exhibited by the individual members of the multiplet, so that singlets possess the highest possible symmetry within a lamina shape. In structures where symmetry is totally absent, there exist no symmetry multiplets and hence all slices in a given (hkl) would then be classified as symmetry singlets.

The implications of the existence of symmetry doublets on the crystal growth process, later called symmetry roughening, were encountered for the first time by Hartman and Heynen (1983).

5.3.2. *Choice of Cell for Computational Purposes*

Since the PBC direction is defined as a primitive lattice translation $[uvw]$, these indices are mutually prime integers with reference to a primitive cell. If we were to adopt a nonprimitive cell for operational purposes, we would be introducing some unnecessary complications. First, we should admit fractional as well as integer indices $[uvw]_{nonprim}$. Second, for every obtained chain direction with integer indices $[uvw]_{nonprim}$ we should examine the corresponding primitive indices $[u'v'w']_{prim}$ and accordingly accept (reject) the PBC in $[uvw]_{nonprim}$ if the $[u'v'w']_{prim}$ indices are mutually prime (integer multiples). This implies that a primitive cell is the appropriate choice for carrying out a computational F-face analysis.

5.3.3. *Search Procedures for Growth Layers*

It is physically evident that a growth layer must at any rate have a lamina shape, satisfying the space group selection conditions, stoichiometry and flatness. In the PBC context, this class includes F, S or K slices. In the absence of additional stipulations, the above specification of the chemical composition of a growth layer remains unrelated to its orientation. We saw that the PBC theory imposes an additional requirement on a candidate growth layer, namely: it is not enough for a growth layer to be just a lamina, it must be especially an F slice, as opposed to an S or K slice.

There exist no physical or chemical grounds for assuming the layer boundaries to be geometrically planar cut surfaces. Consequently, the above definition of slice does not impose any restrictions on the geometrical shape of the boundaries, emphasizing that a slice could even have strongly undulating boundaries.

In what follows, we will outline the implications of that F-slice postulate for the search procedure for finding growth layers. Methods used to determine the growth layers from the structure roughly come down to the following steps:

- *Step 1.* Apply a search procedure to extract all possible slice candidates, and retain those satisfying the crystallographic selection conditions.

In this step the searches for face orientations and ionic/molecular compositions of slices would at first glance seem to be decoupled. It would seem intuitively obvious to split step 1 into two substeps, since one ought to (1a) first find an available reservoir of face orientations likely to contain slices, before (1b) examining each face in detail. However, the number of possible faces in a structure required in step (1a) is limited only by the imagination and patience of the researcher, amounting to a discrete infinity.

The PBC theory unifies the specifications of orientation and ionic/molecular composition of a growth layer – now postulated to be an F slice – into a single definition. Then the two search procedures (1a) and (1b) become inextricably interwoven into one. It is noteworthy that the number of face orientations ensuing from this – restricted or enriched – notion of "F slice" turns out to be sensibly limited. The intuitive concept of a continuous layer boundary, be it planar or undulating, as long as it avoids cleaving the molecules, is less well suited for use as a computational instrument because we are dealing with a crystal structure that is discrete. In order to cast the *F* slice search of step 1 into an algorithmic procedure, the intuitive notion of "boundary" may be replaced by the rigorous notion of "network", which implies the existence of a graph, outlined in Section 5.4.

- *Step 2.* Having constructed a collection of *F* slices in various orientations, the ultimate goal is to find out which slice in a given orientation is viable as, or likely to become, the actual growth layer – in other words, to isolate the slice considered "energetically optimal" in the context of the prevailing experimental circumstances.

The case of growth out of vapour or a weakly interacting solution is straightforward. The optimal *F* slices are structurally determined by maximizing $|E^{\text{slice}}|$.

However, in the case of growth out of the melt or a strongly interacting solution, the optimal *F* slices are the most strongly interacting with the solution. It is empirically known that, in crystals growing out of a polar solvent, the polar faces dominate, whereas growth out of a nonpolar solvent enhances the nonpolar faces (Berkowits-Yellin 1985). Out of all the available surface compositions, the solvent in fact "chooses" the substrate with which to interact. A lucid example is the case of hydrated compounds growing out of aqueous solutions (Van der Voort and Hartman 1991); see also Section 5.4.4. At the interface, the substrates activated by the liquid terminate in water molecules at crystallographic positions or at solvent accessible sites. In both cases the *F* slices adjacent to the solution are asymmetric layers, quite different from the usually symmetric and energetically optimal theoretical *F* slices.

Various effects take place to stabilize polar faces. Polar faces are unstable because the electrostatic potential diverges at the surface. The dipole moment perpendicular to the surface becomes neutralized by polarization of bonds within the slice, change of ionic charges, redistribution of ions, or complete surface reconstruction (Van der Voort and Hartman 1990b).

5.3.4. The Indeterminacy of the Overall Sign of the Face Indices (hkl) and its Bearing on Polar Habits

An important point worth mentioning is that a distinction between the opposite faces (hkl) and $(\bar{h}\bar{k}\bar{l})$ is impossible by construction. The difference in behaviour between opposite faces cannot be found by structural considerations. In complete consistency with the discussion of the previous section, where the interaction between the crystal surface and the adjoining fluid was left out of consideration, this indeterminacy holds also for all energy quantities. In the framework of the Hartman–Perdok theory, the actual distinction between (hkl) and $(\bar{h}\bar{k}\bar{l})$ has an environmental, not a structural, origin. The distinction must be attributed to the interaction of the crystal surface with the adjoining fluid. Consequently, in the case of a growth layer with symmetrically unrelated boundaries, the structural morphology cannot determine which of these two boundaries will occur at the interface. That question can only be resolved by calculating the solid–fluid interaction or by experiment. Therefore, a polar habit cannot be treated as a manifestation of the structural morphology, but as an environmentally induced habit modification. Identifying the composition of the growth front out of all possible PBC-theoretical alternatives in the presence of a strongly interacting solution is a matter of surface chemistry for the specific case under study. Surface chemistry is not considered subject to prior computation, except fortuitously for a restrictive class of some simple Ising-type cases.

5.4. Graph-Theoretic Formulation of the PBC Theory

5.4.1. *Crystal Structure Viewed as a Network*

A graph is defined as a collection of points and links, so the concept of a graph can be employed in a useful way to represent a structure. When the molecules or ions or groups of atoms that may not be cleaved at the surface are abstracted to points or vertices, and the crystallization bonds between them are abstracted to links, the result is a graph capturing the bonding pattern of the structure.

That graph is expressed by an adjacency matrix like the one symbolically presented in equation (5.1). If, moreover, a set of coordinates x, y, z is associated with each point, one obtains an embedding of the graph in three-dimensional space, which is the structure itself. The graph representing the crystal structure is infinite but periodic. Therefore a finite – and often small – portion of it, the so-called "basic crystal graph", suffices to store the complete information content of the bonding pattern in the entire system.

A network is a graph with the property that between every possible pair of points in the network there exists at least one path connecting the two points. Graphs possessing this so-called connectedness property are also called one-component graphs (Harary 1969; Biggs 1974). The graphs featuring in the PBC theory are networks.

Leaving inessential molecules like zeolitic water out of consideration, the graph representing the infinite periodic structure is indeed a network. Otherwise we would be dealing with a superposition of more than one independent structure, each structure corresponding to a different component within the graph, which is an unphysical hypothetical case. The basic crystal graphs, PBCs and F slices, being subgraphs of the infinite structure, are also of necessity networks. Taking the embedding into account, we may further conclude that the basic crystal graph is a three-dimensional network of finite extent; the PBC is a network infinite in one dimension; and the F slice is a network infinite in two dimensions.

The graph-theoretic approach to the derivation of F slices in this section is combinatorial rather than geometrical. It is independent of geometrical details like choice of cell or specific values of cell parameters and position coordinates.

We are, however, constrained to employ a primitive cell as already stated, for the purely logistic reason that the computation of PBCs and F slices can only be effectively implemented with reference to primitive cells.

The only essential feature is the existence of periodicity, implying the existence of a unit cell. The basic data comprise a list of labelled points and a list of associations (crystallization bonds) between the points. The labels are sets of three integers because the periodicity is in three dimensions; the labels are interpreted as cell indices. In the following inessential geometrical details, such as transformation between primitive and conventional cells, position coordinates, etc., are sketched in the figures, while the primitive-cell indices are reported only when necessary.

5.4.2. *The Basic Crystal Graph Expressed as a Network, Illustrated by ADP-type Structures*

The basic crystal graph is a finite subgraph of the graph of the entire structure, and is so defined as to contain the complete bonding information stored in the original graph. The practical significance of the basic crystal graph lies in the fact that the necessary data can easily be confined in a limited and manageable matrix. The periodicity of the structure makes such a size reduction possible.

If the basic crystal graph is restricted to an operational minimum, it should contain all points in the unit cell, with all links between them, and all points in neighbouring cells as well as cells further afield, linking with the points of the unit cell.

For the purpose of illustration, there now follows a description of the combinatorial properties of ADP-type structures. The interested reader is referred to Strom et al. (1995*b*) and Aquilló and Woensdregt (1984, 1987) for a detailed specification of the conventional $I\bar{4}2d$ cell and the primitive cell of $NH_4H_2PO_4$ (ADP).

The primitive cell of ADP contains two NH_4 and two PO_4 groups with four hydrogens serving as hydrogen bonds between the oxygens of the PO_4 groups. The growth units are considered to be the individual ionic groups, and their mass centers are denoted by P_i, N_i ($i = 1, 2$), shown in Fig. 5.2(*a*), while Fig. 5.2(*b*) shows the content of the conventional (Aquilló and Woensdregt 1984, 1987) cell at the atomic level.

The likely candidates for the bonds in the first coordination sphere fall into three categories (Fig. 5.2(*a*)). First, the ionic bonds labeled f and g by Aquilló and Woensdregt (1984, 1987) are beyond doubt crystallization bonds (Hartman 1956*b*, 1959) and should be taken into account. Second, the hydrogen bonds labeled e by Aquilló and Woensdregt (1984, 1987) are almost certainly weak bonds, which as will be shown later play a negligible role in the morphology (Strom et al. 1995*b*). Third, the feasibility of the ionic bonds labeled h (Hartman 1956*b*, 1959) (not to be confused with the weak hydrogen bonds e) has been subject to dispute. The h category was not discussed by Aquilló and Woensdregt (1984, 1987). The h bond strength is somewhat lower than, but comparable to,

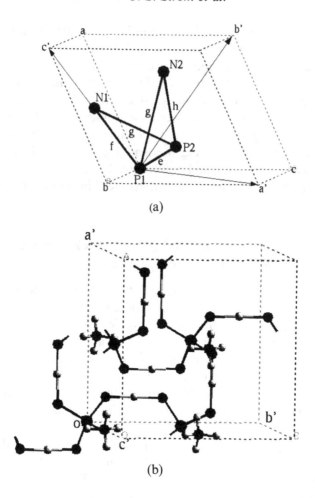

(a)

(b)

Fig. 5.2. (a) Conventional unit cell of ADP (Aquilló and Woensdregt 1984, 1987) defined by axes $\mathbf{a}'$, $\mathbf{b}'$, $\mathbf{c}'$, having twice the volume of the primitive unit cell (Strom et al. 1995b) defined by axes $\mathbf{a}$, $\mathbf{b}$, $\mathbf{c}$. The primitive portion of the unit cell comprises molecules P_1, P_2, N_1 and N_2 reduced to mass centers. The bonds of type f, g, e and h within the unit cell are shown. (b) The atomic structure of the conventional unit cell.

the f and g bond strengths. Nevertheless, in Hartman's opinion (private communication, 1995) the h bond may not be viable as a crystallization bond, because of its specific location with respect to other bonds in the immediate vicinity; put differently, the actual formation of the h bond during crystallization was questionable, despite the respectable value of the h bond strength. It should be added that the ionic f, g, h bonds, as well as the weak e bonds, occur via hydrogen atoms. Section 5.6 provides conclusive evidence of the irrelevance of the e bond, but *only* in comparison with the $\{f, g\}$ set of bonds.

The adjacency matrix representing the graph of ADP-type structures has $A_{+0} = A_{-0} = A_{++} \equiv 0$ submatrices, where, as before, the subscripts $+$, $-$, 0 refer to cations, anions, and neutrals, respectively. ADP would be strictly ionic if the anion–anion hydrogen e bonds were ignored. Submatrix A_{--} contains the hydrogen e bonds, submatrix $A_{-+} = A_{+-}^T$ contains the ionic f, g, and h bonds. The bonding information of the entire structure is confined in the following infinite adjacency submatrix $A = A_{-+}A_{--}$, the rows of which consist of all P_i and columns of all P_i and N_j molecules in the structure:

$$A = \begin{array}{c} P_1^0, P_2^0 \\ P_1', P_2', \ldots, \infty \end{array} \begin{array}{cccc} N_1^0, N_2^0 & N_1', N_2' \ldots, \infty & P_1^0, P_2^0 & P_1', P_2' \ldots, \infty \\ A_{-+}^{00} & A_{-+}^{0\,\prime} & A_{--}^{00} & A_{--}^{0\,\prime} \\ A_{-+}^{\prime 0} & A_{-+}^{\prime\prime} & A_{--}^{\prime 0} & A_{--}^{\prime\prime} \end{array} \, ,$$

where the superscript 0 refers to molecules in the 0-cell, and the primes refer to all remaining cells. As a result of the existing periodicity, the bonding information is recovered from the finite submatrices A_{-+}^{00}, A_{--}^{00}, and further (small) nonzero portions of the infinite submatrices, such as $A_{-+}^{0\,\prime}$, $A_{--}^{0\,\prime}$. These are given below, with the various molecules accompanied by the cell indices (having paradigmatic significance), with reference to the primitive cell:

$$A_{\text{ADP}} = (A_{-+}^{00} \quad A_{--}^{00} \quad A_{-+}^{0\,\prime} \quad A_{-+}^{0\,\prime}) \tag{5.4}$$

The intracell bonds are given by

$$A_{-+}^{00} = \begin{array}{c} \\ P_1^0 \\ P_2^0 \end{array} \begin{array}{cc} N_1^0 & N_2^0 \\ f & g \\ g & h \end{array} \, , \quad A_{--}^{00} = \begin{array}{c} \\ P_1^0 \\ P_2^0 \end{array} \begin{array}{cc} P_1^0 & P_2^0 \\ 0 & e \\ e & 0 \end{array} \, . \tag{5.5}$$

The intercell bonds are given by:

$$A_{-+}^{0\,\prime} = \begin{array}{c} \\ P_1^0 \\ P_2^0 \end{array} \begin{array}{ccccccccccccc} N_1 & N_1 & N_1 & N_1 & N_1 & N_1 & N_2 & N_2 & N_2 & N_2 & N_2 & N_2 & N_2 \\ {}[001] & [011] & [\bar{1}00] & [0\bar{1}0] & [\bar{1}\bar{1}0] & [\bar{1}\bar{1}\bar{1}] & [010] & [011] & [\bar{1}00] & [\bar{1}10] & [\bar{1}0\bar{1}] & [\bar{1}\bar{1}0] & [\bar{1}\bar{1}\bar{1}] \\ h & 0 & h & h & f & h & 0 & 0 & g & 0 & g & 0 & g \\ g & g & g & 0 & 0 & 0 & f & h & f & h & h & 0 & 0 \end{array} \, ,$$

$$A_{--}^{0\,\prime} = \begin{array}{c} \\ P_1^0 \\ P_2^0 \end{array} \begin{array}{cccccc} P_1 & P_1 & P_1 & P_2 & P_2 & P_2 \\ {}[010] & [011] & [111] & [0\bar{1}0] & [0\bar{1}\bar{1}] & [\bar{1}\bar{1}\bar{1}] \\ 0 & 0 & 0 & e & e & e \\ e & e & e & 0 & 0 & 0 \end{array} \, . \tag{5.6}$$

Formally, the matrix elements are 1 or 0 designating presence or absence of a link. For clarity the bond types are used above instead of 1. The basic crystal graphs associated with the various alternative bond sets are obtained by appropriate assignments of f, g, h, $e = 1$ or 0 in equations (5.5) and (5.6). In the $\{f, g, e\}$ bond set $f = g = e = 1$ and $h = 0$ in the A_{ADP} matrix, which is shown in Fig. 5.3.

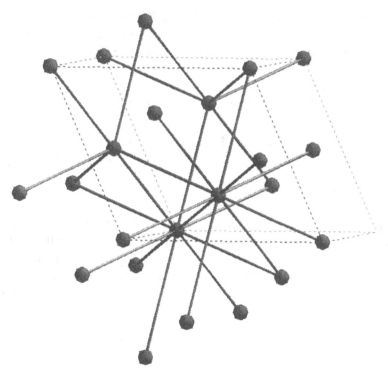

Fig. 5.3. Schematic of the basic crystal graph of ADP corresponding to the $\{f, g, e\}$ bond set, consisting of points in the primitive 0-cell and bonds within the cell, bonds between 0-cell points, and points in any other cells linking with points in the 0-cell.

5.4.2.1. *General Remark on Adopting Crystallization Bonds*

We may at this point entertain the implications of replacing the notion of bond formed during crystallization with the notion of attraction, regardless of how far apart the interacting particles may be. For structures with short-range forces, the basic crystal graph would expand considerably, and might in a limited number of cases remain within manageable proportions. For structures with long- or infinite-range forces – the latter category includes ionic crystals – the basic crystal graph would become enormously large or formally infinite. For example, in the specific case of ADP, if we were to consider that in principle all attractive interactions should qualify as links in the crystal graph, then the matrices $A_{-+}^{0\prime}$ and $A_{--}^{0\prime}$ would have to become infinite in principle, and unmanageable in practice. Let us leave aside for the moment the practical computational complications that would then arise (see Section 5.4.3). The inevitable prospect of requiring an infinite subgraph to store the bonding information of a periodic structure can hardly be considered theoretically appealing. Furthermore, one of the practical consequences of working with an enormously large basic crystal graph is that the method available for a first-principles derivation of F slices, based on processing the adjacency matrix as explained in Section 5.4.3 would become unworkable.

5.4.3. *Graph-Theoretic Derivation of F Slices in Terms of Networks*

From now on we concentrate on deriving F slices, because they play a central role in flat-face growth. However, it is in principle possible to compute PBCs, and methods have been developed to that effect (Strom 1980, 1981).

Theoretically, F slices could be obtained by combining PBCs, but in practice this route is unnecessarily complicated. Since PBCs are bundles of chains, in order to construct them, simpler "direct chains" must first be found and then combined. There emerge many more PBCs than direct chains. Subsequently the PBCs must be (re)combined to get the F slices. The first combinatorial procedure can be circumvented by adopting an alternative definition of F (sub)slice. The new definition in terms of networks replicates in reality all characteristics supplied by the PBC theory, and rests on the concept of "direct chain" (Strom 1980, 1981, 1985). In the remainder of this section face and cell indices refer to the primitive cell.

A "direct chain" (Strom 1980) is simply a path between any two translationally related points, subject to the condition that the two endpoints should be the only points belonging to the path that are related by a lattice translation. Since we are working in a three-dimensional embedding, a direction labeled by three indices $[uvw]$ can be uniquely assigned to a direct chain. Thus, if the endpoints of the path are molecule M in some initial and final cell, the chain direction is the difference in cell translations between the initial and final endpoints, $[uvw]_{\text{chain}} \equiv [uvw]_{\text{final}}^{M} - [uvw]_{\text{initial}}^{M}$.

A definition of F slice equivalent to the one given in Section 5.3 is: an F slice is a network including, possibly among other molecules (see below), a set of (at least two) direct chains in intersecting coplanar directions, satisfying the well known conditions of stoichiometry – in applying the stoichiometry criterion only one of the two endpoints in each direct chain is taken into account – and flatness.

The familiar properties of F slices follow from this definition. First, the face orientation of the F slice is unique and obtained in the obvious way from any two of the intersecting chain directions. Second, the average thickness of an F slice is at most d_{hkl}, and is determined by the number of translationally unique points, or growth units, composing the slice. This turns out to be equal to the interplanar distance or fraction thereof, Rd_{hkl}, where R is the fraction of the unit-cell content present in the slice. F subslices are characterized by $R < 1$. F subslices could form (overlapping) parts of a larger F (sub)slice. F subslices arise because direct chains and PBCs exist containing a fraction of the molecules in the unit cell (see, for example, Fig. 5.8). It is possible to have several slices associated with different values of $R \leq 1$ in the same face (hkl).

The points (molecules) and links (bonds) included in the combined direct chains form the *core* of the F (sub)slice (Strom 1985). In addition to a core, an F (sub)slice may also contain irreducible borders. The edge sets of the irreducible borders do not overlap with the edge sets of the direct chains. The simplest form of irreducible borders is dangling bonds. The crucial role played by irreducible borders when the solid–fluid interaction causes habit modification is illustrated below in Fig. 5.9, by the example of gypsum.

Concentrate for a moment on F (sub)slices possessing irreducible borders. By definition, the core of such an F (sub)slice contains a part of the unit cell, is a network, and satisfies flatness. If it also happens to be stoichiometric, then it is itself an F subslice; otherwise it can be simply characterized as a core.

Formally, the composition (that is, the points and links) of an F slice is expressed by the adjacency matrix of the combination of the constituent direct chains. Physically, an F *slice cell* is a more meaningful entity to use. It is defined as a possible choice of a building block, which, when repeated by all lattice translations parallel to the face, generates the F slice under consideration.

Now, the only techniques necessary to derive the F slices from first principles are the following:

1. Generate all valid direct chains in the structure by finding all paths; every obtained direct chain is characterized by its composition and its direction.
2. Form all possible combinations of direct chains and then (obviously) discard the combinations violating flatness.
3. Apply an irreducible border search while preserving connectedness as well as the flatness condition. This search involves multiple applications of the path finder algorithm, to be defined in Section 5.4.3.1, with the origin at each point in the core and with available vertices the complement of the core (that is the unit-cell content excluding the core).
4. Discard from the final collection all networks violating stoichiometry, bearing in mind that networks with thickness equal to the full interplanar distance, that is, with $R = 1$, are by definition stoichiometric.

Step 4 may be applied repeatedly, should it so happen that different stoichiometric prescriptions are conceivable for one and the same structure. This entire procedure depends critically on step 1, which we next proceed to outline heuristically.

5.4.3.1. *Computing Direct Chains in Terms of Paths*

For further details of the material here the reader is referred to the original works (Strom 1980, 1981, 1985).

A direct chain C is a path between translationally related endpoints,

$$C_n[u_N v_N w_N] = \{i[u_0 v_0 w_0] \rightarrow j[u'v'w'] \rightarrow k[u''v''w''] \rightarrow \cdots \rightarrow i[u_f v_f w_f]\}, \quad (5.7)$$

where $i \neq j \neq k$, and point i is the initial and final point in the primitive cells $[u_0 v_0 w_0]$ and $[u_f v_f w_f]$, respectively. C_n is the nth chain belonging to the class characterized by the label $[u_N v_N w_N]$. From equation (5.7), $[u_N v_N w_N] = [u_f v_f w_f] - [u_0 v_0 w_0]$, the difference between the final and initial labels. Obviously, the net difference $[u_N v_N w_N]$ is the direction of the chain. The net label $[u_N v_N w_N]$ is subject to the following restrictions: $[u_N v_N w_N] \neq [000]$, so that the path is not a closed loop; and $[u_N v_N w_N] \neq m[uvw]$ (with m some integer), in other worlds, the chain direction indices are mutually prime.

A path-finding algorithm is applied to the adjacency matrix of the basic crystal graph under question, say equation (5.4), using back-tracing according to the following steps.

1. A path begins with some vertex i in the 0-cell and continues with the first encountered link in the adjacency matrix, say, $i[000]–j[u_1 v_1 w_1]$, $i \neq j$.
2. Vertex j is referred to the 0-cell, and the first encountered link with $j[000]$, say $j[000]–k(u_2 v_2 w_2)$, where $i \neq j \neq k$, is registered. This procedure is continued as long as possible before entering the next step.
3. The process terminates when the initial vertex is reencountered, say via the link $k[000]–i[u_3 v_3 w_3]$.

This example leads to the following direct chain by propagating the translation indices through the above progression of links (i is the initial point in the 0-cell):

$$\begin{aligned}
\{i \to j[u_1 v_1 w_1] &\to k[u_1 + u_2, v_1 + v_2, w_1 + w_2] \\
&\to i[u_1 + u_2 + u_3, v_1 + v_2 + v_3, w_1 + w_2 + w_3]\}.
\end{aligned} \tag{5.8}$$

The general form of the chain direction indices when the initial point is in the 0-cell is

$$[uvw] = [u_1 + u_2 + \cdots, v_1 + v_2 + \cdots, w_1 + w_2 + \cdots].$$

Chains with zero or multiple indices are discarded.

The process of step 2 gets repeated as many times as possible by deleting each time the last registered link before restarting with step 1. The procedure ends when the adjacency matrix becomes exhausted. The chain directions are, of course, not limited to the cell translations present in the basic crystal graph, but can extend in the infinite graph of the structure. Finally, the direct chains and, by implication, also the F slices are determined without restrictions by this method, which is exhaustive (and exhausting!).

5.4.3.2. *Computing F Slices in Terms of Direct Chains*

Having been constructed, the direct chains need to be combined in order to produce the molecular compositions of the F slices. A possible F slice in (hkl) includes at least two nonparallel direct chains, $\{\{C_m[u_M v_M w_M] + C_n[u_N v_N w_N]\} + \text{IR}\}$, where IR (irreducible border) means points and links not belonging to either chain, and also not belonging to the combination after the bonds are completed.

The distinction between core and irreducible border depends on the extent of the chosen coordination sphere. As the coordination sphere is gradually extended beyond its meaningful bounds, some that have been previously classified as irreducible borders become incorporated in the core, while in the meantime new irreducible borders appear.

The face indices are determined from the chain direction indices, by $\pm(hkl) = [u_M v_M w_M] \times [u_N v_N w_N]$. So the chain and face orientations are obtained in the output, rather than being specified in the input. Furthermore, these are not obtained geometrically. They follow by propagating labels in a combinatorial formulation. The final labels have the aforementioned unambiguous geometrical interpretation. It is a trivial matter to test every candidate F slice for the conditions of "flatness" and stoichiometry explained earlier, and maintain the ones satisfying these conditions.

5.4.4. *Illustration: Solution-Activated F Slices with Irreducible Borders in Gypsum*

Disregarding irreducible borders is an undesirable restriction. The most widespread manifestation of irreducible border, the set of dangling bonds, is an essential feature of the surface structure, especially of ionic compounds. Hydrated compounds growing out of aqueous solutions offer another interesting illustration of the importance of the irreducible borders. There the role played by F slices with irreducible borders shows at the same time how the solution selects which F slices should become activated.

In studying the habit of gypsum (calcium carbonate dihydrate) growing out of aqueous solutions, Van der Voort and Hartman (1991) considered that in the case of a hydrated structure the water molecules play a double role: Some water molecules function as structure particles occupying crystallographic positions at the interface, while other water molecules act as solution particles occupying random positions in the solution, and hydrating the available surface by occupying accessible positions. Therefore, the relevant F slices in the case of an aqueous solution are either F slices terminating in water molecules at crystallographic positions or F slices terminating at solvent-accessible positions at the surface.

Figure 5.4 shows the F slices that are relevant with reference to vacuum, and with reference to the aqueous solution for the (020) and (120) faces. The former are primarily of the core type, whereas the latter terminate in crystallographic water in the form of dangling bonds. The F slices terminating in crystallographic water need no dehydration because those water molecules belong to the structure, and do not need to be removed in order to allow the surface to grow. In contrast, the F slices that have solvent-accessible positions become hydrated by water molecules belonging to the solution. Subsequently these surfaces need to become dehydrated in order to grow; that process of dehydration slows down the growth of these surfaces and hence increases their morphological importance.

The difference in the need for dehydration between these two kinds of F slices explains the discrepancy between the theoretical habit, Fig. 5.5(a), and the observed habit of gypsum obtained from a pure aqueous solution, Fig. 5.5(b). Van der Voort and Hartman (1991) provide additional evidence supporting their explanation. In the case of growth out of an impure aqueous solution containing strongly interacting organic molecules, the difference between crystallographic and noncrystallographic water positions at the interface diminishes. Then, indeed, the observed growth habit resembles the theoretical one, Fig. 5.5(a).

5.4.5. *Specialization to Ionic Structures Illustrated by Structures of the Barium Sulphate Type*

The numbers of direct chains, and especially PBCs, can become rapidly unmanageable; moreover, their combinations grow out of proportion, resulting in a profusion of valid *F* slices. This holds particularly for ionic structures, where the growth units are individual ions, reaching many ions in a single unit cell. The computational *F* slice construction just outlined then becomes too lengthy because of its simple enumerative nature. Efficiency is vastly improved by replacing a single simple, but lengthy, procedure by two shorter procedures at the cost of increased complexity (Strom 1981). This improvement becomes possible for *strictly* ionic cases by taking advantage of the bipartite nature of the crystal graph.

As a first procedure, the positive or negative ions are compressed out of the ionic graph, by squaring the adjacency matrix *A*, equation (5.4). We have:

$$A = \begin{pmatrix} & \{+\} & \{-\} \\ \{+\} & A_{++} & A_{+-} \\ \{-\} & A_{-+} & A_{--} \end{pmatrix}, A^2 = \begin{pmatrix} & \{+\} & \{-\} \\ \{+\} & A_{+-}A_{+-}^{\mathrm{T}} & 0 \\ \{-\} & 0 & A_{+-}^{\mathrm{T}}A_{+-} \end{pmatrix}.$$

It is known (Harary 1969; Biggs 1974; Strom 1981) that the matrix element $(A^2)_{ij}$ equals the number of ways, including 0, in which ion i can be connected to ion j via two-bond links. When every nonzero element of the square block-diagonal matrix $A_{+-}A_{+-}^{\mathrm{T}}$ (or of $A_{+-}^{\mathrm{T}}A_{+-}$) is replaced by a 1 and the diagonal elements by 0, there arises a square symmetric matrix B, containing cation–cation (or anion–anion) connections. Two ions are considered linked in B if they are bonded to at least one and the same oppositely charged ion in A. Thus, the compressed matrix B can be interpreted as the adjacency matrix of a fictitious graph containing only cations (or only anions), much smaller than the original ionic graph. It is of the form which can be contrasted with equation (5.4):

$$B = \begin{pmatrix} B_{++}^{00} & B_{++}^{0\prime} \\ B_{++}^{\prime 0} & B_{++}^{\prime\prime} \end{pmatrix}, \text{ or with } + \text{ changed to } -. \tag{5.9}$$

The superscript 0 means ions in the unit cell, and the prime means ions in other cells. The upper portion of B suffices to store the complete information on bonding.

The algorithm of the previous subsection can be applied to B identically. Although the resulting networks are not the sought ionic F slices, they can be used as generators to obtain the ionic F slices in the second procedure that follows. The ionic F slices are produced by injecting in the F-slice generators

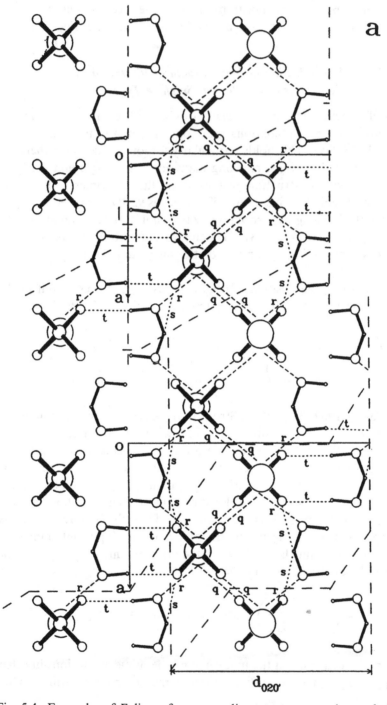

Fig. 5.4. Examples of *F* slices of gypsum adjacent to vacuum (upper half) and water solution (lower half), terminating in crystallographic water molecules in the form of dangling bonds. Large circles: Ca; medium circles: C; small circles: O; points: H. Dashed lines indicate slice boundaries. (*a*) (020); (*b*) (120). (Reprinted from Van der Voort and Hartman (1991), with permission from Elsevier Science.)

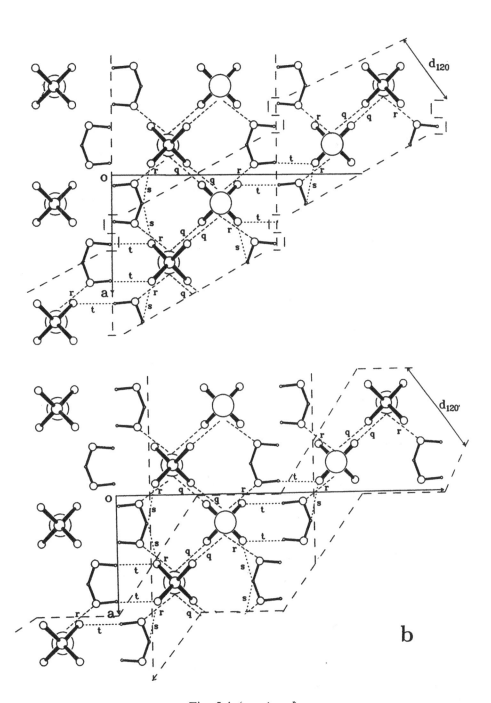

Fig. 5.4 (*continued*)

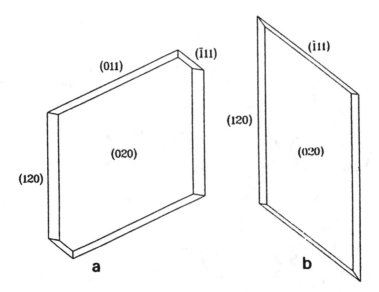

Fig. 5.5. Morphology of gypsum. (*a*) Structural habit, also resembling the growth habit obtained from aqueous solutions containing strongly interacting organic impurities. (*b*) Habit obtained from pure aqueous solutions. (Reprinted with permission from Van der Voort and Hartman (1991).)

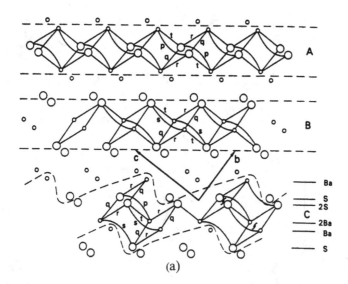

(a)

Fig. 5.6. The (011) *F* slices of categories A, B, and C of barite projected on (100). (*a*) One typical example from each class A, B, and C, showing the surface structure in detail. Small circles: Ba; large circles: sulphate. (Reprinted with permission from Hartman and Strom (1989, p. 505).) (*b*) The thirty-three *F* slices of categories A, B, and C. Beneath each figure the upper number is the attachment energy in kJ mol^{-1}, the lower number the surface energy in mJ m^{-2}, with the oxygen charge equal to -1. (Reprinted from Hartman and Strom (1989, p. 506), with permission from Elsevier Science.)

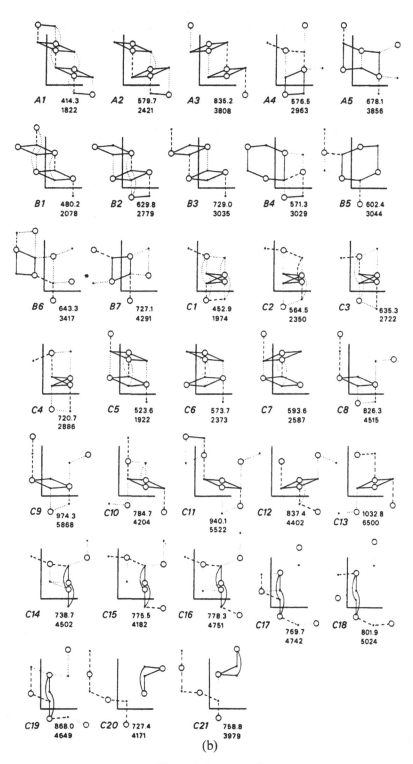

A1 414.3 1822
A2 579.7 2421
A3 835.2 3808
A4 576.5 2963
A5 678.1 3856

B1 480.2 2078
B2 629.8 2779
B3 729.0 3035
B4 571.3 3029
B5 602.4 3044

B6 643.3 3417
B7 727.1 4291
C1 452.9 1974
C2 564.5 2350
C3 635.3 2722

C4 720.7 2886
C5 523.6 1922
C6 573.7 2373
C7 593.6 2587
C8 826.3 4515

C9 974.3 5868
C10 784.7 4204
C11 940.1 5522
C12 837.4 4402
C13 1032.8 6500

C14 738.7 4502
C15 775.5 4182
C16 778.3 4751
C17 769.7 4742
C18 801.9 5024

C19 868.0 4649
C20 727.4 4171
C21 758.8 3979

(b)

Fig. 5.6 (*continued*)

the ionic bonds that were removed in the first procedure. The reader is spared the tedious details (Strom 1981, 1985). Here follows an assessment and an illustration.

The distinction between positive and negative ions is symbolic, so that either can be compressed out of the ionic basic graph. If the unit cell contains equal numbers of cations and anions, compression leads to the same size reduction. The minimum size of the B matrix, implying maximum size reduction, is achieved by removing as many ions as possible in the first procedure. The price to be paid is increased complexity (Strom 1981, 1985) in recovering the desired ionic F slices.

The need to resort to the described compression for ionic structures is by no means exaggerated. A nice example of the profusion of ionic F slices that can occur even in structures with a modest number of ions in the unit cell, is the set of centrosymmetric structures of the barite type, $BaSO_4$. The unit cell contains four cations and four anions, the B matrix for the F face analysis contains a 4×4 submatrix B^{00}. A long-standing controversy in the literature (Dowty 1976; Hamar 1981; Küppers 1982; Tasoni, Riquet, and Durand 1978) about the correct classification and slice compositions of the {011} face of barite was resolved when the present graph-theoretic derivation produced no less than 64 valid nonpolar F slices. These (011) F slices consist of (Hartman and Strom 1989) twelve symmetry singlets, classified in groups A and B; and twenty-one symmetry doublets, classified in groups C and D, where group D is obtained from group C through the symmetry operations. The thirty-three F slices of groups A, B, and C are shown in Fig. 5.6, together with their respective attachment and specific surface energies. It is easily seen that no two of the depicted F slices are symmetrically related, since no two energy quantities have the same value.

5.4.6. On the Profusion of F Slices

From graph-theoretic studies of several structures, we have drawn certain conclusions as regards the profusion of F slices regularly observed in one and the same orientation (*hkl*). Sets of parallel F slices turn out to be partially overlapping growth fronts at the same level or shifted with respect to each other by a fraction of d_{hkl}. (This is especially true if F half-slices occur, alternating to form a pair of full F slices staggered by $\frac{1}{2} d_{hkl}$. See, for example, the half-slices of the pyramidal {011} face of ADP in Fig. 5.9, elaborated further in Section 5.6. The same feature is found in fully fledged PBCs, which can be composites of smaller PCBs containing a portion of the unit cell, see, for example, Fig. 5.8.) The staggering of F slices at steps $\frac{1}{4} d_{hkl}$ or smaller are for larger unit cells a common occurrence. The relation between individual slices is sometimes caused by symmetry, but in many cases the existence of large numbers of F slices cannot be traced to symmetry relations.

Two mechanisms tend to increase the numbers of generated F slices: symmetry and combinatorial inflation. The symmetry multiplicity is reflected in the numbers and multiplicities of the groups of symmetrically related F slices. Combinatorial inflation is related to the notion of the *degree* of a point in a graph, which is defined as the number of links emanating from that point. Points of low degree have little impact on the generation of direct chains. In particular, points of degree 2 can be removed and replaced by a link without affecting at all the direct chains of the structure. On the other hand, the diversity and abundance of direct chains, and by implication also of *F* slices, is caused by the presence of points with high degree. So we see that it is not so much the size of the vertex set (number of growth units) in the unit cell, but the size of the edge set (or the coordination) that could cause combinatorial inflation and ultimately combinatorial explosion.

Larger interplanar distances tend to promote both mechanisms. The reason is that within the boundaries of the thicker slices, the density of bonds per unit (*hkl*) surface area is higher, implying a large number of bonds oblique to the surface. This increases the likelihood of bond formation out of the face, leading to an increase in the number of bond combinations, and a larger number of direct chains parallel to (*hkl*). In contrast, boundaries of thinner slices contain fewer bonds per unit area, more nearly parallel to the surface, leading to fewer bond combinations and fewer direct chains within the slice. The large density of bonds per unit area in thicker slices is also manifest in the existence of irreducible borders (see Fig. 5.9), since these dangling-bond constructions consist mainly of bonds oblique to the surface.

5.4.7. Technical Issues of Geometry, Symmetry, Validation, and Computational Tasks

Because the graph-theoretic *F* slice derivation method is independent of geometry and symmetry, the results obtained hold for all structures sharing the same combinatorial properties. So, rather than individual structures, the method is applicable to families of structures, even with broadly different geometries and symmetries.

Three examples of such combinatorial families of structures which can be subjected to the same graph-theoretic derivation of *F* slices are:

1. The barite type structures (Hartman and Strom 1989; Hartman 1991*a*, *b*) MSO_4 (M = Sr, Pb, Ba), $MCrO_4$ (M = Sr, Pb, Ba, Ra), $MClO_4$ (M = K, NH_4, Rb, Tl, Cs), $MMnO_4$ (M = K, Rb, Cs), MBF_4 (M = Ag, K, NH_4, Rb, Cs).
2. Groups (*a*) and (*b*) of a large number of monoclinic structures classified by Hartman (1991*b*).
3. $NH_4H_2PO_4$ (ADP) and the isomorphous compound potassium dihydrogen phosphate (KDP), both treated in Sections 5.6 and 5.9.
4. Ammonium chloride (mineral sal-ammoniac) shares network characteristics with the halides. It merits a comparative study because it is an unusual halide, owing its peculiar surface properties (Section 5.7) to the ammonium content.

The common features shared by structures belonging to the same combinatorial family are the parameters specifying the first coordination sphere, namely:

the number of growth units in the unit cell; the correspondence between intracell points and links; the cell labels; and correspondence between intercell points and links. This structural information is essential and constitutes the primary data. The remaining structural information constitutes the secondary data, enters at the output stage in the form of output options, and can be varied at will after the *F* slice computation has been basically completed:

1. The stoichiometry condition is enforced simply by discarding the established candidate *F*-slices having nonstoichiometric composition. Consequently, it is a trivial matter to obtain the various *F*-slice sets associated with different stoichiometric combinations in quick succession.
2. The space group selection conditions are enforced by discarding the established candidate *F* slices when the face indices violate these conditions.
3. The overall centrosymmetry may be enforced if there are reasons for doing so, by discarding those *F* slices whose opposite boundaries are unrelated by symmetry.
4. The cell parameters and mass-center coordinates in the unit cell are only necessary for embedding the *F* slices in space.
5. The *F* slices are made visible in structure projections in the system of axes chosen by the user, which would in most cases be the conventional cell.

As for the symmetry, interestingly it forms part of the *output* rather than the *input* information, albeit implicitly. Symmetry, being basically a geometrical concept, has no inherent affinity with a combinatorial approach. This is evidenced by the fact that we have to assign the geometrical interpretation of direction indices to the final chain labels, as we have seen. The symmetry properties of the resulting *F* slices become manifest in structure projections, after embedding in space has been carried out. Symmetry cannot be generically incorporated in the graph-theoretic derivation of direct chains; it can only be superficially imposed by suppressing resulting chains with directions outside a certain portion of the sphere in space.

If no symmetry restrictions are imposed, it can be easily seen that the resulting PBCs and *F* slices exhibit exactly the symmetry of the input bonding pattern. The symmetry of the bonds often coincides with, or is higher than, the space-group symmetry. In that case the direct chains and *F* slices obtained in the output possess symmetry properties at the very least equal to the known space-group symmetry of the structure. The symmetry of bonds can also be lower than the space-group symmetry, as is the case with paraxylene and several other related monoclinic structures investigated collectively in group (a) of Hartman's (1991*b*) article. In that case the obtained chains and *F* slices exhibit symmetry lower than that of the space group; the observed difference in symmetry elements equals precisely the symmetry elements missing in the bonding pattern (Strom 1993).

Symmetry is an important guideline in assessing the correctness of the software code. Obtaining in the output the appropriate symmetry properties, without having introduced them in the input, is the most thorough and reliable way at our disposal for testing correctness. Imposing symmetry restrictions tends to mask flaws in the software and is ill-advised. In order to verify the results, the structure should be treated as triclinic.

An additional test for correctness of the software is available in the case of strictly ionic structures, namely: Identical sets of F slices should ultimately result after compressing either the cations or the anions from the original ionic graph.

5.4.8. *Description of the Software and Technical Specifications*

Program FFACE (Strom 1985, 1998) is a complete and rigorous implementation of the automated graph-theoretic derivation of F slices, meeting the previous input and output specifications. It was developed by Strom between 1979 and 1985, in the Universities of Leiden and Utrecht. It has been fully tested and operational for over a decade, on, for example, early work on weddellite (Strom and Heijnen 1981), and later work on lysozyme crystallization (Strom and Bennema 1997a, b).

FFACE has two versions, for the general ionic cases. The former is applicable to all structures, including strictly ionic and mixed ionic–covalent systems; the latter is an option available to strictly ionic structures, exploiting the computational advantage offered by their bipartite character. In both versions the surface dipoles can be computed if nonzero charges are input, with the available option to restrict the list of F slices to a limiting dipole value or to zero (nonplanar F slices).

FFACE works without restrictions, and the correctness of the code has been repeatedly verified as it replicates in reality the existing symmetry properties. The resulting F slices of both FFACE versions are written in output files in the form of adjacency matrices and slice cells, and can be obtained as printer output. Lists of algebraic expressions for the energy quantities relevant to the broken-bond model, that is slice and attachment energies (Strom 1985; Strom and Bennema 1997a, b), are also output. When bond strengths are specified, additional numerical tables follow.

The computation for a customary run is distributed with respect to the various tasks excluding input/output as follows.

1. A small part of the computation is devoted to finding the direct chains and the formation of network cores. The stoichiometric subset of the network cores forms a subset of the F slices, specifically the F slices that happen to contain no irreducible borders.
2. A larger portion of the computation is consumed by constructing irreducible borders. For complicated structures or faces with large interplanar distances the IRs require an enormous amount of additional computation. Restricting the computation to only core-type networks is a serious handicap for molecular crystals. For ionic crystals such a restriction is absurd, since it would even preclude the existence of dangling bonds.

3. A certain amount of the processing is devoted to sorting and ordering procedures in order to prevent (or postpone) the dreaded combinatorial explosion. This is a situation where the number of combinations rises so rapidly that the available amount of computer memory and the processing time become determining factors. By excluding illegal or unviable combinations before they are formed, and by ordering the remainder, the combinatorial explosion can be averted.

4. Finally the ionic formulation is a major complication. However, this task is not essential since ionic structures can be treated by the general case.

It is worth noting that there exists no functional replication of program FFACE. Commercial efforts under way to imitate the above described software, are restricted to attempting task (1). Obviously failure to implement essential tasks such as computing the growth layers with IRs makes commercial morphology (Cerius2 1995) imitation software unsuitable for serious scientific work. The consequences are illustrated by two examples. The growth layers that become activated by the presence of solution in the case of gypsum would be overlooked since they contain IRs. An inadequate network analysis resulting from the neglect of IRs, as done for KDP in Section 5.9, could have contributed to errors (De Vries et al. 1998) in electrostatic surface properties, as pointed out by Strom (1999).

5.5. Energy Computations in Ionic Structures

5.5.1. *Electrostatic Point Charge Model in the PBC Theory*

In the past, the PBC theory has been applied more frequently to ionic than organic substances. The slice and attachment energies of the F slices obtained from PBCs had to be calculated in order to arrive at the relative growth rates in vacuum. In ionic structures, the van der Waals interactions are usually negligible as compared with the Coulomb interactions, or form at any rate a small fixed percentage of the Coulomb contribution. Consequently, the overwhelming majority of the early applications of the PBC theory ignore the van der Waals contributions, dealing exclusively with the calculation of electrostatic energies. This is better known as the "electrostatic point charge model," and is based on the use of formal point charges.

As is known, the Coulomb potential has infinite range, being of the form $1/r$. So the value of the potential at a sampling point arising from an infinite or semi-infinite lattice or an infinite two-dimensional lamina involves the sum of infinitely many terms. If we attempt a straightforward summation, no matter how large a radius we choose away from the sampling point, the result will always amount to a number picked almost randomly from the computer's memory when execution is interrupted. (In that sense, direct summation of Coulomb terms may be put to better use as a procedure for pseudorandom number generation than for computation of crystal energies!)

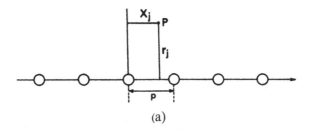

(a)

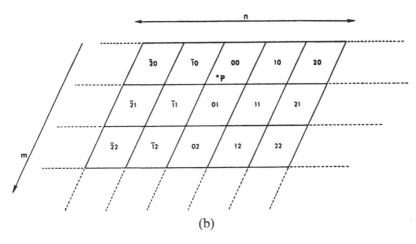

(b)

Fig. 5.7. Illustration of two of the data infrastructures used for Madelung's site potential calculations. (*a*) Ion line or linear array coordinate x, corresponding in text notation to the line spacing a and height z, respectively. (*b*) An ion mesh or rectangular array of equally charged ions. (Reprinted from Hartman (1973*a*, pp. 382 and 383), with permission from the author.)

5.5.1.1. *Madelung Formulation*

In 1918 Madelung provided expressions for evaluating the site potential due to the collective presence of periodic linear and planar point-charge distributions (Madelung 1918; Kleber 1939):

- a one-dimensional array of collinear equally charged ions (with charge Q) parallel to [uvw], with extension $(-\infty, +\infty)$, here called for convenience an "ion line;"
- a two-dimensional array of coplanar equally charged ions (with charge Q) parallel to (hkl), with extension $(-\infty, +\infty) \times (-\infty, +\infty)$, here called for convenience an "ion mesh."

Fig. 5.7 shows an ion line with spacing a and an ion mesh with spacings a, b. Each ion has charge Q. The net charge on a single ion mesh or a single ion line is formally infinite. The charge distributions are expressed as one- and two-dimensional Fourier series, respectively.

One-Dimensional Fourier Series. The expression for the site potential due to an ion line with $N \to \infty$ ions at sampling point P has cylindrical symmetry. It is given in terms of the cylinder radius r and distance z of the sampling point from the closest ion measured along the ion line. The site potential is obtained by either direct integration or solution of Laplace's equation:

$$V(r, z) = \frac{2Q}{a} \log \frac{2a}{r} + \frac{4Q}{a} \sum_{m=1}^{\infty} \cos\left(\frac{2m\pi z}{a}\right) K_0\left(\frac{2m\pi r}{a}\right) + \frac{2Q}{a} \log N, \qquad (5.10)$$

for $r > 0$. For $r = 0$ the following expression holds:

$$V(0, z) = -\frac{Q}{a}\left[\Psi\left(\frac{z}{a}+1\right) + \Psi\left(2 - \frac{z}{a}\right) - \frac{a}{z} - \frac{a}{a-z}\right] + \frac{2Q}{a} \log N. \qquad (5.11)$$

For $r = 0$, $z = 0$ we have:

$$V(0, 0) = (2Q/a)(C + \log N), \qquad (5.12)$$

where K_0 is a Hankel function, Ψ is the logarithmic derivative of the Γ function (Abramowitz and Stegun 1965), and $C = -\Psi(0)$ is Euler's constant. As $N \to \infty$, the potential due to a single charge line has a logarithmic divergence, but the logarithmic term vanishes when the complete contribution to the potential of a charge-neutral collection is calculated. Thus, given a neutral superposition of ion lines, expressions (5.10)–(5.12) converge everywhere.

Two-Dimensional Fourier Series. The expression for the site potential due to a rectangular ion mesh at sampling point P is given in terms of the x and y coordinates measured along the mesh, and height z above the mesh plane:

$$
\begin{aligned}
V(x, y, z) \simeq{} & \frac{4Q}{ab} \sum_{l=1}^{\infty} \sum_{m=1}^{\infty} \frac{\exp(-2\pi k_{lm} z)}{k_{lm}} \cos\left(\frac{2\pi lx}{a}\right) \cos\left(\frac{2\pi my}{b}\right) \\
& + \frac{2Q}{b} \sum_{l=1}^{\infty} \frac{\exp(-2\pi lz/a)}{l} \cos\left(\frac{2\pi lx}{a}\right) \qquad (5.13)\\
& + \frac{2Q}{a} \sum_{m=1}^{\infty} \frac{\exp(-2\pi mz/b)}{l} \cos\left(\frac{2\pi my}{b}\right) - \frac{E}{ab} 2\pi z,
\end{aligned}
$$

where $k_{lm} = \sqrt{l^2/a^2 + m^2/b^2}$. An analogous expression holds for the case of a parallelogram. Equation (5.13) converges poorly for small z and diverges when $z = 0$, owing to the appearance of $z \geq 0$ in the negative argument of the exponential.

Each single ion existing in the unit cell generates one ion mesh parallel to a given lattice plane (*hkl*), and one ion line parallel to a given lattice direction [*uvw*].

Obviously two or more ion lines must be superimposed to produce a one-dimensional charge-neutral assembly in [*uvw*]; two or more ion meshes must be superimposed to produce a two-dimensional charge-neutral assembly in (*hkl*). Charge-neutral compositions need not be stoichiometric, but stoichiometric compositions, such as PBCs and slices, always have zero net charge.

Employment of Madelung's Ion Meshes. To employ Madelung's expression for the site potential due to ion meshes, equation (5.13), one specifies a slice with thickness d_{hkl} to be a composite of as many ion meshes parallel to (*hkl*) as there are ions in the unit cell, in such a way as to obtain a suitable description of the appropriate growth front. The total site potential due to the slice is then the sum of the contributions of the individual ion meshes composing that slice. The interaction energies are obtained by the usual prescription: Evaluate the potential at all ion sites in the unit cell, multiply by the ion charges, and sum appropriately. The advantage offered by Madelung's equation (5.13) is that the slice of interest can be directly expressed by superimposing ion meshes. It is outweighed by the disadvantage caused by the divergence of the site potential as soon as the variable sampling point approaches or lies in *any one* of the mesh planes (thus resulting in $z_k = 0$ for the *k*th mesh). Therefore, equation (5.13) is only valid for calculating interaction energies between the 0-slice and slices parallel to (*hkl*) shifted by an integral multiple of d_{hkl}. Consequently, equation (5.13) will provide the attachment energy and the surface energy, but will fail to provide the slice energy and crystal energy.

Employment of Madelung's Ion Lines. To employ Madelung's expressions (5.10)–(5.12) one must investigate individual PBCs in detail. Since a PBC is either a single chain or a bundle of chains, the site potential due to a PBC follows by repeated application of the Madelung expressions for ion lines. Furthermore, since an *F* slice is a superposition of two PBCs, say in [*uvw*]$_1$ and [*uvw*]$_2$, one may proceed in an obvious way, by summing the resulting expressions for the potentials due to PBCs, in order to arrive at the slice and attachment energies.

Detailed specification of the interaction energies between individual slices is not directly possible. Therefore (surface and) specific-surface energies cannot be computed. However, the drawback of implementing Madelung's ion-line expressions for the computation of energies due to lamina shapes is that one must first decompose the particular slice under study into its constituent PBCs. Subsequently, one must reassemble correctly the site-potential contributions of the various PBCs to replicate the result of the original slice. For that reason, data preparation can be unduly cumbersome. The computational procedure employing Madelung's ion lines has been implemented by Woensdregt (1971). The advantage of the Madelung "ion-line" implementation is that convergence is slow but safe – *convergence to which quantity?*

5.5.1.2. *Issue of Convergence: Madelung versus Ewald*
Formulations

Although in most cases the results turn out to be consistent and in agreement with results obtained by other methods, persistent and systematic inconsistencies turn up for *some* site potentials in *some* faces of *some* structures, for example, fluorite, rutile, corundum, and fluorphlogopite. In his paper entitled "An empirical correction term for the calculation of site potentials with the Madelung method" Hartman (1978*b*) poignantly sums up, "... the crystal structure is divided into parallel slices d_{hkl}. Each slice is divided into parallel identical ionic chains [uvw]. Site potentials calculated for the same (hkl) but different [uvw] have the same values. However, site potentials calculated for the same [uvw] but different (hkl) give inconsistent values."

To find a remedy for the problem we need to enter Hartman's tabernacle of empirical evidence. On page 144 of that paper, Hartman (1978*b*) arrived empirically at a correction term depending on the primitive cell volume, face indices, ion charges, and height of ion meshes above the plane (hkl). For example, in fluorite, some of the necessary correction terms in units e/Å for site potentials are: in face (111) the correction is -0.19169 for a potential value of -1.19319 and for a potential value of $+0.93684$; in face (110) the correction is zero; in face (001) the correction is -0.57507 for a potential value of -0.80984 and a potential value of $+1.32025$. Hartman concludes, "The corrected site potentials agree with those calculated by the Ewald method."

So much for the remedy, but where lies the *cause* of the problem? In the light of large and systematic discrepancies – in the above example of fluorite the corrections can be almost as large as the magnitude of the potential itself – the difficulty could hardly be attributed to algebraic error, computational precision, or inaccuracy resulting from approximation. In Hartman's view, "The discrepancy between the Madelung and Ewald methods is presumably due to an incorrect derivation of the potential ..." In a later publication, Hartman (1982) demonstrated explicitly the necessity for the correction term, arising because the value of the site potential depends on the *order* of summation of the infinite double sums – a clear sign of ambiguity and inadequacy in the process of convergence.

A plausible explanation could be that, although Madelung's derivation is correct, it fails to guarantee consistency in the choice of boundary conditions. The value of the potential at a point can only be determined up to an arbitrary additive constant. The quantity of interest is the potential difference between the sampling point and infinity. Conventionally, we fix that constant by taking the value of the potential at infinity to be zero, but, in the Madelung formulation, a single ion mesh or a single ion line possesses net charge located at infinity. Can we legitimately expect to enforce the boundary condition of zero potential in a location where net charge is present (at infinity)? The physical quantity to which Madelung's expressions in equations (5.10)–(5.12) converge could well be the correct site potential, but accompanied each time by an additive constant depending circumstantially on symmetry conditions. So the various site-potential contributions, to be ultimately summed, could sometimes be implicitly carrying with them inconsistent conventions.

In the light of this evidence, it is easy to see why the Ewald method does not suffer from inherent inconsistencies. In the case of an infinite three-dimensional structure, the Ewald (1921) method is based on two spheres (Tosi 1964); one sphere refers to the direct lattice and the other to the reciprocal lattice. When the Ewald method is adapted to the case of an infinite two-dimensional lamina shape, the analytic closed-form formulation (Heyes 1980, 1981; Heyes and van Swol 1981; Strom and Hartman 1989) involves two disks; one disk refers to the direct lattice and the other to the reciprocal lattice. Under no circumstances is one confronted with the occurrence of net charge at infinity. The Ewald expression for the site potential due to lamina shapes is given in the following.

5.5.2. *Coulomb and van der Waals Interactions*

The expressions for the Coulomb and van der Waals interactions between atoms i and j are given by Strom et al. (1995a):

$$\phi_{ij}^{\text{Coul}} = q_i q_j / r_{ij}, \tag{5.14}$$

$$\phi_{ij}^{\text{LJ}} = -p_{ij}\varepsilon_{ij}\left[\left(\frac{r_{ij}^*}{r_{ij}}\right)^{12} - 2\left(\frac{r_{ij}^*}{r_{ij}}\right)^{6}\right],$$

$$r_{ij}^* \equiv \frac{1}{2}(r_i^* + r_j^*), \quad \varepsilon_{ij} \equiv (\varepsilon_i\varepsilon_j)^{1/2}, \tag{5.15}$$

where r_{ij} is the distance between atoms i and j. The total interaction between two atoms separated by distance r_{ij} is of course $\phi(r_{ij}) = \phi_{ij}^{\text{Coul}} + \phi_{ij}^{\text{LJ}}$.

The physical parameters entering the Coulomb expression are the atomic point charges q_i, which are usually determined by semiempirical or ab initio procedures. The physical parameters entering the Lennard-Jones (L-J) 6–12 expression are atomic potential-well depth ε and equilibrium radius r^*. They can be found in chemical tables of atom properties or are compiled in commonly accessible databases.

The prefactor p_{ij} in the L-J expression acts as a weight depending on the configurational relationship between atoms i and j. Three classes can be distinguished: First, the i–j pair is "nonbonded" if the atoms belong to different molecules, or to the same molecule but are far apart, meaning that they differ by four or more bonds; this case corresponds to $p_{ij} \equiv 1$, that is, the i–j interaction is given full weight. Second, the i–j pair is "bonded" if both atoms i and j belong to the same molecule, but differ by one, two, or three bonds; this case corresponds to $p_{ij} \equiv 0$, meaning that the i–j interaction is discarded, being considered as part of the internal energy of the molecule. The third category comprises atoms i and j belonging to the same molecule but differing by four bonds; the i–j interaction is then known as the "1–4 bonding" case and is given half weight, $p_{ij} \equiv 0.5$.

5.5.3. Habit-Controlling Energy Quantities

Take orientation $(hkl) = (001)$ without loss of generality (Strom and Hartman 1989; Strom et al. 1995a), since transformation of axes can always be found such that $(hkl) \rightarrow (001)$. Define a distance function between atom i in the zero cell and atom j translated to cell $[\lambda, \mu, m]$. Since the face is (001), m is the slice index, and λ and μ are translation indices parallel to (001). The distance between the above two atoms is

$$d_j^i(\lambda, \mu, m) = |\mathbf{R}_i - \mathbf{R}_j - \lambda \mathbf{a} - \mu \mathbf{b} - m\mathbf{c}|, \qquad (5.16)$$

where $\mathbf{R}_i$ ($\mathbf{R}_j$) is the position of atom i (j) in the zero cell and λ, μ, m are integers. The slice energy has two terms, $E_{sl} = E_{sl}^0 + E_{sl}'$:

$$E_{sl}^0(001) = \frac{1}{2} \sum_{i=1, i\neq j}^{N} \sum_{j=1}^{N} \phi_{ij}(d_j^i(0,0,0)) = \sum_{i=1, i<j}^{N} \sum_{j=1}^{N} \phi_{ij}(d_j^i(0,0,0)), \qquad (5.17)$$

$$E_{sl}'(001) = \frac{1}{2} \sum_{\lambda, \mu=-\infty}^{\infty\,'} \sum_{i,j=1}^{N} \phi_{ij}(d_j^i(\lambda, \mu, 0)), \qquad (5.18)$$

where N is the number of atom charges in the unit cell, and the prime ($'$) means that the term $\lambda = \mu = 0$ is excluded from the summation. The interaction energy between the $F(001)$ cell and the mth slice is

$$E_m^0(001) = \sum_{\lambda, \mu=-\infty}^{\infty} \sum_{i,j=1}^{N} \phi_{ij}(d_j^i(\lambda, \mu, m)), \quad m \neq 0 \qquad (5.19)$$

(here the term $\lambda = \mu = 0$ is included in the summation). The attachment energy thus becomes

$$E_{att}(001) = \sum_{m=1}^{\infty} E_m^0(001) = \sum_{m=1}^{\infty} \sum_{\lambda, \mu=-\infty}^{\infty} \sum_{i,j=1}^{N} \phi_{ij}(d_j^i(\lambda, \mu, m)). \qquad (5.20)$$

The dipole, surface and specific surface energies are as defined in equations (5.2) and (5.3b).

5.5.4. *Site Potential in the Ewald Formulation for Lamina Shapes*

The charge density at location $\mathbf{R}$ produced by the existence of charges $\{q_n\}$ located at $\mathbf{R}_n$, $n = 1, \ldots$ has the form $\sum_n \delta(\mathbf{R} - \mathbf{R}_n)q_n$. The Ewald (1921) transformation (Tosi 1964) consists in splitting this expression in two parts, by adding and subtracting a continuous spherically symmetric charge distribution $q_n \sigma(|\mathbf{R} - \mathbf{R}_n|, \eta)$:

$$\sum_n q_n \delta(\mathbf{R} - \mathbf{R}_n) = \left\{ \sum_n q_n \sigma(|\mathbf{R} - \mathbf{R}_n|, \eta) \right\}$$
$$+ \left\{ \sum_n q_n [\delta(\mathbf{R} - \mathbf{R}_n) - \sigma(|\mathbf{R} - \mathbf{R}_n|, \eta)] \right\}. \tag{5.21}$$

The explicit functional form of σ can be Gaussian, exponential, or a square well. The contribution of the first term to the potential is processed in reciprocal space; that of the second term, in direct space. The total potential:

$$U(\mathbf{R}) = U^{\mathrm{dir}}(\mathbf{R}, \eta) + U^{\mathrm{rec}}(\mathbf{R}, \eta) \tag{5.22}$$

is entirely independent of the Ewald sphere radius η.

In order to transform the Ewald method analytically from spherical to lamina shapes (Heyes 1980, 1981; Heyes and van Swol 1981) the charge distribution of the first term in equation (5.21) is Fourier-decomposed in terms of the reciprocal lattice vector $\mathbf{K}$. To arrive at a formulation applicable for a lamina shape instead of a sphere, with orientation $(hkl) \rightarrow (001)$, $\mathbf{K}$, $\mathbf{R}$ and $\mathbf{R}_n$ are decomposed into components $(\mathbf{k}, \mathbf{k}')$, $(\mathbf{r}, \mathbf{z})$ and $(\mathbf{r}_n, \mathbf{z}_n)$, respectively, with $\mathbf{k}, \mathbf{r}, \mathbf{r}_n \parallel (001)$, and $\mathbf{k}', \mathbf{z}, \mathbf{z}_n \perp (001)$.

The reader is spared the tedious details of the ensuing transformation from two spheres to two disk shapes in reciprocal and direct spaces. The resulting site potential of equation (5.22), due to a collection of, not necessarily consecutive, slices with index set $\{m\}$, is (Strom and Hartman 1989):

$$U^{\mathrm{rec}}(\mathbf{R}, \eta) = \frac{2}{A} \sum_{j=1}^{N} q_j \sum_{\lambda,\mu=-\infty}^{\infty}{}' \sum_m \exp[i\mathbf{k} \cdot (\mathbf{r}_j - \mathbf{r})] \exp(im\mathbf{k} \cdot \mathbf{c})$$
$$\times \psi(k, z_j - z + mV/A, \eta)$$
$$+ \frac{2}{A} \sum_{j=1}^{N} q_j \sum_m \psi(0, z_j - z + mV/A, \eta) \tag{5.23}$$
$$- \sum_{\lambda,\mu=-\infty}^{\infty} \sum_m U_0(\eta) \delta_{\mathbf{R},\mathbf{R}_j + \lambda\mathbf{a} + \mu\mathbf{b} + m\mathbf{c}},$$

where the prime on the first sum means $\lambda \neq 0$ or $\mu \neq 0$, and

$$U^{\text{dir}}(\mathbf{R}, \eta) = \sum_{j=1}^{N} q_j \sum_{\lambda,\mu=-\infty}^{\infty} {\sum_{m}}' [(d^j_{\lambda,\mu,m})^{-1} + I(d^j_{\lambda,\mu,m})],$$

$$I(d^j_{\lambda,\mu,m}) = -2\pi \left(\frac{1}{d^j_{\lambda,\mu,\nu}} \int_0^{d^j_{\lambda,\mu,m}} x^2 \sigma(x, \eta) \mathrm{d}x + \int_{d^j_{\lambda,\mu,m}}^{\infty} x\sigma(x, \eta) \mathrm{d}x \right) \qquad (5.24)$$

$$d^j_{\lambda,\mu,m} \equiv |\mathbf{R} - \mathbf{R}_j - \lambda\mathbf{a} - \mu\mathbf{b} - m\mathbf{c}|,$$

where the prime means that terms with $\mathbf{R} = \mathbf{R}_j + \lambda\mathbf{a} + \mu\mathbf{b} + m\mathbf{c}$ are excluded from the summation; $j = 1 \ldots N$ enumerates the charges in the unit cell of volume V, and A is the area on (001). A note on the atomic position vectors $\mathbf{R}_j$ is now in order. They are given by the unit cell coordinates plus appropriate lattice translations that determine the slice boundaries; in other words, the lattice translations inherent in the $\mathbf{R}_j$ determine the composition of the *slice cell*, as explained in Section 5.4.3.

The electric field at site $\mathbf{R}$ is obtained by differentiating the above expressions (Strom and Hartman 1989).

5.5.5. Description of the Data

The analytic closed-form expressions of the Ewald formulation enable the exact computation of the electrostatic potential and electric field vector at any point in space, that is, in the bulk, on or away from the surface, whether they coincide with atomic positions or not. Since these expressions are specifically developed for lamina shapes, the Coulomb and van der Waals interactions of any arbitrary collection of slices can be readily obtained (Strom and Hartman 1989). A stacking of infinitely many consecutive slices can model a half-infinite structure. Other collections of slices can readily be devised to model, for example, stacking faults or other experimental situations.

The conditions of validity of these computations are as follows. The crystal surface is in contact with vacuum, so its interaction with any adjoining substance is neglected. Therefore, the most pronounced surface relaxation or reconstruction effect that is caused by the adjoining fluid is excluded from the treatment. The less important surface relaxation in vacuum, which some structures are known to undergo, is also neglected. In other words, the computed laminas are assumed to be the same at the surface as in the bulk. The molecules are assumed to be rigid. This means that the atomic positions are not allowed to vary and that internal molecular degrees of freedom cannot be accommodated. Finally, the molecular charge distributions are modeled by an appropriate set of partial atomic charges. These charges depend on the choice of force field. In the absence of guidelines to decide which force fields should be employed, one can repeat the computation using a number of force fields, and subsequently compare the theoretical results with experiment as far as possible.

The required data are classified in two categories:

1. *Structural or bulk-related data.* Atomic coordinates are obtained from databases. One or several partial atomic charge sets are obtained by semiempirical Hamiltonian methods developed for the purpose. Lennard-Jones potential depth and equilibrium distance are obtained from databases. The atoms belonging to a growth unit are specified, forming a so-called mass center, that is, a cluster of one or several atoms considered to be permanently bonded to each other; this is usually a molecule. If relevant, hydrogen atoms serving as hydrogen bonds are specified, together with all alternative hydrogen bond distributions in the bulk, to be explained in the note below.

2. *Orientation-dependent data.* The cell indices of the various mass centers or growth units provide information on the particular growth front under study, specifying the slice cell as indicated in Section 5.4.3. (These input cell indices are applied to all the atoms in the cluster defining a growth unit, since by definition a growth unit may not be cleaved.) It should be stressed that the orientational data specify in general a layer satisfying the definition of *slice*, which may – but need not – satisfy the stronger condition of *F slice*.

5.5.5.1. *Note on the Data in Category 1 Concerning Hydrogen Bond Distributions*

Hydrogen bonds are points in the structure graph with degree 2 (Section 5.4.6). Hydrogen atoms functioning as bonds between *a pair* of molecules are not permanently attached to either molecule. So, in order to determine the alternative distributions, these hydrogen atoms could be treated as individual mass centers. However removing them from the crystal graph and replacing them with links, as is done in ADP, does not lead to any irrevocable loss of combinatorial information. The various hydrogen-bond distributions at the surface can be readily generated from the bulk information.

It is noteworthy that all of the above described data are physical quantities. The absence of adjustable parameters leads to model-independent results.

5.5.6. *Description of the Software and Technical Specifications*

Programs SURFENERGY and SURFGRID (Strom and Hartman 1989; Strom 1998) are composed of the SURFPOT modules developed by Strom between 1985 and 1988 at the University of Utrecht. These modules perform the aforementioned basic Ewald computations and meet the specifications outlined in the previous paragraph. They have been extensively tested and fully operational for almost a decade. In addition to energy computations of *F* slices (Hartman and Strom 1989; Strom et al. 1995*a*, 1997), they have been used for studying habit modifications due to surface interaction with aqueous solutions, by considering the response of the dipole moment of the water molecule to the electric field caused by the lattice, see for example Van der Voort (1991) and Van der Voort and Hartman (1988, 1990*a*, 1991).

Symmetry properties do not enter the input of these modules, but are reflected in the output.

The packages composed of the SURFPOT modules are operational on PCs in an OS2 (and MS DOS) environment and are characterized by: general applicability for all types of crystals, not just molecular crystals, employing van der Waals and Coulomb contributions, and delivering exact results; total absence of any model-dependent, or ad hoc, or adjustable parameters, whether explicitly specified or internally generated; excellent convergence properties within at most 20 Å (so that bulk or semi-infinite-structure properties can be readily obtained). The accuracy and model independence offered is particularly suited for studying polar surfaces.

SURFGRID is a related package, composed of SURFPOT modules, suitable for use as a tool in assessing the influence of the mother phase on the morphology. Preliminary output includes results for contour plots of the electrostatic potential and electric field vector caused by an infinite stacking of laminas, sampled on a series of grids; the points on these grids represent the closest approach between the van der Waals radii of the *F* slice under study and the radius of the particles of the solution.

SURFENERGY can handle in a single run per (*hkl*) multiple alternative partial atomic charge sets and multiple atomic slice compositions, selected from an input reservoir. The final output consists of the slice, attachment, crystal (= constant), dipole, supplementary-slice, surface, and specific-surface energies of the Coulomb, van der Waals, and total interactions for every charge set (equations (5.3*a,b*)). The final output is written to a file and to the printer. Optional intermediate output may be obtained of the individual interaction energies between the slices.

Some SURFENERGY runs on ADP are reported in the next section. The primitive unit cell contains sixteen atoms. Each molecular slice composition corresponds to six atomic compositions, that is, six ways of distributing hydrogen bonds at the surface. A typical ADP energy run includes six charge sets and about forty atomic slice compositions.

5.6. Case Study 1: Structural Morphology of ADP-type Structures

In the following sections the *F*-face analysis of ADP- and KDP-type structures is elaborated. The energy quantities of the ADP surfaces take account of the alternative hydrogen-bond distributions at the surface and utilize various atomic point-charge assignments. The constructed theoretical growth habit diverges from the experimentally observed one, indicating the strong effect of environmentally induced habit modification.

5.6.1. *F-face Analysis*

As a result of the presence of the P—P *e* bonds, ADP must be considered as a mixed structure rather than a strictly ionic one. For that reason, the *F* slices were

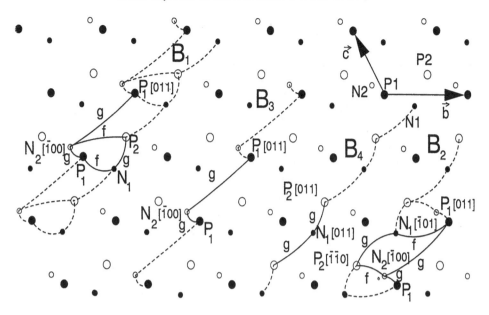

Fig. 5.8. Schematic (indices in the primitive cell) of some parallel PBCs of ADP, projected on a plane parallel to the common direction. It shows that PBCs like B_1 and B_2 could be bundles of smaller chains which are themselves PBCs, like B_3 and B_4. These PBCs lack axial symmetry, having dipole moment components perpendicular to their directions. Large open circles, P_1; large filled circles, P_2; small open circles, N_1; small filled circles N_2. (Reprinted from Strom et al. (1995b, p. 148), with permission from Elsevier Science.)

computed using the general method, meaning the first version of FFACE. The adjacency matrix A_{ADP} of equation (5.4) was used as input. For the first run the $\{f, g\}$ bonds were taken into account by letting $f = g = 1$ and $h = e = 0$ in A_{ADP}; in the second run, the $\{f, g, e\}$ bonds were considered by changing to $e = 1$; Fig. 5.3 is an exact pictorial representation of this second input.

One more item entering the input data of ADP concerns the stoichiometric combinations of molecules, which are: $\{P_1, N_1\}$, $\{P_1, N_2\}$, $\{P_2, N_1\}$, $\{P_2, N_2\}$, and any further combinations thereof.

Here we illustrate some features, like irreducible borders and symmetry relations, of the F slices obtained from the $\{f, g\}$ bond set and draw some conclusions from a comparative analysis of the results obtained from the $\{f, g\}$ and $\{f, g, e\}$ bond sets of ADP (Strom et al. 1995b). We limit attention to symmetrically unique face orientations. (Quantitative results are expressed in terms of networks (Strom et al. 1995b, Appendix B).)

Some of the PBCs obtained from the $\{f, g\}$ bond set are shown in the structure projection of Fig. 5.8, illustrating the composite nature of PBCs as bundles of chains or smaller PBCs. All the F slice compositions obtained from the $\{f, g\}$ bond set are shown in the structure projections of Fig. 5.9. In the upper right-hand corner the single additional F slice resulting from the $\{f, g, e\}$ bond set is

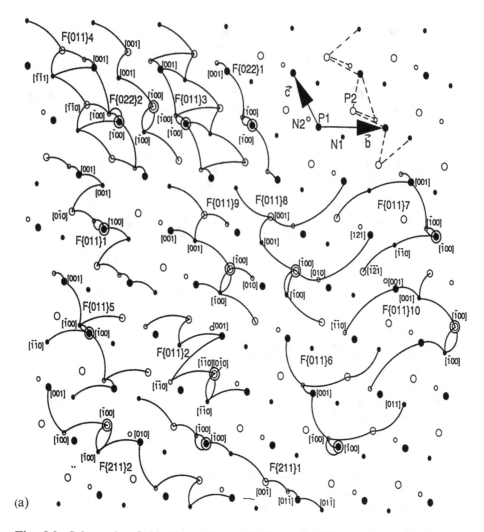

(a)

Fig. 5.9. Schematic of the $\{f, g, e\}$-based F slices of ADP; symbols as in Fig. 5.8. Translation indices are given for the primitive cell, in the conventional form. (*a*) The eleven $\{011\}$ and two $\{022\}$ F slices are perpendicular to the projection plane, the two $\{211\}$ F slices are oblique. The slice indicated by dashed lines in the upper right-hand corner becomes disconnected when the e bond is removed. (*b*) The nine $\{020\}$ F slices are perpendicular to the projection plane and the two $\{031\}$ slices are oblique. (Reprinted from Strom et al. (1995*b*, p. 150), with permission from Elsevier Science.)

shown with dashed lines. Note the widespread lack of any symmetry relation between opposite boundaries of each individual F slice.

The upper left-hand corner of Fig. 5.9(*a*) shows how two half $\{022\}$ F slices can alternate to form two full $\{011\}$ F slices, which are very compact consisting of only cores, having no irreducible borders. The $\{011\}$ F slices numbered 5–10 have irreducible borders of the form of single and double dangling bonds. In $\{020\}$ there exists one compact symmetry singlet, $F\{020\}1$ consisting of a core,

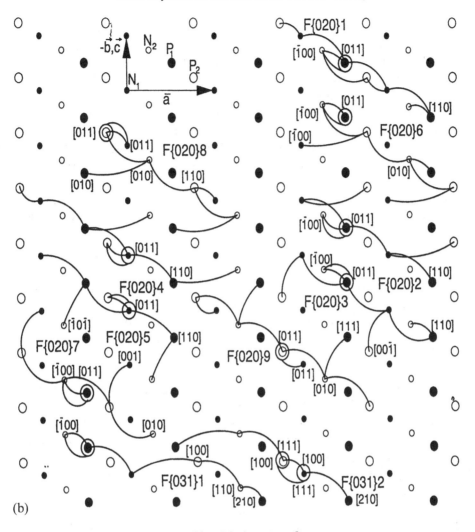

Fig. 5.9 (*continued*)

the remaining *F* slices being pairwise symmetric and having dangling bonds. (Note that the {020} *F* slices are not half-slices but full slices in the primitive cell.) In a projection exactly perpendicular to {211} (in the figure the plane of {211} is oblique to the plane of the paper), it would be possible to see that the {211} *F* slices form a symmetry doublet. An analogous conclusion is reached for {031} in Fig. 5.9(*b*). The symmetry properties of the *F* slices, as well as the theoretical implications of the pairwise-symmetry of the {031} and {211} doublets are touched on in Sections 5.8 and 5.9.

A brief summary of the {*f*, *g*}-based treatment (Strom et al. 1995*b*) against the background of earlier results (Aquilló and Woensdregt 1984, 1987) is now given. The form {110} contains otherwise valid networks, but is forbidden by the selection conditions of the space group. The form {112} is classified as an *S* form in both the earlier (Aquilló and Woensdregt 1984, 1987) and recent

(Strom et al. 1995b) studies. Forms {220} and {224} emerged in the recent study as S forms. Some additional F slices emerged in the F forms predicted earlier, namely {020}, {011}, and {022}. Also three new F forms emerged, {031}, {211}, and {21$\bar{1}$}, containing two F slices each. Form {031} was earlier classified as S; forms {211} and {21$\bar{1}$} were earlier classified as K.

The following numbers of full F slices emerged from {f, g} in symmetrically unique faces: nine in (020), ten in (011), two in (211) (or (21$\bar{1}$)), and two in (031), amounting to a total of twenty-four. There are two F half-slices in (022). The analysis was repeated after adding the e bond to the {f, g} set. Not surprisingly, the {f, g, e} bond set produced many more PBCs than the {f, g} bond set (Strom et al. 1995b). However, the {f, g, e} bond set produced ultimately the same set of F slices as {f, g}, except for a single additional F slice, shown in the upper right-hand corner of Fig. 5.9(b).

The impact of adding the weak e bond to the {f, g} set can be explained as follows: The massive increase in the number of new PBCs is a direct result of the increase in the coordination number. Each cation has coordination six with reference to the {f, g} set of bonds, but coordination ten with reference to the {f, g, e} set of bonds. However, the additional PBCs did not produce any new F-face orientations or F-slice compositions, contrary to what one might have expected. Instead, the effect of these additional PBCs was merely to add more bonds to the F slices already obtained from the {f, g} set. The main effect of adding the e bond was to incorporate the irreducible borders, seen in the F slices of Fig. 5.9, into the cores of the already existing F slices.

Two conclusions may be drawn. The first conclusion (Strom et al. 1995b) confirms the intuitive statement (Aquilló and Woensdregt 1984, 1987) that the e bond is indeed negligible with respect to the {f, g} set of bonds. The second conclusion concerns a property of the internal structure of the ADP F slices, confirming the dependence of the irreducible borders on the choice of coordination sphere, as explained in Section 5.4.3.1.

Another conclusion concerning the ADP structure follows from the F-face analysis using the {f, g, h} bond set. The h bond need not be negligible with respect to the {f, g} bond set. It results in a more extensive list of F-slice compositions, as well as F-slice orientations, and is responsible for the F character of the {112} form, as discussed further in the next section.

In closing this section, we have furnished examples for the statements made earlier (Section 5.4.2) regarding the choice of crystallization bonds. There we anticipated that overspecifying the first coordination sphere would lead to model-independent results, while underspecifying it would lead to inadequate results.

5.6.2. Energy Computations of the ADP Slices

The primitive portion of the unit cell on the atomic scale is shown in Fig. 5.2(a) and the atomic structure of the conventional unit cell, as defined by Aquilló and Woensdregt (1984, 1987), is shown in Fig. 5.2(b). The fractional axial atomic

Table 5.1. *Lennard-Jones atomic parameters of ADP (force field CHARMm); $|\varepsilon|$ is the magnitude of the potential well depth in kcal/mol, and r^* is the equilibrium distance in Å*

Atom	$|\varepsilon|$	r^*	Atom	$|\varepsilon|$	r^*
P	0.1	1.8	N	0.15	1.65
O	0.1521	1.55	H, HB	0.0498	0.8

Table 5.2. *Partial atomic charges in units of the elementary charge e, obtained by various ab initio and semiempirical mechanics procedures*

Atom	A	B	C	D	E	F
P	0.7444	0.7058	1.64	2.52	1.2234	1.6645
O	−0.4651	−0.4813	−1.16	−1.38	−1.0659	−1.1663
N	−0.7476	−0.8052	0.9976	−0.0944	0.0642	0.2383
H	0.2589	0.2999	0.0006	0.2736	0.244	0.1906
HB	0.414	0.4125	1	1	1	1

Note: Charge equilibration Q_{eq} (force field CHARMm) method with periodic boundary conditions: *A*, *B* (Rappé and Goddard 1991); *C–F*, MOPAC 6 with geometry optimization using ESPD charges; *C*, PM3 Hamiltonian; *D*, AM1 Hamiltonian; *E*, MNDO; *F*, MINDO 3 (Stewart 1995).

coordinates were taken from Khan and Baur (1973). The remaining input data for phosphorous (P), oxygen (O), nitrogen (N), hydrogen bonded with the nitrogen (H) and hydrogen serving as hydrogen bond between the oxygens (HB) are reported in Tables 5.1 (Lennard-Jones parameters) and 5.2 (atomic charges). It should be noticed that the atomic partial charge sets *C–F* use normalized values, because the hydrogen bond charge is fixed to $+1$, and hence these point charges have unduly high values.

There are six possible ways of distributing the four hydrogen-bond hydrogens, two at a time, between the two PO_4 groups to produce two H_2PO_4 groups. In other words, the number of *atomic* slice configurations is a factor six larger than the number of *molecular* slice configurations.

Program SURFENERGY (Strom and Hartman 1989; Strom 1998) performed energy calculations on the atomic configurations of the following slices: the

Table 5.3. *Sum of Coulomb and van der Waals energy quantities in kcal/mol of the two {211} pairwise symmetric ADP F slices (F1 and F2) using charge set A, with six surface hydrogen-bond distributions (HBD1–HBD6) per slice (Slice, attachment and supplementary-slice energies are listed to one significant figure less than surface and crystal energies)*

		Slice	Attachment	Supplementary slice	Surface	Crystal
F1	HBD1	−967.315	−58.571	−81.187	−1083.483	−1048.502
	HBD2	−945.884	−26.369	−102.637	−997.996	−1048.521
	HBD3	−969.265	−56.636	−79.252	−1080.705	−1048.517
	HBD4	−977.220	−70.709	−71.315	−1116.202	−1048.535
	HBD5	−970.901	−67.382	−77.617	−1100.912	−1048.518
	HBD6	−984.634	−63.323	−63.298	−1111.172	−1048.563
F2	HBD1	−983.511	−64.391	−64.997	−1113.608	−1048.508
	HBD2	−976.312	−61.972	−72.207	−1098.198	−1048.519
	HBD3	−983.575	−64.343	−64.949	−1111.215	−1048.524
	HBD4	−962.409	−63.498	−86.114	−1086.672	−1048.523
	HBD5	−939.971	−32.282	−108.550	−1001.742	−1048.521
	HBD6	−967.937	−57.961	−80.578	−1081.805	−1048.515

twenty-four slices with maximum thickness d_{hkl} summarized in Section 5.4 and shown in Fig. 5.9 (twenty-three of these have F character based on the $\{f, g\}$ bonds, one is S based on the $\{f, g\}$ bonds, but F based on the $\{f, g, e\}$ bonds); the planar slice (112) having S character on the basis of $\{f, g\}$, but F character on the basis of the $\{f, g, h\}$ bond set; a randomly chosen planar slice in (332) that is not an F slice either on the basis of $\{f, g\}$ or $\{f, g, h\}$. (In this section the face indices refer to the conventional I cell.)

In total, we computed Coulomb and van der Waals contributions for the slice, attachment, crystal (= constant), dipole, supplementary-slice, surface, and specific-surface energies of $(24 + 2 = 26)$ slices, times six hydrogen bond distribution schemes, times six charge sets, giving 936 slices in all! Table 5.3 shows the added van der Waals and Coulomb contributions to the slice, attachment, supplementary-slice, surface, and crystal energies, for only charge set A of the pairwise symmetric (211) F slices. The twelve configurations presented in that table refer to two F slices with six hydrogen-bond surface distributions per F slice. The crystal energy is calculated independently for each surface configuration, so the constancy of that tabulated quantity reflects the accuracy of the computation.

A provisional analysis leads to the following conclusions: the various partial atomic charges supplied by molecular mechanics give entirely consistent results;

Table 5.4. *Illustration of theoretical habit construction of ADP, based on optimal attachment energies*

$\{hkl\}$	R_{hkl}^{in}	R_{hkl}^{cr}	IMI
$\{011\}$	25	*	
$\{020\}$	42	35.24	0.161
$\{112\}$	59	28.90	0.510
$\{031\}$	70	33.45	0.522
$\{211\}$	86	43.21	0.497
$\{332\}$	86	45.13	0.475

Note: Dimensionless input central distances R_{hkl}^{in}, critical central distance R_{hkl}^{cr} and implicit morphological importance (IMI) (Strom 1979). An asterisk indicates the form occurring on the theoretical growth habit.

the different hydrogen-bond distributions of the slices of the various faces induce dipole moments perpendicular to the face, causing the dipole energy and the other energy quantities to fluctuate, sometimes substantially. The van der Waals contributions to the energies in charge sets A and B are extremely small compared to the Coulomb; in the last four charge sets, they are entirely negligible – however, these charges are considered less meaningful as a result of the afore-mentioned normalization.

The behaviour of the (112) slices with respect to energy is comparable to the behaviour of the rest of the slices. All slices except for (332) have negative attachment energies, indicating attraction, regardless of the hydrogen-bond distribution. However, in the (332) slices, the sign of the attachment energy is sensitive to the choice of hydrogen-bond distribution. Although the attachment energy in (332) is chiefly negative, positive attachment energies indicating repulsion are occasionally encountered, accompanied by large magnitudes of the dipole moment perpendicular to the face, indicating polar surface instability.

Restricting attention to the slice compositions that would be most stable in vacuum – minimizing the dipole moment perpendicular to the given face – we obtain the following minimum magnitudes of attachment energy in kcal/mol for each considered face: $\{011\}$, 25; $\{020\}$ 42; $\{112\}$ 59; $\{031\}$ 70; $\{211\}$ 86; $\{332\}$ 86.

5.6.3. *Construction of the Theoretical Growth Form of ADP Considering van der Waals and Coulomb Interactions with Several Partial Atomic Point-Charge Sets*

The results of the energy computations of ADP illustrated in Section 5.6.2 are compiled in Table 5.4.

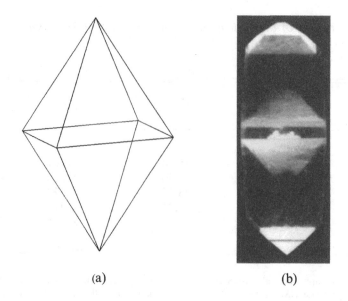

(a)   (b)

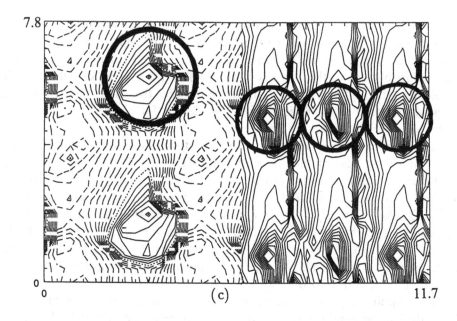

(c)

Fig. 5.10. (*a*) Theoretical growth habit of ADP. (*b*) Photographs of ADP crystals grown out of aqueous solution by Y. S. Wang (1983–1993). (*c*) Left: equipotential curves at locations accessible to a particle with radius 2.0 Å on an undulated surface on the KDP prism. Positive, zero and negative potential indicated by solid, dotted and dashed lines respectively. Right: equifield curves at locations accessible to a particle with radius 1.4 Å. Adsorption sites of an anion at a potential maximum and water molecules at field maxima in the upper portion.

The theoretical growth habit follows when the theoretical growth rates are taken proportional to the minimum magnitudes of the attachment energy of the most stable slice compositions in each face. The input growth rates R_{hkl}^{in} are shown in Table 5.4 for charge set A.

The theoretical growth habit is constructed using as input the cell parameters, point-group symmetry, and forms {hkl}, accompanied by input central distances R_{hkl}^{in}. Whether or not a face appears on the growth form depends on the geometrical characteristics of the adjoining faces and is expressed by the critical central distance R_{hkl}^{cr}. The implicit morphological importance IMI $\equiv R_{hkl}^{cr}/R_{hkl}^{in}$ (Strom 1979) extends the description of a growth habit beyond the forms actually present. It is the fraction of the input central distance in excess of the critical value. When IMI > 1, the face is present on the growth form; $0 <$ IMI < 1 shows how close the face is to the threshold of appearance at IMI $= 1$. The IMI parameter is a well defined quantity, the value of which can be determined under all circumstances (Strom 1979, p. 186).

The theoretical growth form construction is carried out by program CRYSTALFORM (Strom 1979, 1998), the input and numerical output of which are shown in Table 5.4 (program CRYSTALDRAW provides drawings of crystals at equal volumes). Figure 5.10(a) shows that the pyramidal form {011} dominates the structural habit, whereas the prism in Fig. 5.10(b) dominates the habit resulting from solution growth. This discrepancy is attributed to the habit modification caused by the interaction between crystal surface and aqueous solution. Figure 5.10(c) shows adsorption sites of (left) a KDP anion and (right) water molecules on the KDP prism surface covering 3×2 primitive cells.

5.7. Case Study 2: Gel-Induced Fractal Habit of Ammonium Chloride via Pseudo-flat Surface Reconstruction

The morphology of ammonium chloride illustrates the predictive power of the PBC theory. A pseudo-F face is predicted to play a dominant role in the habit by triggering reconstruction. That same mechanism is also expected to function as an agent of epitaxial nucleation. This effect, called Solution Induced Reconstructive Epitaxial Nucleation (SIREN), explains not only the important morphological features of the mineral, but also the fractal morphology of ammonium chloride as a gel-induced habit modification.

5.7.1. *Flat and Pseudo-flat Surfaces*

The NH_4Cl structure under study (Vainstein 1956) crystallizes in the space group $Pm\bar{3}m$ with one formula unit $Z = 1$ in the asymmetric portion of the unit cell. In this work we use $a = 3.8756$ Å (International Tables for X-ray Crystallography 1969). The N and Cl atoms are located at $(0, 0, 0)$ and $(\frac{1}{2}, \frac{1}{2}, \frac{1}{2})$, respectively. The four H atoms belonging to the NH_4 molecule surround N in tetrahedral

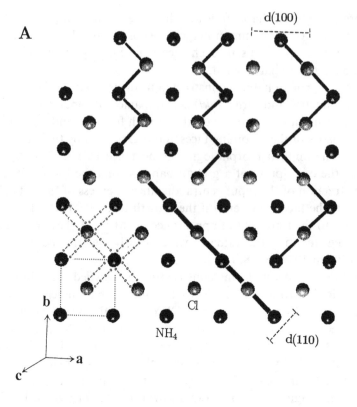

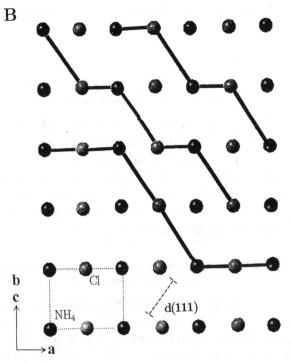

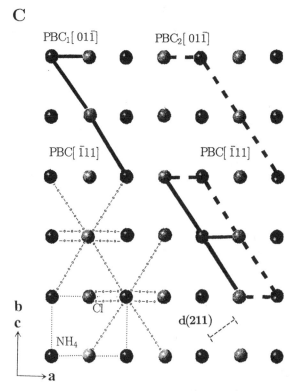

Fig. 5.11. Structure projections showing the bonds in the first coordination sphere and the growth layers of the NH_4Cl. (A) Singlet dipoleless F slice d_{110}, doublet of polar F slices d_{100}, and a possible reconstructed F slice $2d_{100}$, projected on (001). (B) Doublet of polar F slices d_{111} and a possible reconstructed F slice $2d_{111}$ projected on (01$\bar{1}$). (C) Pseudo-F slices d_{211} and the reconstructed F slice $2d_{211}$ projected on (01$\bar{1}$).

coordination. Vainstein (1956) gives positions of H at (0.146, 0.146, 0.146), in which case the multiplicity of the space group generates eight hydrogen positions corresponding to each ammonium. The distribution of the four hydrogen atoms surrounding each ammonium must be taken as the eight symmetry-generated sites occupied statistically. The nominal atomic charges are $q_{Cl} = -1$, $q_N = -3$, and $q_H = +\frac{1}{2}$ (instead of $q_H = +1$) for each of the statistically distributed hydrogens.

Clearly the growth units of NH_4Cl are the NH_4 cations, represented by their mass centers at the N positions, and the Cl anions.

The crystallization bonds follow from a histogram of the number of N—Cl pairs versus separation distance. The number of cation–anion pairs reaches a peak of eight at a distance $d = 3.5152$ Å and a first minimum equal to zero at $d \sim 3.52$ Å, which we take as the radius of the first coordination sphere (see Fig. 5.11). (The second and third coordination spheres are marked by peaks of twelve and six pairs corresponding to zero minima at radii $d \sim 5.8$ Å and $d \sim 8.2$ Å, respectively.)

Table 5.5. *Coulomb energies of the F faces of ammonium chloride in the point-charge model in which N is surrounded by eight statistically distributed H atoms*

F face	E_{sl} $(e^2 \text{Å})$	E_{att} $(e^2 \text{Å})$	E_{dip} $(e^2 \text{Å})$	E_{suppl} $(e^2 \text{Å})$	E_{sp} $(e^2/\text{Å}^3)$	$d_{\perp}$ $(e \text{Å})$
(110)	-7.6785	-0.08162	0	-0.08162	-0.3690	0
(100)	-7.3835	$+0.02869$	-0.4053	-0.3766	-0.4877	±1.9378
(111)	-7.6204	-0.00464	-0.1351	-0.1397	-0.2939	±1.1188

Note: Point charges taken as $q_N = -3$, $q_H = +0.5$, $q_{Cl} = -1$. e is the elementary charge. Slice E_{sl}, attachment E_{att}, dipole E_{dip} supplementary-slice $E_{suppl} \equiv E_{cr} - E_{sl}$, and specific surface E_{sp} energies. The constant crystal energy is $-7.7601\ e^2 \text{Å}$.

PBCs are found in directions $\langle 100 \rangle$, $\langle 110 \rangle$, and $\langle 111 \rangle$. Forms $\{100\}$, $\{110\}$, $\{111\}$, and $\{211\}$ contain intersecting PBCs. Of these, the first three faces are flat as they contain F slices, whereas $\{211\}$ contains only pseudo-F slices. Infinitely many other faces contain either no PBCs at all (K character) or PBCs in only one direction (S character).

Figure 5.11 shows (A) the hexahedron $\{100\}$ F slices and the dodecahedron $\{110\}$ F slices in a structure projection on (001) and (B) the octahedron $\{111\}$ F slices in a structure projection on $(01\bar{1})$. The singlet $\{110\}$ is dipoleless. $\{100\}$ and $\{111\}$ have pairwise symmetric F slices, the so-called doublet configurations. The F slices within each doublet have equal, oppositely directed surface dipole moments. Table 5.5 lists energy quantities in the electrostatic point-charge model. The van der Waals contributions turned out to be negligible and were left out of consideration.

In Fig. 5.11(C) it is easy to see in a structure projection on $(01\bar{1})$ why the would-be networks in face $\{211\}$ could never materialize as valid growth layers. There are three PBCs parallel to $\{211\}$: $\text{PBC}\langle\bar{1}11\rangle$ is the chain $NH_4[000] \rightarrow Cl[\bar{1}00] \rightarrow NH_4[\bar{1}11]$; $\text{PBC}_1\langle 01\bar{1}\rangle$ is the chain $NH_4[000] \rightarrow Cl[00\bar{1}] \rightarrow NH_4[01\bar{1}]$, while $\text{PBC}_2\langle 01\bar{1}\rangle$ is the chain $NH_4[000] \rightarrow Cl[\bar{1}0\bar{1}] \rightarrow NH_4[01\bar{1}]$.

The combination $\text{PBC}\langle\bar{1}11\rangle \times \text{PBC}_1\langle 01\bar{1}\rangle$ involves a lattice translation $[10\bar{1}]$ between indispensible Cl anions. Similarly, the combination $\text{PBC}\langle\bar{1}11\rangle \times \text{PBC}_2\langle 01\bar{1}\rangle$ involves a lattice translation $[100]$ between indispensible Cl anions. Removal of any of the indispensible Cl ions leaves the network disconnected, whereas, on the other hand, maintaining the out-face translations results in loss of the "flatness" characteristic. It is also easy to see from Fig. 5.11(C) that if a 2×1 reconstruction were to occur on the $\{211\}$ face, a highly stable, dipoleless, planar, and compact $2d_{211}$ F slice would arise, with an enormous energy gain as a result. Therefore, such reconstruction is considered plausible on theoretical grounds, both in vapour and in solution.

5.7.2. *Theoretical Predictions*

It can be seen from Fig. 5.11 and Table 5.5 that the dodecahedron is the most compact, energetically optimal (unreconstructed) growth form. The polar cubic and octahedral forms have similar surface structures and similar energy properties. Both have attachment energies close to zero, owing to the instability of the polar surfaces. {100} has higher dipole energy than {111}, to the extent of resulting in a positive amount of attachment energy, indicating repulsion between the growth layer and the existing structure.

The 2×1 reconstructed dipoleless F slices in Fig. 5.11(A) and (B) are combinations of the two polar {100} F slices and the two polar {111} F slices. If reconstruction were to occur, it would lead to similar undulating configurations $2d_{100}$ and $2d_{111}$. The polar F slices {100} and {111} also show similarities as regards the interaction between slice boundaries and polar particles. The opposite boundaries of both slices are characterized by NH_4 molecules on one side and Cl on the other. It remains to be seen whether the NH_4 side or the Cl side interacts more strongly with the polar particles of the present experiment, that is, water molecules. Because of their similar bonding patterns at the surface, these faces should stabilize by similar mechanisms.

The PBC theory gives the following morphological predictions:

- The dipoleless dodecahedral form {110} should dominate in growth out of a nopolar medium.
- Furthermore, owing to the enormous energy gain achieved by the 2×1 reconstruction on {211}, that pseudo-F face could be widespread, regardless of the growth conditions.
- In growth out of a polar solvent, the polar forms {100} and {111} should dominate at the cost of the dodecahedron, by interacting with charged or polar particles – in the present case under study, the polar water molecules in the gel. The growth rates of these faces would be reduced by the need to dehydrate the surfaces at the solvent-accessible sites.
- Because of the similarity in bond structure and energy properties, both of these polar forms should feature prominently, unless some other mechanism operates to mask one of the forms in the presence of the other.

5.7.3. *Experimental Confirmation*

The Groth (1906) Atlas summarizes the global morphological features of the ammonium chloride mineral as follows: the dodecahedron {110} face dominates in growth out of sublimation; the less polar octahedron {111} face dominates in growth out of aqueous solutions; the more polar {100} face dominates in growth out of ionic solutions; finally, {211} is the most important form, occurring in almost all growth conditions. Indeed, observation confirms the prediction.

Figures 5.12 and 5.13 show fractal features of ammonium chloride grown out of gel. The dodecahedron, either by itself or in combination with the cube or

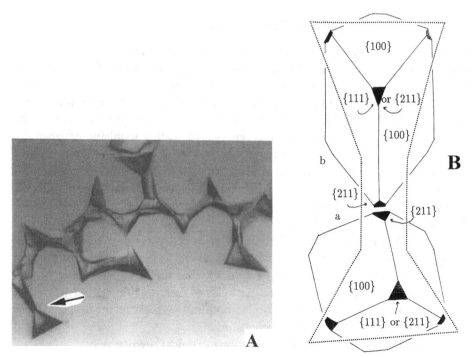

Fig. 5.12. (A) Optical image of fractal growth pattern of ammonium chloride, where {100} dominates and {111} is inhibited by continuous nucleation taking place via reconstruction on pseudo-F {211} surfaces. The arrow points to the geometrical details of the region of nucleation between crystallites a and b, drawn in (B). Crystallite b is elongated along the preferred growth direction ⟨111⟩, that is, the axis of the {211} trigonal pyramid. Dotted lines isolate the regions of crystallization visible in the photograph.

octahedron, does not match the impressions provided by the experimental growth forms.

The nucleation and twinning patterns observed in Figs. 5.13(A) and (B) show uninterrupted periodicity and consistency with respect to crystallographic orientation, as explained by M. Wang et al. (1998a, b) and Strom et al. (1999). The crystallographic orientation of the NH₄Cl crystallites were analyzed after classifying them in types a and b, shown in Figs. 5.12 and 5.13. The experimental results show that the a and b crystallites alternate strictly in Fig. 5.13, and more loosely in Fig. 5.12. Periodic alternating elongation of the b-type crystallites occurs consistently along the direction of preferred growth. The lower portions of the crystallites in both the optical and atomic force microscopy (AFM) images are absent, owing to mechanical constraints. The major form in Fig. 5.12 is the hexahedron, whereas the major form in Fig. 5.13 is the octahedron.

The theoretical discussion of the previous section gave a satisfactory explanation as to why the dipoleless, most compact, and energetically optimal dodecahedral form {110} does not occur, either prominently by itself or otherwise in combination with some other major form. The absence of {110} simply confirms the known properties of the gel: that it contains water molecules that have dipole moments.

We will show that the fractal morphology is foreseen by the theoretical mechanism of solution-induced reconstructive epitaxial nucleation (SIREN). We will argue as follows. First, in the crystallites of the hexahedral modification of Fig. 5.12, the morphology is dominated by {100}, while {111} is inhibited by nucleation. Second, in the crystallites of the octahedral modification of Fig. 5.13, the morphology is dominated by {111}, while {100} is inhibited by nucleation. Third, we will provide evidence that the same nucleation mechanism operates in both cases and is triggered by the 2×1 reconstruction on the pseudo-F {211} surfaces. In both cases the axis of the {211} pyramidal faces is the direction of preferred growth and crystallite elongation.

The preferred direction of growth in Fig. 5.12 is $\langle 111 \rangle$. The orientations of the crystallites in the photograph show clearly that the nucleation plane cannot be {111}. The geometrical situation is shown in Fig. 5.12(B). The geometrical details are replicated when the nucleation planes are taken as the opposite faces of the {211} trigonal pyramid. Thus the (211) face of, say, crystallite a and the $(\bar{2}\bar{1}\bar{1})$ face of crystallite b form the $2d_{211}$ double layer composed of the two d_{211} pseudo-F slices, indicated by the thick solid and dashed bonds at the top of Fig. 5.11(C).

By analogy, the preferred direction of growth in Fig. 5.13 is $\langle 100 \rangle$. The geometrical details of the AFM images are not consistent with a nucleation plane along {100}, but with two faces of the tetragonal pyramid {211} belonging to crystallites a and b. As shown in Fig. 5.13(C), the nucleated faces are rotated about the pyramidal axis relative to each other in such a way that the direction $[\bar{1}02]$ of one of the crystallites coincides with $[01\bar{1}]$ of the other partner crystallite. The figure also indicates how nucleation evolves, producing either mirror twinning or a form of single branching. This phenomenon can be explained in detail (M. Wang et al. 1998a, b; Strom et al. 1999) on the basis of the structural properties of anisotropic nucleation-limited aggregation.

5.7.4. *Conclusion and Future Work*

In conclusion, in the case of ammonium chloride the PBC theory provided a gratifying prediction of the global mineral habit found in nature, as well as the agent triggering epitaxial nucleation in gel-grown fractal crystallites. Surface reconstruction by means of the {211} pseudo-flat faces is a strong and stable mechanism, operating independently of the properties of the mother phase. The emergence of fractal features in the case of growth out of gel are attributed to the local action of the supersaturation arising from the presence of the gel.

As regards the fractal habit, more detailed experimental work is necessary to investigate why in the case of Fig. 5.12 it is the cubic face that dominates while the octahedral is inhibited by {211} whereas in Fig. 5.13, on the other hand, the {211} pyramid enables the octahedral modification to dominate by inhibiting the cubic form. Another task should be to investigate why, in the octahedral

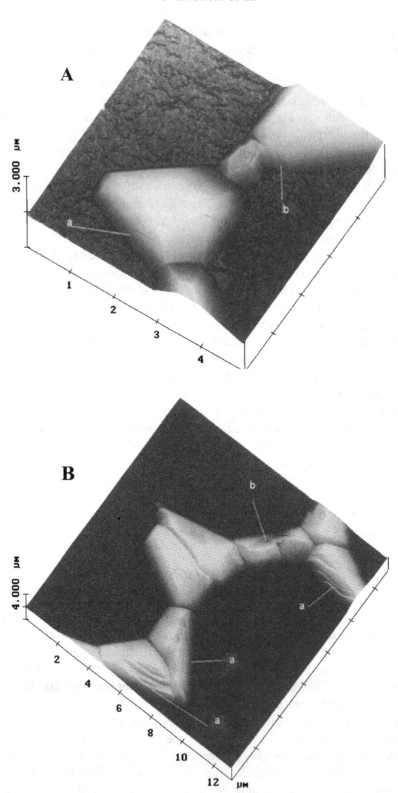

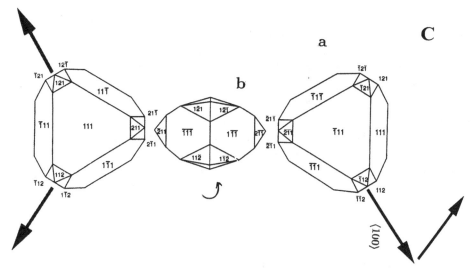

Fig. 5.13. (A–B) AFM images of octahedral habit modification of ammonium chloride, where {111} dominates and {100} is inhibited by reconstructive nucleation on pseudo-*F* {211} surfaces. Crystallite types *a* and *b* are discernible. (A) Region of nucleation between crystallite *a* and elongated crystallite *b*. (B) Twinning and branching in the fractal pattern. (C) Geometrical details of the nucleation and pattern formation of (A) and (B). Preferred growth direction ⟨100⟩ is the axis of the {211} tetragonal pyramid. Nucleation occurs on opposite {211} faces after rotation along that axis.

modification, the epitaxial nucleation is accompanied by a rotation about the axis of the pyramid, whereas in the cubic case we have continuous epitaxial nucleation. We suspect that both problems may be treated by studying the interfacial behaviour of the hydrogen atoms surrounding the ammonium at the surface and their interaction with the water molecules of the gel.

5.8. Interfaces and Roughening

5.8.1. *Introduction on Roughening Models*

The principle integrating the preceding models for the interface between crystal and mother phase with kinetic crystal-growth models is the concept of surface roughening or thermal roughening (Bennema 1993; Bennema and van der Eerden 1987). This means that below a critical temperature T^R a surface with an orientation (hkl) is flat on an atomic scale and also on a macroscopic scale. Such a surface will grow by a layer mechanism (spiral growth or 2-D nucleation). Above the critical temperature T^R a surface with orientation (hkl) is rough on an atomic scale and, macroscopically, a rounded-off face without a specific orientation (hkl) will result. In this case, growth occurs by means of a continuous mechanism rather than a layer mechanism. Whether a specific orientation (hkl)

has a T^R equal to zero ($T^R = 0$ K) or larger than zero ($T^R > 0$ K) depends on whether or not for the orientation (hkl) a connected net (network) (see Section 5.4) can be identified. In order to disconnect a connected net, a positive cleavage or cut energy is needed along all directions [uvw] parallel with the net. This implies that for all directions parallel to the connected net, the edge energy is positive. If in only one direction [uvw] (or more directions) of a supposedly connected net the edge (free) energy equals zero, such a (nonconnected) net will have no roughening transition; this can be put formally as $T^R = 0$ K. Such a face will not grow with a layer mechanism; only growth with a continuous mechanism would be possible. For a connected net $T^R > 0$ K and growth proceeds via a layer growth mechanism, provided $T < T^R$ (T is the actual temperature).

5.8.2. *First-Nearest-Neighbour Models versus Long-Range Models*

Recently surveys have appeared on an integrated theory consisting of the statistical mechanical roughening transition theory and the crystallographic morphological Hartman–Perdok theory (Bennema 1993, 1995, 1996; Bennema and van der Eerden 1987).

The Hartman–Perdok theory is, first of all, a theory based on chains involving links between first-nearest-neighbour growth units, characterized by proximity, and referred to as crystallization bonds in Sections 5.3 and 5.4.

As discussed in Section 5.5, for ionic crystals the relevant energy quantities, with E_{hkl}^{att} and E_{hkl}^{slice} as the most important, are calculated with sophisticated formalisms implemented in software for ionic point charge models, taking account of both short- and long-range interactions up to infinity. The employed Madelung or Ewald summation techniques show appropriate convergence properties. So, in the framework of the present chapter, the following paradoxical situation occurs:

- On one hand, for the calculation of E_{hkl}^{att} and E_{hkl}^{slice} all short- and long-range interactions are used; while
- On the other hand, the classification into F, S and K faces is based essentially on the assumption that local short-range events dominate crystallization, having as outcome the first coordination sphere.

The experimental situation shows that in the case of crystal structures with dominant first-nearest-neighbour van der Waals or covalent interactions, as well as in the case of ionic interactions, only F faces determined by first-nearest-neighbour bonds occur. Researchers viewing this situation as paradoxical face the challenge of finding an explanation in the context of modern morphological theories. In such theories the concept of connected net or F slice discussed earlier plays a key role.

5.8.3. *Actual Bond Energies at the Interface*

In the original Hartman–Perdok theory the links defining the periodic bond chains can be interpreted as bond energies Φ_i^{ss}; these may be viewed as energies resulting from a thought experiment where growth units in the crystal structure are brought together from positions infinitely far apart. As shown in previous papers (Bennema 1993, 1995, 1996; Bennema and van der Eerden 1987), actual bond energies occurring at the interface between crystal and fluid mother phase are expressed by

$$\Phi_i = \Phi_i^{sf} - \frac{1}{2}(\Phi_i^{ss} + \Phi_i^{ff}), \tag{5.25}$$

where the labels sf, ss, ff refer to the energies of solid–fluid, solid–solid and fluid–fluid bonds, respectively. As is well known, the Φ_i^{ss} bond energies can be calculated. However, the Φ_i^{sf} and Φ_i^{ff} bond energies are unknown. In order to estimate Φ_i, the so-called proportionality relation is often assumed implicitly:

$$\Phi_1 : \Phi_2 : \cdots : \Phi_i \cdots = \Phi_1^{ss} : \Phi_2^{ss} : \cdots : \Phi_i^{ss} \cdots. \tag{5.26}$$

Employing equation (5.26) implies that, in principle, the Hartman–Perdok morphology and the mother-phase morphology become equal. Equation (5.26) may be in the first instance justified on basis of general physical principles (Bennema 1993).

5.8.4. *Roughening*

5.8.4.1. *Thermal Roughening*

For the Solid On Solid (SOS) model of the (001) surface of a simple Kossel crystal with one bond energy Φ, having the form of equation (5.25), the dimensionless roughening temperature θ^R resulting from renormalization theory and inspired by Monte Carlo simulations (Liu and Bennema 1995; Grimbergen et al. 1998a) of the face (001) is defined as:

$$\theta_{001}^R = (2kT/\Phi_{str})_{001}^R. \tag{5.27}$$

In equation (5.27) T is the absolute temperature and k the Boltzmann constant. For a (more or less complex) 2-D connected net (hkl) the dimensionless order–disorder temperature or Ising temperature θ^C resulting from the theory of Onsager (1944) and a generalized Onsager theory for (more) complex nets according to Rijpkema et al. (1983) is defined as:

$$\theta_{hkl}^C = (2kT/\Phi_{str})_{hkl}^C. \tag{5.28}$$

Here Φ_{str} is the strongest bond energy of the connected net, which in principle, depending on the adopted conventions, is also the strongest bond of the crystal graph.

Within the framework of the theory also a dimensionless temperature θ can be defined as:

$$\theta = 2kT/\Phi_{\text{str}}, \tag{5.29}$$

where Φ_{str} is the bond energy of the strongest bond of the crystal graph. The use of Φ_{str} and the factor 2 in equations (5.28) and (5.29) is just a matter of convention. In practice, θ^R_{hkl} and θ^C_{hkl} do not differ that much (Bennema 1993; Bennema and van der Eerden 1987), so

$$\theta^R_{hkl} \approx \theta^C_{hkl}. \tag{5.30}$$

It would be desirable to calculate θ^R_{hkl} for complex connected nets occurring in a complex crystal structure. This is in practice impossible. So, instead of calculating the real SOS roughening transition, the θ^C_{hkl} according to Rijpkema et al. (1983) (see Bennema 1993, 1995, 1996; Bennema and van der Eerden 1987) can be calculated.

The theoretical concepts of roughening transition in an SOS (-like) surface and an order–disorder phase transition within a connected net can be summarized as follows:

$$\text{if}\quad \theta < \theta^C_{hkl}(\approx \theta^R_{hkl}),\qquad \gamma_{hkl} > 0;\qquad \text{if}\quad \theta \geq \theta^C_{hkl}(\approx \theta^R_{hkl}),\quad \gamma_{hkl} = 0. \tag{5.31}$$

In equation (5.31) γ_{hkl} is the free energy of a step within the connected net (hkl). The physical implication of equation (5.31) is that below a certain roughening temperature, a crystallographic face parallel to (hkl) remains in essence flat on an atomic scale. Above this roughening temperature, it will be atomically rough. Macroscopically, a rough face shows up as rounded-off face without any orientation (hkl).

Using the theory of Rijpkema et al. (1983) (as presented in Bennema (1993) and Bennema and van der Eerden (1987)) for each connected net that is a planar connected net (without crossing bonds), θ^C_{hkl} can be calculated exactly (within the precision of the numerical solution of a matrix equation). Where connected nets have a few crossing bonds, an upper and a lower bound of θ^C_{hkl} may be calculated (Rijpkema et al. 1983). If connected nets become very complex (consisting, for example, of two or more connected subnets), the calculation of θ^C_{hkl} may become impossible; see the work on apatite of Terpstra, Rijpkema, and Bennema (1986).

In the experimental reality, actual roughening temperatures T^R_{hkl} or order–disorder transition temperatures T^C_{hkl} ($\approx T^R_{hkl}$) are measured or estimated. This implies that in inequalities (5.31) θ^R, θ^C, and θ have to be replaced by T^R, T^C and T. In recent research T^R_{hkl} ($\approx T^C_{hkl}$) was estimated from experimental data. From this and from the theoretically calculated θ^C_{hkl}, Φ_{str} could be calculated (equation (5.28)). From this value and the proportionality relation all Φ_is of the crystal graph could be calculated. This opens up a whole new field of research concerning the complete theoretical calculation of Φ_is at the interface (hkl) (Liu 1993; Liu and Bennema 1995).

It follows from this section that the Hartman–Perdok crystallographic morphological theory and the statistical mechanical theory of roughening transition or order–disorder phase transition of Rijpkema et al. come together in the concept of connected net.

5.8.4.2. *Symmetry Roughening*

It has been shown that, quite often, pairs of connected nets occur that can be transformed into each other by two classes of symmetry elements (Grimbergen et al. 1998*a*; Meekes et al. 1998). Such symmetry elements are mirror planes or glide planes or twofold, fourfold, and sixfold axes that are oriented perpendicular to the plane (hkl) and similar symmetry elements of order 2 that are parallel to (hkl). By combining such pairs, a stepped profile with an edge energy equal to zero may be obtained. This effect is complementary to the well established BFDH law, for which the effective growth layer is reduced due to such symmetry elements, since the surface energy of the symmetrically related connected nets is the same while the edge (free) energy remains non-zero (Hartman 1973*a*, 1978*a*).

From equation (5.31) it can be concluded that if the edge (free) energy of a pair of connected nets turns out to be zero in some direction [uvw] parallel to a face (hkl), the roughening temperature equals zero. This will be called symmetry roughening in what follows. Although the theory of symmetry roughening was developed for first-nearest-neigbour models, it may also be applied to crystal structures with long range forces. Symmetry roughening can be of either macroscopic or microscopic nature. Macroscopic symmetry roughening corresponds to thermal roughening at absolute zero (0 K). Microscopic symmetry roughening offers a special case for which the orientation (hkl) stays macroscopically flat but is microscopically rough. This results in increased growth rates for a macroscopically flat face (Grimbergen et al. 1998*a*, *b*).

5.8.4.3. *Kinetic Roughening*

Besides the basic integrating principle of thermal roughening and the derived principle of symmetry roughening, another type of surface roughening can be derived from thermal roughening. This is the so-called kinetic roughening (Bennema 1993). Kinetic roughening is defined for a surface, with orientation (hkl) bounded by a connected net, that is growing below its roughening temperature T_{hkl}^{R}. As the supersaturation increases, the size of critical 2-D nuclei decreases down to a critical driving force for crystallization, at which the critical 2-D nucleus attains the size of a few growth units. Above this critical driving force, the crystal face (hkl) grows as a rough, rounded-off crystal face. As the supersaturation decreases beyond the critical driving force, the surface starts to grow again as a flat surface with orientation (hkl). This phenomenon has been observed for the faces {110} and {20$\bar{1}$} of naphthalene crystals growing from a toluene solution by Jetten et al. (1984) and Human et al. (1981).

Interestingly, a special kind of kinetic roughening has been recently observed. The so-called Rough–Flat–Rough (RFR) kinetic phase transition was observed on the {110} faces of paraffin growing from a hexane or toluene solution (Liu et al. 1992, 1994). The RFR transition is caused by a saturation temperature just below the roughening temperature. It can be shown that, just below the roughening temperature, the faces {hkl} under consideration are rough. As the supersaturation increases, the faces {hkl} become flat; as it increases again, the faces become rough again (Bennema 1993; Liu et al. 1992, 1994).

The experimental data supporting the concept of (thermal or kinetic) roughening refer to organic crystals. However, thermal roughening has also been observed for ionic crystals (Sawada and Shichiri 1989). Moreover, the step free energy of an orientation (hkl) can also become zero at a finite driving force as a result of the presence of multiple connected nets that are not symmetrically related, but give rise to a very small step free energy at zero driving force (Grimbergen et al. 1998a, b). This we call pseudo symmetry roughening. An extensive theoretical and computer-based study on kinetic roughening has been done by Van Veenendaal et al. (1998).

5.8.4.4. Sublayer Roughening

It will be shown in a forthcoming paper that in the case of a connected net (hkl) consisting of two or more connected subnets stacked upon each other, and above a certain critical driving force for crystallization, growth via a 2-D nucleation mechanism by means of a connected subnet becomes rate determining (Grimbergen et al. 1998a). Below this critical driving force, growth via a 2-D nucleation mechanism occurs by means of full connected net layer, having thickness d_{hkl} according to BFDH (Bravais, Friedel, Donnay, and Harker). Theoretically, it may well be that the connected subnet is either thermally roughened or kinetically roughened. In this case roughening occurs either at lower temperatures or at lower driving forces, as compared to the quantities corresponding to the full connected net (Grimbergen et al. 1998b).

Habit modifications due to half–whole or whole–half layer transitions are dependent on whether the growth process is reaction controlled or diffusion controlled. Their interdependence on supersaturation or electrolyte concentration has been reported by Van der Voort and Hartman (1990a) and Heijnen (1985).

A connected net often consists of two or more connected subnets. This holds, for example, for the pyramidal faces {011} of ADP, as will be shown in Section 5.9. When in contact with the mother phase, each of the subnets shows its own θ_{hkl}^{R} and broken bond slice energy E_{hkl}^{slice}.

In practice, it may be impossible to observe the change from a full- to a half-layer growth mechanism. The reason is that when the half-layer growth mechanism is rate determining, it is accompanied by a very high driving force to fill the second half of the nucleating layer. Thus, it may be impossible to observe experimentally genuine half-layer growth by 2-D nucleation or a spiral mechanism (Grimbergen et al. 1998b).

5.8.5. *Molecular Dynamics Studies of the Fluid Part and Prediction of Morphology*

One way to obtain the necessary data for a theoretical calculation of the bond energies in the form of equation (5.25) is to carry out Molecular Dynamics (MD) simulations of the fluid part of the interface between the crystal and the mother phase. An example is presented in recent articles on the morphology of urea by Boek et al. (1992, 1994). Better theoretical predictions of morphologies could employ:

1. Wulff plots based on the proportionality of the growth rate of faces (*hkl*), R_{hkl}, and the broken-bond attachment energy expressed in bond energies given by equation (5.26)
2. Wulff plots based on expressions for R_{hkl} in the dependence of the driving forces derived from kinetic theories, in which as well as the kinetics the actual bond energies $(\Phi_i)_{hkl}$ play a key role.

It is obvious that option 2 leads to the more reliable prediction of growth forms (see work on urea by Liu et al. (1995*a, b*)). It does not seem possible yet to apply the approach of Boek et al. (1992, 1994) to ionic crystals, but in the longer term this ought to be possible as computers with ever-increasing capacity become available.

5.8.6. *Statistical-Mechanical Theories on Roughening Transition and Long-Range Interaction*

In the following, we will limit the discussion to thermal roughening. From the discussion on roughening, basic principles on the integrated roughening-transition–Hartman–Perdok theory follow:

- Connected nets are obtained from crystal graphs, essentially based on first-nearest-neighbour interactions.
- Φ_i bond energies given by equation (5.25) are assumed to maintain the same ratio as the bond energies according to equation (5.26).
- Actual $(\Phi_i)_{hkl}$ bonds for a given face (*hkl*) and connected net in (*hkl*) can be calculated from observed roughening phenomena; or
- The real bond energies $(\Phi_i)_{hkl}$ occurring at the surface can be calculated from the structures of the fluid part of the interface by carrying out MD or other studies.

It would be worth while generalizing the theory based on first-nearest-neighbour bonds and the concept of roughening transition in order to make these methods applicable to ionic crystals with long-range attractive and repulsive forces. This is, however, at present questionable for the following reasons:

- As already noted, if all short- and long-range forces are taken into account, a partitioning of a crystal graph in connected nets becomes practically impossible and even meaningless theoretically, as explained in Section 5.8.2. Handling connected nets based on first nearest neighbours is feasible, but when next nearest neighbours enter the picture, the numerous crossing bonds that may occur make calculations of θ_{hkl}^C prohibitive.

- In the first-nearest-neighbour bond models a proportionality relation like equation(5.26) has to be taken into account in order to carry out summations based on the Madelung or Ewald technique to calculate E_{hkl}^{slice}, E_{hkl}^{att}, and θ_{hkl}^C.

5.8.7. *On the Problem of Short- and Long-Range Interactions*

In order to compare the role of the bond energies in the first coordination sphere with the role of the bond energies of the subsequent coordination spheres in crystal graphs of ionic crystals growing from an ionic solution, we start with the generalized bond energy given by equation (5.25). It will be shown that, in taking equation (5.25) into account, a distinction has to be made between bond energies Φ_i in the first (fc) and next (nc) coordination spheres. We will now focus our attention on first the next-nearest-neighbour interactions and then the first-nearest-neighbour interactions.

5.8.7.1. *Next-Nearest-Neighbour Interactions*

Since in the mother phase solute and solvent ions do not have fixed positions as they do in crystals, and solvent molecules usually have dipole (for example H_2O molecules) or quadrupole moments, the magnitude of sf and ff bond energies may in some cases fall off rapidly with distance as a result of screening. Thus, it is reasonable to assume that $\Phi_{i,\text{nc}}^{\text{sf}} \approx \Phi_{i,\text{nc}}^{\text{ff}} \approx 0$, leading to

$$\Phi_{i,\text{nc}} \approx -\Phi_{i,\text{nc}}^{\text{ss}}. \tag{5.32}$$

The expression (5.32) is not strictly correct, because the interface bounded by a (connected) net, at least close to the roughening temperature, consists in a statistical sense, on average over a large number of configurations, of one half of solid cells (growth units) and one half of fluid cells. The fluid cells may have a dielectric constant that is rather close to the dielectric constant of the bulk of the mother phase. It may be reasonable to assume that the dielectric constant of the mixed solid–fluid layer $\bar{\varepsilon}$ can be approximated by some average of the dielectric constant of the solid phase ε_c and the dielectric constant of the mother phase ε_f. Thus, we can assume that the dielectric constant of a mixed solid–fluid interface $\bar{\varepsilon}$ is given by

$$\varepsilon_c < \bar{\varepsilon} < \varepsilon_f. \tag{5.33}$$

In the case of water, ε_f has the extremely high value $\varepsilon_f \simeq 80$, and in general, ε_f varies from 10 to this value of 80. Looking at the bulk of the crystal, it is reasonable to assume that for the nc bonds ε will be equal to the bulk of the crystal ε_c. For alkali halide crystals ε_c varies from 2 to 15. So for alkali halide crystals growing from an aqueous solution $10 < \bar{\varepsilon} \leq 80$.

In general the above range of values gives a good estimate of $\bar{\varepsilon}$, suggesting that, by comparing equation (5.32) with the above range, the actual values of the long range interaction energies may be obtained as:

$$\Phi_{jk} = -\Phi_{jk}^{ss}/\varepsilon = -q_j q_k/\bar{\varepsilon} r_{jk}. \qquad (5.34)$$

This Φ_{jk} may be an order of magnitude lower than the Φ_{jk}^{ss} that is referenced to vacuum. (In equation (5.34) q_j and q_k are the positive or negative charges on the interacting ions j and k separated by the distance r_{jk}.) It follows from equation (5.34) that for next-nearest-neighbour interactions the proportionality condition of equation (5.26) is valid provided $\bar{\varepsilon}$ does not depend on r_{jk}.

5.8.7.2. *First-Nearest-Neighbour Interactions*

Concerning $\Phi_{i,\mathrm{fc}}$ we now have

$$\Phi_{i,\mathrm{fc}} = \Phi_{i,\mathrm{fc}}^{sf} - (\Phi_{i,\mathrm{fc}}^{ss} + \Phi_{i,\mathrm{fc}}^{ff}), \qquad \text{with} \quad \Phi_{i,\mathrm{fc}}^{sf} \neq \Phi_{i,\mathrm{fc}}^{ff} \neq 0 \qquad (5.35)$$

and, since $\varepsilon_0 = 1$,

$$\Phi_{i,\mathrm{fc}}^{ss} = q_j q_k/r_{jk}. \qquad (5.36)$$

If equations (5.35) and (5.36) are compared, the following reasons to distinguish fc bond energies and nc bond energies can be given: first, the dielectric constant for fc bond energies is probably close to $\varepsilon_0 = 1$ and for nc bond energies about $\bar{\varepsilon}$. Secondly Φ_i^{sf} and Φ_i^{ff} rapidly fall off with distance.

Note that in equation (5.36) $\bar{\varepsilon}$ is no longer present and hence can no longer cause Φ to become an order of magnitude smaller than the Φ_{jk}^{ss}. So the contribution of the $\Phi_{i,\mathrm{fc}}^{ss}$ bonds to the overall bond energy (equation (5.34)) may be an order of magnitude larger than the corresponding contribution to the overall bond energy Φ_i (equation (5.25)) for nc bonds. This much higher bond energy is now, however, compensated by the $\Phi_{i,\mathrm{fc}}^{sf}$ and $\Phi_{i,\mathrm{fc}}^{ff}$ bond energies, which no longer equal zero. On the basis of our present state of knowledge and lack of an adequate theoretical treatment of the interface between mother phase and ionic crystals, we distinguish for the sake of argument two extreme cases below: In equation (5.37a) the sum of the fc bonds, $\Phi_{i,\mathrm{fc}}$, is much larger than the sum of the nc bonds; while in equation (5.37b) the reverse extreme holds:

$$\sum_{i(\mathrm{fc})} \Phi_{i,\mathrm{fc}} \gg \sum_{i(\mathrm{nc})} \Phi_{i,\mathrm{nc}}, \qquad (5.37a)$$

$$\sum_{i(\mathrm{fc})} \Phi_{i,\mathrm{fc}} \ll \sum_{i(\mathrm{nc})} \Phi_{i,\mathrm{nc}}. \qquad (5.37b)$$

In the case of equation (5.37a), we are back to a first-nearest-neighbour model and the whole Hartman–Perdok roughening theory described above applies (Bennema 1993, 1995, 1996; Bennema and van der Eerden 1987; Rijpkema et al. 1983). In the case of equation (5.37b), integration of the Hartman–Perdok and a generalized Onsager theory according to Rijpkema et al. (1983) is not possible.

In this case, however, the mean-field Jackson (1958) model, to be described in the next section, is applicable.

We note that both expression (5.34) for nc bonds and expression (5.36) for fc bonds, the bond energies are reduced by an order of magnitude. Now, according to the calculations of Hartman and coworkers, typical values of slice energies are $E_{hkl}^{\mathrm{slice}} \sim -1000$ kcal/mol. So if $\bar{\varepsilon}$ is taken equal to a few tens, the broken-bond slice energy $E_{\mathrm{interface}}^{\mathrm{slice}}$ will be a few tens of kcal/mol. A typical E_{hkl}^{slice} may correspond to say four to eight growth units. Thus, the energy per growth unit will become say 5 to 10 kcal/mol. Such values seem reasonable because with $RT \approx$ 0.6 kcal/mol for room temperature and $RT \approx 2$ kcal/mol for $T \approx 1000$ K, the Φ_i values are around a few RT. These are normally the bond energies and temperatures at which growth takes place.

Now, as a result of contact with an ionic melt or solution, the lowering of bond energies can be considered as a screening effect. The screening effect of lowering the nc bond energies results because the inserted dielectric constant is not that of vacuum ε_0 but a kind of average $\bar{\varepsilon}$ between vacuum and mother phase. However, taking a constant average $\bar{\varepsilon}$ for the solid phase ε_s and ε_f for the fluid phase is a crude ad hoc estimate. It is well known that dielectric constants may depend on whether the surface is below or above its roughening temperature, and may vary from face to face.

5.8.8. *Mean Field Theory and Ionic Bond Energies*

It follows from the previous section that an "exact" Hartman–Perdok roughening transition theory can be applied to a planar first-nearest-neighbour crystal graph (without crossing bonds). This approach is then also possible for a first-nearest-neighbour model for the ionic-crystal case mentioned earlier. It is, however, also possible to integrate a mean-field interface model with the occurrence of long-range electrostatic (point-charge) interactions. This logical integration will be described below.

Starting with the crude mean-field approach first introduced by Jackson (1958), consider a purely solid two-dimensional crystal (Bennema 1993) consisting of xM square solid blocks or cells mixed with a purely fluid two-dimensional crystal consisting of $(1-x)M$ fluid blocks of the same size and shape. Then, as a result of the mixing process, the change in free energy ΔF_{mix} with reference to the purely solid and fluid crystals is given by:

$$\Delta F_{\mathrm{mix}} = \Delta H - T\Delta S. \tag{5.38}$$

Introduce the dimensionless function:

$$f = \Delta F_{\mathrm{mix}}/MkT = \alpha x(1-x) + x \ln x + (1-x)\ln(1-x), \tag{5.39}$$

$$\alpha_{hkl} = \sum (\Phi_i)_{hkl}/kT. \tag{5.40}$$

Because of the competition between the entropy-of-mixing term $x \ln x + (1 - x)\ln(1 - x)$ and the energy term $\alpha x(1 - x)$, a plot of the function f versus the fraction of solid cells x exhibits, for sufficiently high values of α, two minima separated by a maximum. As α decreases, the minima shift towards the centre until finally only one minimum survives at a critical point.

The case of two minima separated by a maximum corresponds to a solid phase or solid domains and a fluid phase or fluid domains, and that of one minimum corresponds to a mixed solid–fluid phase. According to Jackson's original idea, the first case can be interpreted as corresponding to a flat surface consisting of flat domains separated from fluid domains by steps and the second case to a rough face. It is noteworthy that, in the mean-field Jackson (1958) model, the critical value α^{cr} corresponding to what may be called the roughening transition is $\alpha^{cr} = 2$. As shown by Bennema (1993), the Jackson model can be easily generalized to all crystal structures with all kinds of chemical compositions. The Jackson (1958) mean-field or regular-solution model can be applied to ionic F slices with all interactions from the fc to all nc bonds (up to infinity).

Madelung or Ewald calculations can provide the actual quantity $\alpha_{hkl} = \widetilde{E}_{hkl}^{slice}/kT$, where $\widetilde{E}_{hkl}^{slice}$ is defined as the sum of all the Φ_i within the F slice under consideration. The Φ_i have again the form of equation (5.26). Note that $\widetilde{E}_{hkl}^{slice}$ is positive, but $\widetilde{E}_{hkl}^{slice,ss}$ is negative. Because of the proportionality relation (5.26) Madelung or Ewald summations can be applied to calculate $\widetilde{E}_{hkl}^{slice}$. If the fc bonds are not adequate, a correction is introduced to obtain a suitable value for the actual $\widetilde{E}_{hkl}^{slice}$:

$$\widetilde{E}_{hkl}^{slice} = \Delta \widetilde{E}_{hkl}^{slice} + \widetilde{E}^{slice,ss}/\bar{\varepsilon}, \quad \Delta \widetilde{E}_{hkl}^{slice} = \sum_{i(fc)}(\Phi_{i,fc} - \Phi_{i,fc}^{ss}/\bar{\varepsilon}). \quad (5.41)$$

The first-nearest-neighbour bond model and hence the integrated Hartman–Perdok roughening theory are valid for $\Delta \widetilde{E}_{hkl}^{slice} \gg \widetilde{E}^{slice,ss}/\bar{\varepsilon}$.

The dimensionless roughening temperatures cannot be calculated in the mean-field Jackson model. However, owing to the proportionality relation, the sequence of α values coincides with the sequence of E_{hkl}^{slice} values, leading to a reliable assessment of the order of the morphological importance (MI). In this way a statistical mechanical interpretation is assigned to the relative MI of various faces.

5.8.9. *The Success of the First-Nearest-Neighbour Bond Model*

Our present state of knowledge of the growth of ionic crystals from ionic solvents, or solvents with dipolar molecules like water, does not allow us to predict whether the appropriate model is a first-nearest-neighbour crystal graph, a next-nearest-neighbour model, or a mixture of the two.

A strict first-nearest-neighbour bond approach to the garnet structures provided an explanation for the morphology of synthetic and natural garnets, including the dependence on chemical composition of the facets occurring on spheres of single crystals (Tolksdorf and Bartels 1981), the anisotropy of growth fronts, and the interlacing on {110} facets. The success on the garnet structure (Bennema et al. 1976) suggests that the first-nearest-neighbour model indeed works (Cherepanova et al. 1989, 1992a,b). This success may also be attributed to two properties of the garnet structure:

- For synthetic garnets Fe—O, Al—O, and perhaps Y—O bonds are to a high extent covalent and only partially ionic.
- Each positive cation in the garnet structure is surrounded by a sphere of negative oxygen atoms, giving rise to a strong screening effect within a highly symmetrical cubic crystal structure.

Boutz and Woensdregt (1993) explained the morphology of garnets with E_{hkl}^{att} values calculated by employing Madelung calculations in a point-charge model.

In the future the results of the nearest-neighbour bond and next-nearest-neighbour bond approaches have to be compared systematically, the former to be carried out in the context of a computerized F-face (or PBC) analysis. In the past, the first-nearest-neighbour bond approach was successfully applied to a series of ionic crystals (Bennema and Van der Eerden 1987; Bennema 1993).

In conclusion, with an eye to the future, the role played by surface reconstruction in ionic crystals is most worthy of note. The compelling influence exerted by pseudo-flat surface reconstruction on the morphology of ammonium chloride in Section 5.9 is by no means an isolated occurrence. A recent study on the morphology of cesium halide crystals grown from the vapour phase shows the relevance of reconstructed surface phases, giving rise to so-called Disordered Flat (DOF) phases and resulting in morphological transitions as a function of the driving force for crystallization and of the temperature (Grimbergen et al. 1998c).

5.9. Case Study 3: Interface Behaviour and Surface–Fluid Interactions of ADP and KDP

5.9.1. *Introduction to ADP and KDP*

Large single crystals of the isomorphous salts ammonium dihydrogen phosphate (ADP) and potassium dihydrogen phosphate (KDP) have been grown in large quantities for industrial purposes. During and after World War II, large numbers of these crystals have been produced because of their piezoelectric properties (Jaffe and Kjellgren 1994). Both crystals became well known model compounds for the investigation of the fundamental problems associated with crystallization processes.

At present, KDP is often used as an optical frequency doubler (Loiacono et al. 1983). In laser technologies this application requires the growth of large, high quality single crystals. A bottleneck for the production of these crystals lies in the growth of the {020} faces. These prismatic faces usually grow very slowly and are

Table 5.6. *Positions of molecular mass centres in the I cell in fractional axis coordinates and bonds defining the crystal graph*

Mass centre	X	Y	Z	Mass centre	X	Y	Z	Label	Bond
P_1	0.00	0.00	0.00	N_1	0.00	0.00	0.50	f	P_1—N_1
P_2	0.50	0.00	0.25	N_2	0.00	0.50	0.25	g	P_1—N_2; P_2—N_1
P_3	0.50	0.50	0.50	N_3	0.50	0.50	0.00	h	P_1—N_3
P_4	0.00	0.50	0.75	N_4	0.50	0.00	0.75	e	P_1—P_2

very sensitive to impurities and inclusion formation. This problem will be discussed below.

5.9.2. Crystal Structure and Bonds of the First Coordination Sphere

In Sections 5.3 and 5.4 the crystal graph of ADP was introduced, and treated from a graph-theoretical point of view on the basis of a primitive cell. In this section we will treat the crystal graph of ADP based on the body-centered or I cell. The space group is $I\bar{4}2d$, with cell parameters $a = 7.50$ Å $(= b)$ and $c = 7.55$ Å $(\alpha = \beta = \gamma = 90°)$. In Table 5.6 the molecular mass centres of $H_2PO_4^-$ (P) and NH_4^+ (N) in the I cell are listed. The selection rules for (hkl) of the ADP crystal structure are given by: $h + k + l = 2n$; $2k + l = 2n + 1$ or $4n$; and $2h + l = 4n$ $(l = 2n)$.

In Fig. 5.14(a) the f and g bonds already discussed in Section 5.4 are drawn. We have omitted the hydrogen bonds e because the discussion in Section 5.6 showed that they are not important. However, a closer examination (Section 5.6) leads to the conclusion that, although the e bonds are negligible in the context of the $\{f, g, e\}$ bond set, they need not be negligible in the context of the $\{f, g, h, e\}$ bond set.

Table 5.6 lists the important bonds f and g, the weaker diagonal bond h, and the omitted bond e for the sake of completeness. The f, g and h bonds are drawn in Fig. 5.14(b).

5.9.3. Derivation of the Connected Nets Based on f, g and h Bonds

All the stoichiometric nets presented in this section were obtained by a computer program and, according to the notation of Section 5.4.3.2, are of the restricted form $\{\{C_m[u_M v_M w_M] + C_n[u_N v_N w_N]\}\}$ and classified as purely core-type F slices.

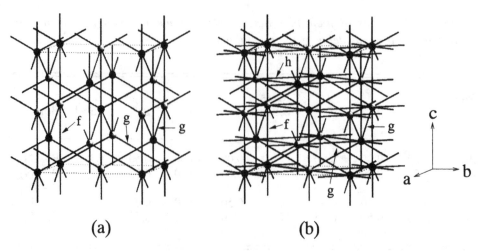

Fig. 5.14. The crystal graph of ADP based on (a) f and g bonds and (b) f, g and h bonds. For convenience, a few bonds are labelled; the remaining bonds follow by symmetry. Mass centers: black spheres, PO_4; gray spheres, NH_4.

Table 5.7. *Connected nets based on f, g and f, g, h bonds*

Form	d_{hkl} (Å)	Number of nets based on $\{f, g\}$	Number of nets based on $\{f, g, h\}$
{011}	5.321	4	6
{022}	2.661	2	6
{020}	3.750	1	9
{112}	3.075	—	9
{121}	3.065	2	6
{220}	2.652	—	1
{031}	2.373	2	2
{130}	2.372	—	4
{013}	2.386	—	6
{123}	2.013	—	2
{004}	1.888	—	1

For the ADP structure the connected nets based on f, g and h bonds are presented in Table 5.7. For all forms $\{hkl\}$ for which core-type connected nets were found to satisfy the selection rules, Table 5.7 lists the interplanar distance d_{hkl}, as well as the numbers of connected nets based on the bond sets $\{f, g\}$ and $\{f, g, h\}$.

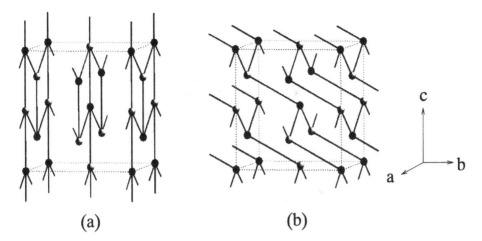

Fig. 5.15. (*a*) Connected nets (020) and (*b*) connected subnets $(022)_{1,2}$ shown within the bounds of the *I* cell.

5.9.4. *Pictorial Presentation of Some Connected Nets*

5.9.4.1. *Nets Derived with f and g Bonds Taken into Account*

In this section we will focus on nets for which the $\{f, g\}$ bond set suffices. It can be immediately seen from Fig. 5.14(*a*) that the crystal graph can be partitioned into equal parallel connected nets (020), being highly symmetrical and having low dipole moment, as presented in Fig. 5.15(*a*).

In orientation (011) the crystal graph can be partitioned into connected nets $(011)_{1,2}$. Each single (011) connected net consists of two connected subnets $(022)_{1,2}$ (Fig. 5.15(*b*)).

Two full-layer connected nets, $(011)_1$ and $(011)_2$, can be identified as $(011)_1$ consisting of the connected subnet $(022)_1$ on top of connected subnet $(022)_2$, and the connected net $(011)_2$ where now $(022)_2$ is on top of $(022)_1$. In Fig. 5.16(*a*) the $(011)_1$ connected net is seen from above at a slightly oblique direction. The honeycomb structure of $(011)_1$ consists of connected sublayer $(022)_1$ on top of $(022)_2$. In Fig. 5.16(*b*) the same connected net is seen from the side. The connected nets $(011)_{1,2}$ will have high dipole moments in the absence of surface reconstruction, adjacent to vacuum.

In addition, the connected nets $(011)_3$ and $(011)_4$ were found. They are connected subnets $(022)_1$ and $(022)_2$, linked, respectively, by two growth units N and P spanned over the hexagons of these nets. In Fig. 5.16(*c*) and (*d*) one alternative is seen from above and edge-on.

The connected nets $(121)_{1,2}$ and $(031)_{1,2}$ were first identified by Strom et al. (1995*b*). If we assume that their broken bonds react in the same way with the mother phase, they can be demonstrated to undergo the symmetry roughening discussed in Section 5.8 (Grimbergen et al. 1998*a*; Meekes et al. 1998). If that assumption does not hold, one of the two nets may dominate. If the roughening

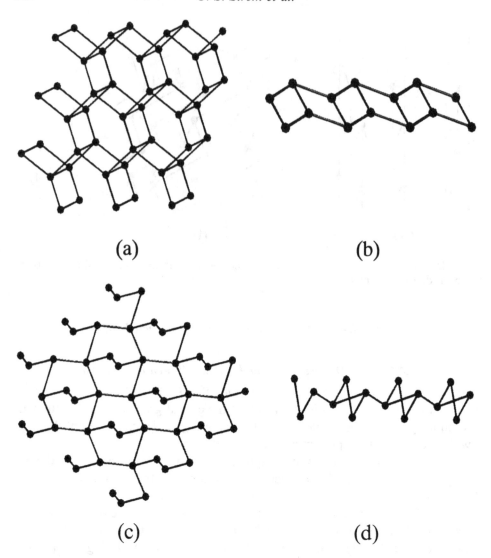

(a)   (b)

(c)   (d)

Fig. 5.16. The connected nets (a) $(011)_1$ from the top; (b) $(011)_1$ edge-on; (c) $(011)_3$ from the top; (d) $(011)_3$ edge-on.

temperature of the dominant connected net is lower than the growth temperature, faces {121} may appear.

By complete analogy, if the pair of connected nets $(031)_{1,2}$ react in the same way with the mother phase, they will lead to symmetry roughening; otherwise one of them may be dominant. The pairs of connected nets $(121)_{1,2}$ (Fig. 5.17(a), (b)) and $(031)_{1,2}$ (Fig. 5.17(c), (d)) are very similar.

5.9.4.2. *Nets Derived with f, g and h Bonds Taken into Account*

Taking the h bond additionally into consideration results in the crystal graph of Fig. 5.14(b), which can be partitioned into highly symmetrical core-type con-

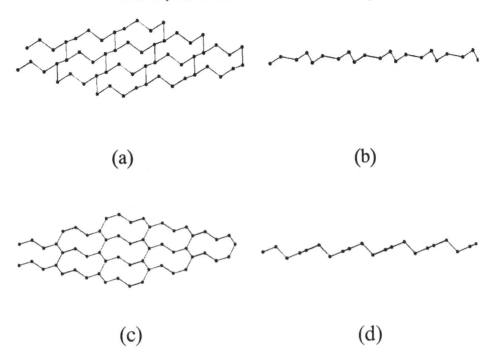

(a)

(b)

(c)

(d)

Fig. 5.17. The connected nets: (*a*) (121)$_1$ from above; (*b*) (121)$_1$ edge-on; (*c*) (031)$_1$ from above; (*d*) (031)$_1$ edge-on.

nected nets with a very low dipole moment. The nets (220) are shown in Fig. 5.18(*a*). The nets (004) shown in Fig. 5.18(*b*) are weak because they consist of only *h* bonds. In Fig. 5.18(*c*) and (*d*) the single connected net (112) can be seen from above and edge-on, respectively.

The introduction of the *h* bond now results in eight alternative core-type connected nets for (020) (see Table 5.7); the same holds true for (112). The occurrence of these eight alternatives for the two connected nets (020) and (112) can be explained as follows. The top view of the nets (020) and (112) shows two N and two P units per unit cell or mesh area. Each of these N or P groups can be shifted over a distance $[\frac{1}{2}\frac{1}{2}\frac{1}{2}]$, giving rise to a new stoichiometric net consisting of *h* bonds, thus producing four alternative connected nets. The two N or two P points can also be shifted by the opposite translation $[\frac{\bar{1}}{2}\frac{\bar{1}}{2}\frac{1}{2}]$, yielding four more alternative connected nets, bringing the total up to $1 + 8 = 9$ connected nets. The eight additional nets can be classified in symmetrically related pairs that could lead to symmetry roughening.

If, for whatever reason, interaction with the mother phase does not lead to symmetry roughening, the eight alternatives may well occur, but they will be suppressed by the singlet that has a higher magnitude of E_{hkl}^{slice}. As usual, in both cases {020} and {112}, the planar highly symmetric and stable (that is, with a low dipole moment) singlet connected nets will survive.

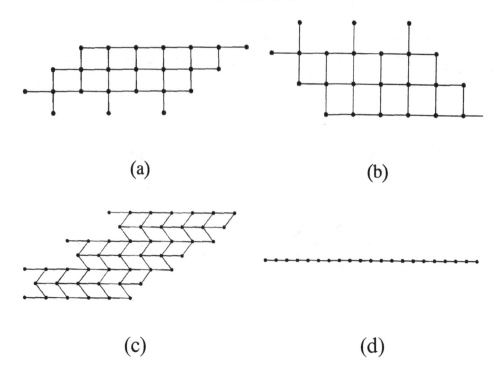

(a)   (b)

(c)   (d)

Fig. 5.18. The connected nets (*a*) (220) from the top; (*b*) (004) from the top; (*c*) (112) from the top; (*d*) (112) edge-on.

Aquilló and Woensdregt (1984, 1987) found from visual inspection the connected nets $(011)_{1,2}$, $(022)_{1,2}$ and (020) by taking *f* and *g* bonds into account, and in addition (112) and (220) by taking *f*, *g* and *h* bonds into account.

5.9.4.3. *Influence of Trivalent Cations on Growth Mechanism and Morphology*

The theoretical treatment of Section 5.6 showed that the *F* slices {011} and {020} are compact and have the lowest magnitude of E_{hkl}^{att} in each face. The discrepancy between theoretical and observed growth forms (Fig. 5.10) was also reported in Section 5.6.

The absence of the {020} faces can be explained by the strong interaction of trivalent cations with the prism faces, as will be argued next.

Belouet, Dunia, and Petroff (1974) were able to show that the segregation coefficient of Fe^{3+} and Cr^{3+} ions in the prismatic sectors increases at low supersaturation. From growth rate versus supersaturation measurements Mullin et al. (1970) deduced that a certain minimum supersaturation is required for growth to take place on the {020} faces of ADP. Subsequently Davey's in-situ study on the kinetics of high macrosteps (of the order of microns) on {020} ADP (Davey and Mullin 1975) reported that no macrosteps could be observed below a certain supersaturation.

These two observations pose an intriguing problem to researchers in the field of crystal growth mechanisms. The question whether growth is really blocked below a certain supersaturation can only be answered by looking in situ at the behaviour of very low steps. Dam et al. carried out in situ experiments on the influence of about ten parts per million (10 ppm) of Cr^{3+} or Fe^{3+} on the propagation of steps on the prism and pyramidal faces of KDP (Dam et al. 1984; Dam and van Enckevort 1984). It was found that, when ultrapure solutions were used, the prism and pyramidal faces grew with roughly equal growth rates. This would produce a habit approximating the one predicted in Fig. 5.10(*a*). However, when small traces of Cr^{3+} or Fe^{3+}, up to 10 ppm, were added to the solution, the growth spirals stopped rotating. This stands in clear contrast to the rotating growth spirals of the pyramidal faces, which are much less sensitive to traces of trivalent cations.

The previous PBC analysis can explain this observation. Figure 5.15 shows, in a projection of the KDP structure normal to the [100] direction, the distribution of the twofold axes, seen end-on. The connected nets $(011)_1$ and $(011)_2$ and (020) are recognized in the projection edge-on. In contrast to the highly symmetrical (020), the two alternative nets $(011)_{1,2}$ are strongly asymmetrical and polar. They will consequently have an electric dipole moment perpendicular to the face (011). It can immediately be seen from Fig. 5.15 that foreign positive cations with a high charge and small radius will show a strong interaction with the negative $H_2PO_4^-$ ions of the (020) surface.

If it is assumed for the time being that the K^+ ions are on the outside of the pyramidal face (this corresponds to the connected net $(011)_2$), we can explain why the traces of the trivalent ions like Fe^{3+} and Cr^{3+} have much less influence on the pyramidal faces. Thus, we arrive at the conclusion that the elongated shape of ADP and KDP crystals growing under normal conditions at a temperature of 30 to 40°C and a pH of about 4 (and a not very pure solution) must be attributed to the special interaction of trivalent cations with the highly symmetrical prism faces. This explains why the prismatic faces {020} occur on the actual growth form. In fact, recent in situ experiments using synchrotron radiation have shown that, indeed, the K^+ ions are on the outside of the pyramidal faces (De Vries et al. 1998).

5.9.4.4. *Results of Sphere Experiments and Step-Height Measurements on Pyramidal Faces*

In the past, polished single crystals of ADP and KDP having spherical shape were exposed to slightly supersaturated solutions (Bennema 1965; Dam et al. 1980; Verheijen et al. 1996). Sphere experiments of high precision and sophistication have been carried out on ADP crystals by Verheijen et al. (1996). The most remarkable observation in all three papers was the occurrence of a system of bands on spheres of ADP having the point group symmetry 222 instead of the $\overline{4}2m$ point group symmetry of the crystal structure of ADP as determined from diffraction experiments. For the details we refer to the original papers on the

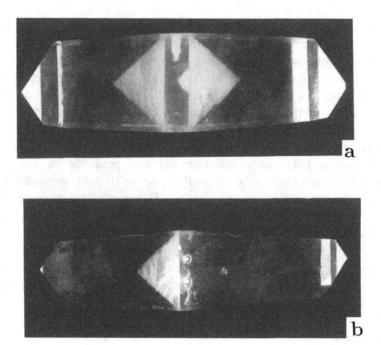

Fig. 5.19. Aqueous solution grown crystals of (a) ADP (Y. S. Wang 1983–1993) and (b) KDP (Y. S. Wang et al. 1992).

ADP hypomorphism (Bennema 1965; Dam et al. 1980; Verheijen, Vogels, and Meekes 1996). The forms of the facets observed are: on the ADP spheres, {011}, {020}, {112}, {11$\bar{2}$}, and {110}; on KDP spheres, {011}, {020}, {112}, {11$\bar{2}$}, and within a band of facets perhaps also a weak {031} form.

On spheres of ADP and KDP, besides faces {020} and {011} parallel to strong nets, the appearance of faces parallel to the highly symmetrical singlet connected nets {112}, {11$\bar{2}$}, and {220} is worthy of note. The {112} net is stronger than the {220} net, explaining why the {220} faces of KDP may grow above their roughening temperature and hence do not show up on the sphere. It remains to be seen whether or not the {031} faces of KDP occurring in bands are genuinely flat.

De Yoreo (private communication 1998, to be published) has measured the step height of growth spirals of KDP crystals using atomic force microscopy (AFM) to be 5.0 Å, in exact agreement with the height of a full layer {011} or double-connected subnet {022}, for which $d_{011} = 5.092$ Å.

5.9.5. Comparison of Theoretical and Experimental Growth Forms of ADP and KDP

Photographs of ADP (Y. S. Wang, 1983–1993) and KDP (Y. S. Wang et al. 1992) are shown in Fig. 5.19. The discrepancy between the theoretical and the actually observed habits of ADP stands in sharp contrast to the reasonable degree of resemblance found between the theoretical habit of KDP reported by Hartman (1956*b*) and Fig. 5.19(*b*), the difference being a somewhat more elongated experimental form.

Aquilló and Woensdregt (1984, 1987) (AW) have carried out extensive calculations on the morphology of ADP based on classical Hartman–Perdok analysis. These authors explain the observed elongated habit by assuming that the pyramidal faces grow with half layers (022). With the theoretical and experimental knowledge currently at our disposal, the following arguments can be put forward against the presuppositions of AW.

- It can be shown that in the case where a connected net is composed of two connected subnets, for example $(011)_{1,2}$, it is only above a certain supersaturation that face (011) may grow with "half layers" $(022)_1$ or $(022)_2$. Thus, growth of the pyramidal faces by means of layers $(022)_1$ and $(022)_2$ at rather low supersaturations is implausible (if we consider here the growth conditions of large crystals).
- According to the recent AFM measurements of De Yoreo (private communication) the pyramidal faces grow with $(011)_1$ or $(011)_2$ slices.
- In situ observations show that small traces of trivalent cations block the growth of the prism faces at normal temperatures.

Even though the present work is of a preliminary nature, the results are expected to lead to an improved explanation of the observed prismatic elongation of ADP. However, the work has not progressed to the point of quantitative determination of the environmentally induced habit modification.

5.10. Conclusions

We have outlined the fundamental developments in computational prediction of ionic structures, including both the traditional PBC theory and as the more recent development on roughening transitions.

The traditional PBC theory, developed by Hartman and Perdok, applies to energy-dominated crystallization processes. It prescribes a morphological analysis in two stages.

In the first stage, the basic structural morphology predicts the growth habit obtained in growth out of vapour or out of a mother phase interacting weakly and/or isotropically with the crystal surface. This stage consists of two steps: The first step comprises a graph-theoretic treatment to compute the chemical compositions and face orientations of all possible substrates, while the second step involves accurate, analytic, and adjustable-parameter-free computations of the various long-range Coulomb energy quantities. The successful predictions of crystal habits in cases where the environmentally induced modification effects are small indicate that crystal morphology is basically a manifestation of the periodic bond chains existing in the structure.

In the second stage, when growth occurs out of a polar or strongly interacting mother phase, the actually observed growth habit is determined as an environmentally induced habit modification imposed on the basic structural morphology. In the modified habit, the mother phase may even activate surface configurations that are different to the surface configurations featuring in the structural habit. The environmentally induced modification effects can be determined from considerations of surface chemistry. These approaches are of necessity qualitative or semiquantitative, since surface chemistry is not subject to prior computation. Nevertheless, they have led to a reasonable degree of success, a host of unresolved issues notwithstanding.

In computing energy quantities, or interactions between the solid and given spatial particle distributions in the fluid, no approximations are necessary, provided we are satisfied to simulate the atomic charge distributions by a sufficiently large number of point charges, the values of which are reliably assigned, and can obtain high-quality parameters for van der Waals interactions.

A number of illustrations and detailed case studies on ADP-type structures and ammonium chloride have been included in support of the theoretical treatment.

The major shortcoming of the traditional PBC theory is its inability to account for thermodynamics-dominated crystallization processes. The well known theory of roughening transitions has been developed in the context of Ising-type systems. In their general formulation Ising-type systems are not expected to capture the full range of physics applicable to ionic processes. One domain of ionic structures where Ising-type formulations can prove fruitful arises when strong screening effects are operating, since then the physical processes are dominated by the short-range phenomena. A recent extension of the theory of roughening transitions to ionic structures has been outlined in this chapter. In any case, recent developments extending the Hartman–Perdok theory, by taking account of multiple connected nets, have offered deeper insight into the thermodynamics of crystal surfaces.

Finally, with regard to ionic structures, potent theoretical methods lending themselves as effective computational instruments for simulating the behaviour, that is, conformational and spatial distributions, of both dilute and dense electrolytic solutions are at present not in sight. The substantial progress already achieved thus far primarily concerns molecular structures or specific and very simple ionic cases. So, despite our best endeavours, the work reported in this chapter is but a simple prelude to the task that still remains to be accomplished.

Acknowledgments

We are much indebted to Professor Y. S. Wang for providing us with the photographs of his grown ADP and KDP crystals in Figs 5.10 and 5.19.

C.S.S. thanks the Netherlands Technology Foundation (STW) for partial financial support during the early stages of draft preparation and the Netherlands Foundation for Research in Astronomy (ASTRON) in Dwingeloo for use of graphics facilities. R.F.P.G. is supported by the Netherlands Technology Foundation (STW). The work of M.W. was supported by the National Science Foundation of China (project no. 19425007), the Chinese Committee of Science and Technology, the Qiu Shi foundation of Hong Kong, and the Dutch Ministry of Science and Education.

References

Abramowitz, M. and Stegun, I. A. (1965). *Handbook of Mathematical Functions.* New York: Dover.

Aquilló, M. and Woensdregt, C. F. (1984). *Journal of Crystal Growth*, **69**, 527–36.

Aquilló, M. and Woensdregt, C. F. (1987). *Journal of Crystal Growth*, **83**, 549–59.

Belouet, C., Dunia, E., and Petroff, J. F. (1974). *Journal of Crystal Growth*, **23**, 243–52.

Bennema, P. (1965). *Zeitschrift für Kristallographie*, **121**, 312–14.

Bennema, P. (1993). In *Handbook of Crystal Growth*, Vol. 1, ed. D. T. J. Hurle, Chapter 7, pp. 477–581. Amsterdam: North-Holland.

Bennema, P. (1995). In *Science and Technology of Crystal Growth*, ed. J. P. van der Eerden and O. S. L. Bruinsma, Chapter 4.1, pp. 149–64. Dordrecht: Kluwer Academic Publishers.

Bennema, P. (1996). *Journal of Crystal Growth*, **166**, 17–28.

Bennema, P. and van der Eerden, J. (1987). In *Morphology of Crystals*, Part A, ed. I. Sunagawa. Tokyo: Terra.

Bennema, P., Giess, E. A., and Weidenborner, J. E. (1976). *Journal of Crystal Growth*, **42**, 41–60.

Berkowits-Yellin, M. (1985). *Journal of the American Chemical Society*, **107**, 8239–53.

Biggs, N. (1974). *Algebraic Graph Theory*, p. 13. Cambridge: Cambridge University Press.

Boek, E. S., Briels, W. J., van der Eerden, J., and Feil, D. (1992). *Journal of Chemical Physics*, **96**, 7010–18.

Boek, E. S., Briels, W. J., and Feil, D. (1994). *Journal of Chemical Physics*, **98**, 1674–8.

Boutz, H. H. H. and Woensdregt, C. F. (1993). *Journal of Crystal Growth*, **134**, 325–36.

Cerius2 (1995). Molecular Simulations, Cambridge (CHARMm 22.0).

Cherepanova, T. A., Didriksons, G. T. D., Bennema, P., and Tsukamoto, K. (1989). In *Morphology and Growth of Crystals*, ed. I. Sunagawa, pp. 163–75. Tokyo: Terra.

Cherepanova, T. A., Bennema, P., Yanson, Yu. A., and Tsukamoto, K. (1992a). *Journal of Crystal Growth*, **121**, 1–16.

Cherepanova, T. A., Bennema, P., Yanson, Yu. A., and Vogels, L. P. J. (1992b). *Journal of Crystal Growth*, **121**, 17–32.

Dam, B. and van Enckevort, W. J. P. (1984). *Journal of Crystal Growth*, **69**, 306–16.

Dam, B., Bennema, P., and van Enckevort, W. J. P. (1980). In Extended Abstracts, 6th International Conference on Crystal Growth (Moscow, 1980), Vol. 4, pp. 18–23.

Dam, B., Polman, E., and van Enckevort, W. J. P. (1984). *Industrial Crystallization*, 84, eds. S. J. Jancic and E. J. de Jong, pp. 97–102. Amsterdam: Elsevier.

Davey, R. J. and Mulin, J. W. (1975). *Journal of Crystal Growth*, **29**, 45–8.

De Vries, S. A., Goedtkindt, P., Bennett, S. L., Huisman, W. J., Zwanenburg, M. J., Smilgies, D.-M., de Yoreo, J. J., van Enckevort, W. J. P., Bennema, P., and Vlieg, E. (1998). *Physical Review Letters*, **80**, 2229–32.

De Yoreo, J. J., Land, T. A., and Dair, B. (1994). *Physical Review Letters*, **73**, 1938–41.

Donnay, J. D. H. and Harker, D. (1937). *American Mineralogist*, **22**, 446–67.

Dowty, E. (1976). *American Mineralogist*, **6**, 448–53.

Ewald, P. P. (1921). *Annalen der Physik (Leipzig)*, **64**, 253–71.

Grimbergen, R. F. P., Meekes, H., Bennema, P., Strom, C. S., and Vogels, L. J. P. (1998a). *Acta Crystallographica* A, **54**, 491–500.

Grimbergen, R. F. P., Bennema, P., and Meekes, H. (1998b). *Acta Crystallographica* A, in press.

Groth, P. (1906). *Chemische Kristallographie*, pp. 182–4. Leipzig: Wilhelm Engelmann.

Hamar, R. (1981). PhD Thesis, Grenoble.

Harary, F. (1969). *Graph Theory*, Chapters 12 and 13. Reading, MA: Addison Wesley.

Hartman, P. (1956a). *Acta Crystallographica*, **9**, 569–72.

Hartman, P. (1956b). *Acta Crystallographica*, **9**, 721–7.

Hartman, P. (1958a). *Acta Crystallographica*, **11**, 365–9.

Hartman, P. (1958b). *Acta Crystallographica*, **11**, 459–64.

Hartman, P. (1959). *Neues Jahrbuch für Mineralogie Monatshefte*, pp. 73–81.

Hartman, P. (1969). The dependence of crystal morphology on crystal structure. In *Growth of Crystals*, Vol. 7, ed. N. N. Sheftal, pp. 3–18. New York: Consultants Bureau.

Hartman, P. (1973a). In *Crystal Growth: An Introduction*, ed. P. Hartman, pp. 367–401. Amsterdam: North-Holland.

Hartman, P. (1973b). In *Estratto dai Rendiconti della Societá Italiana di Mineralogia e Petrologia*, **XXIX(5)**, 153–71.

Hartman, P. (1978a). *Canadian Mineralogist*, **16**, 387–91.

Hartman, P. (1978b). *Zeitschrift für Kristallographie*, **147**, 141–6.

Hartman, P. (1979). *Fortschritte der Mineralogie*, **57**, 127–71.

Hartman, P. (1980a). *Journal of Crystal Growth*, **49**, 157–65.

Hartman, P. (1980b). *Journal of Crystal Growth*, **49**, 166–70.

Hartman, P. (1982). *Zeitschrift für Kristallographie*, **161**, 259–63.

Hartman, P. (1987). In *Morphology of Crystals*, Part A, ed. I. Sunagawa, pp. 269–319. Tokyo: Terra.

Hartman, P. (1991a). *Journal of Crystal Growth*, **114**, 497–9.

Hartman, P. (1991b). *Journal of Crystal Growth*, **110**, 559–70.

Hartman, P. and Bennema, P. (1980). *Journal of Crystal Growth*, **49**, 145–56.

Hartman, P. and Chan, H.-K. (1993). *Pharmaceutical Research*, **10**, 1052–8.

Hartman, P. and Heynen, W. M. M. (1983). *Journal of Crystal Growth*, **63**, 261–4.

Hartman, P. and Perdok, W. G. (1955a). *Acta Crystallographica*, **8**, 49–52.

Hartman, P. and Perdok, W. G. (1955b). *Acta Crystallographica*, **8**, 521–4.

Hartman, P. and Perdok, W. G. (1955c). *Acta Crystallographica*, **8**, 525–9.

Hartman, P. and Strom, C. S. (1989). *Acta Crystallographica*, **79**, 502–12.

Heijnen, W. M. M. (1985). *Neues Jahrbuch für Mineralogie Monatshefte*, pp. 357–9.

Heyes, D. M. (1980). *Physics and Chemistry of Solids*, **41**, 281–93.

Heyes, D. M. (1981). *Journal of Chemical Physics*, **74**, 1924–9.

Heyes, D. M. and van Swol, F. (1981). *Journal of Chemical Physics*, **75**, 5051–8.

Human, H. J., van der Eerden, J. P., Jetten, L. A. M. J., and Odekerke, J. G. M. (1981). *Journal of Crystal Growth*, **51**, 589–600.

International Tables for X-ray Crystallography (1969). Volume I, Birmingham: Kynoch Press.

Jackson, K. A. (1958). In *Liquid Metals and Solidification*, pp. 174–86. Metal Park, OH: American Society of Metals.

Jaffe, H. and Kjellgren, R. F. (1994). *Discussions of Faraday Society*, **5**, 319–22.

Jetten, L. A. M. J., Human, H. J., Bennema, P., and van der Eerden, J. P. (1984). *Journal of Crystal Growth*, **68**, 503–16.

Khan, A. A. and Baur, W. H. (1973). *Acta Crystallographica*, **B29**, 2721–6.

Kleber, W. (1939). *Neues Jahrbuch für Mineralogie Geologie Palaentologe* Beil. Bd. **A75**, 72–86.

Küppers, H. (1982). *Zeitschrift für Kristallographie*, **159**, 86–96.

Liu, Xiang-Yang (1993). *Journal of Chemical Physics*, **98**, 8154–9.

Liu, Xiang-Yang and Bennema, P. (1995). *Current Topics in Crystal Growth Research*, **2**, 451–534.

Liu, Xiang-Yang, Bennema, P., and van der Eerden, J. P. (1992). *Nature*, **356**, 778–80.

Liu, Xiang-Yang, Knops, H. J. F., Bennema, P., van Hoof, P., and Faber, J. (1994). *Philosophical Magazine*, **70**, 15–21.

Liu, Xiang-Yang, Boek, E. S., Briels, W. J., and Bennema, P. (1995*a*). *Nature*, **374**, 342–5.

Liu, Xiang-Yang, Boek, E. S., Briels, W. J., and Bennema, P. (1995*b*). *Journal of Chemical Physics*, **103**, 3747–54.

Loiacono, G. M., Zola, J. J., and Kostecky, G. (1983). *Journal of Crystal Growth*, **62**, 545–56.

Madelung, E. (1918). Das elektrische Feld, *Physikalische Zeitschrift*, **XIX**, 524–63.

Meekes, H., Bennema, P., and Grimbergen, R. F. P. (1998). *Acta Crystallographica*, **54**, 501–10.

Mullin, J. W., Amatavivadhana, A., and Chakraborty, M. (1970). *Journal of Applied Chemistry*, **20**, 153–9.

Niggli, P. (1923). *Zeitschrift für Kristallographie*, **58**, 490–513.

Onsager, L. (1944). *Physical Review*, **65**, 117–49.

Rappé, A. K. and Goddard, W. A. (1991). *Journal of Physical Chemistry*, **95**, 2358–70.

Rijpkema, J. J. M., Knops, H. J. F., Bennema, P., and van der Eerden, J. P. (1983). *Journal of Crystal Growth*, **61**, 295–305.

Sawada, T. and T. Shichiri (1989). In *Morphology and Growth Unit of Crystals*, ed. I. Sunagawa, pp. 117–29. Tokyo: Terra.

Stewart, J. J. P. (1995). U.S. Air Force Academy, Colorado Springs, distributed by QCPE, Bloomington, IN, USA.

Strom, C. S. (1979). *Journal of Crystal Growth*, **46**, 185–8.

Strom, C. S. (1980). *Zeitschrift für Kristallographie*, **153**, 99–113.

Strom, C. S. (1981). *Zeitschrift für Kristallographie*, **154**, 31–43.

Strom, C. S. (1985). *Zeitschrift für Kristallographie*, **172**, 11–24.

Strom, C. S. (1993). F Face Analysis of Paraxylene. Internal report, University of Nijmegen.

Strom, C. S. (1998). SeriousToo Crystal Morphology Software, P.A. 30002, 1000 PC, Amsterdam, The Netherlands; FFACE (1980–1985) and SURFPOT (1989) modules (for-profit or proprietary use to be arranged directly with the author).

Strom, C. S. (1999). Preprint, Comment on: Surface atomic structure of KDP crystals in aqueous solution: an explanation of the growth shape, to be submitted to *Physical Review Letters*.

Strom, C. S. and Heijnen, W. M. M. (1981). *Journal of Crystal Growth*, **51**, 534–40.

Strom, C. S. and Hartman, P. (1989). *Acta Crystallographica*, **A45**, 371–80.

Strom, C. S. and Bennema, P. (1997*a*). *Journal of Crystal Growth*, **173**, 150–8.

Strom, C. S. and Bennema, P. (1997*b*). *Journal of Crystal Growth*, **173**, 159–66.

Strom, C. S., Grimbergen, R. F. P., Hiralal, I. D. K., Koenders, B. G., and Bennema, P. (1995*a*). *Journal of Crystal Growth*, **149**, 96–106.

Strom, C. S., Vogels, L. J. P., and Verheijen, M. A. (1995*b*). *Journal of Crystal Growth*, **155**, 144–55.

Strom, C. S., Leusen, F. J. J., Geertman, R. M., and Ariaans, G. J. A. (1997). *Journal of Crystal Growth*, **171**, 236–49.

Strom, C. S. et al (1999). Morphology of fractal gel-grown ammonium chloride NH$_4$Cl and solution-induced reconstructive epitaxial nucleation on pseudo-flat surfaces, to be published.

Tasoni, D., Riquet, J. P., and Durand, F. (1978). *Journal of Crystal Growth*, **A34**, 55–60.

Terpstra, R. A., Rijpkema, J. J. M., and Bennema, P. (1986). *Journal of Crystal Growth*, **76**, 494–500.

Tolksdorf, W. and Bartels, I. (1981). *Journal of Crystal Growth*, **54**, 417–24.

Tosi, M. (1964). *Solid State Physics*, **16**, 1–120.

Vainstein, B. K. (1956). *Trudy Inst. Krist. Akad. Nauk SSSR*, **12**, 18–24.

Van der Voort, E. (1991). *Journal of Crystal Growth*, **110**, 653–61.

Van der Voort, E. and Hartman, P. (1988). *Journal of Crystal Growth*, **89**, 603–7.

Van der Voort, E. and Hartman, P. (1990a). *Journal of Crystal Growth*, **104**, 445–50.

Van der Voort, E. and Hartman, P. (1990b). *Journal of Crystal Growth*, **106**, 622–28.

Van der Voort, E. and Hartman, P. (1991). *Journal of Crystal Growth*, **112**, 445–50.

Van Veenendaal, E., Van Hoof, P.J.C.M., Van Suchtelen, J., Van Enckevort, W.J.P., and Bennema, P. (1998). *Surface Science*, **417**, 121–38.

Verheijen, M. A., Vogels, L. J. P., and Meekes, H. (1996). *Journal of Crystal Growth*, **160**, 337–45.

Wang, M., Liu, Xiao-Yong, Strom, C. S., Bennema, P., van Enckevort, W. J. P., and Ming, N.-B. (1998a). Fractal aggregations at low driving force with strong anisotropy, *Physical Review Letters*, **80**, 3089–92.

Wang, M., Liu, Xiao-Yong, Strom, C. S., Bennema, P., and Ming, N.-B. (1998b). Anisotropic nucleation-limited aggregation: the role of alternating epitaxial nucleations on fractal growth, submitted to *Physical Review E*.

Wang, Y. S. (1983–1993). Crystallization experiments on ADP. Internal report, Department of Solid State Chemistry, University of Nijmegen.

Wang, Y. S., Zheng, M. N., Bennema, P., Liu, Y. S., Zhu, R., Ye, G. F., Brian, J., and van Enckevort, W. J. P. (1992). *Journal of Physics D: Applied Physics*, **25**, 1616–18.

Woensdregt, C. F. (1971). Internal report, Geological and Mineralogical Institute, University of Leiden.

6

The Solid-State Structure of Chiral Organic Pharmaceuticals

G. PATRICK STAHLY AND STEPHEN R. BYRN

Chirality, or handedness, is a pervasive element in our perception of the universe. From the obvious differentiation of left- and right-handed gloves to the hard-won understanding of different physiological effects resulting from mirror-image molecular stereoisomers, chirality significantly influences human life. It is interesting to consider that there is often an apparent difference in the chirality of a gross structure versus the building blocks from which it is composed. Take people as examples. We exhibit bilateral symmetry in the whole, having a mirror plane which can generate our left side from our right (or vice versa). Thus, we have a left half and a right half which cancel each other, leaving each person achiral. However, on the molecular level, enzymes, which control the chemical reactions leading to body construction and function, are composed of only the levorotary stereoisomers of the amino acids. This chapter deals with the similar relationship between the chirality of molecules and the nature of the forms they attain in the solid (crystalline) state, particularly as it relates to organic compounds utilized as drugs.

6.1. Stereoisomerism

The development of the theory of molecular stereoisomerism arose from observations of the chirality of observable structures. A series of important discoveries in France were critical to this process. Hemihydrism, the existence of nonsuperimposable crystal forms, was noticed in quartz crystals by the mineralogist Haüy in 1801 (Fig. 6.1) (Haüy 1801). Shortly thereafter (1809) plane-polarized light was discovered by Malus, a physicist. In 1812 Biot found that quartz crystals rotate the plane of polarized light (optical rotation), the direction of rotation varying with the crystal chosen. Within the next few years Biot showed that organic liquids and solutions could also rotate polarized light, and realized that the phenomenon could be of either crystalline or molecular origin. The

Molecular Modeling Applications in Crystallization, edited by Allan S. Myerson. Printed in the United States of America. Copyright © 1999 Cambridge University Press. All Rights reserved. ISBN 0 521 55297 4.

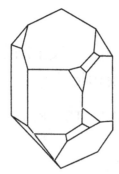 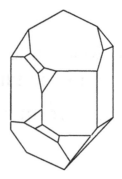

Fig. 6.1. Quartz crystals that are hemihedral. These structures are mirror images that cannot be superimposed by any amount of rotation.

relationship between hemihydrism and the direction of optical rotation in quartz was recognized by the British astronomer Herschel in 1822 (Herschel 1822).

During this period scientists worldwide were struggling to understand the way in which atoms were aggregated to form materials of varying physical and chemical properties. The regular three-dimensional structures exhibited by crystals influenced the French chemist Laurent, who in 1837 proposed in his doctoral thesis that atoms were arranged in regular polyhedra. He supposed that the geometries of substances were responsible for their intrinsic properties, and believed as well that substances with the same crystal form should exhibit the same optical activity (Salzberg 1991).

Laurent proposed to Louis Pasteur that he attempt to confirm this experimentally by a study of tartaric acid. The result was the seminal study of sodium ammonium tartrate. Pasteur was able to separate, crystal-by-crystal, sodium ammonium tartrate into two non-superimposable types. This process, called resolution by triage, was possible because these crystals contain hemihedral faces observable under the microscope. On finding that solutions of each crystal form exhibited optical rotation in opposite directions, Pasteur deduced that the three-dimensional, mirror-image structures observable in crystals must also occur in molecules themselves.

Since that event the preponderance of effort in organic stereochemistry has focused on the molecular level. Bonding theory evolved and the three-dimensional shapes attained by molecules were described. Three-dimensional molecular structure was found to greatly influence the chemical and physical properties of materials. Molecules that have the same atoms and bonding relationships, but which differ in three-dimensional structure, are called stereoisomers. Stereoisomers may be classified as either enantiomers or diastereomers.

Enantiomers are related to each other in that they are nonsuperimposable mirror images. Another term which is used to describe objects, including molecules, which are not superimposable on their mirror images is chiral. An achiral (not chiral) object is superimposable on its mirror image; a chiral object is not. Molecules capable of existing as enantiomers include those containing chiral

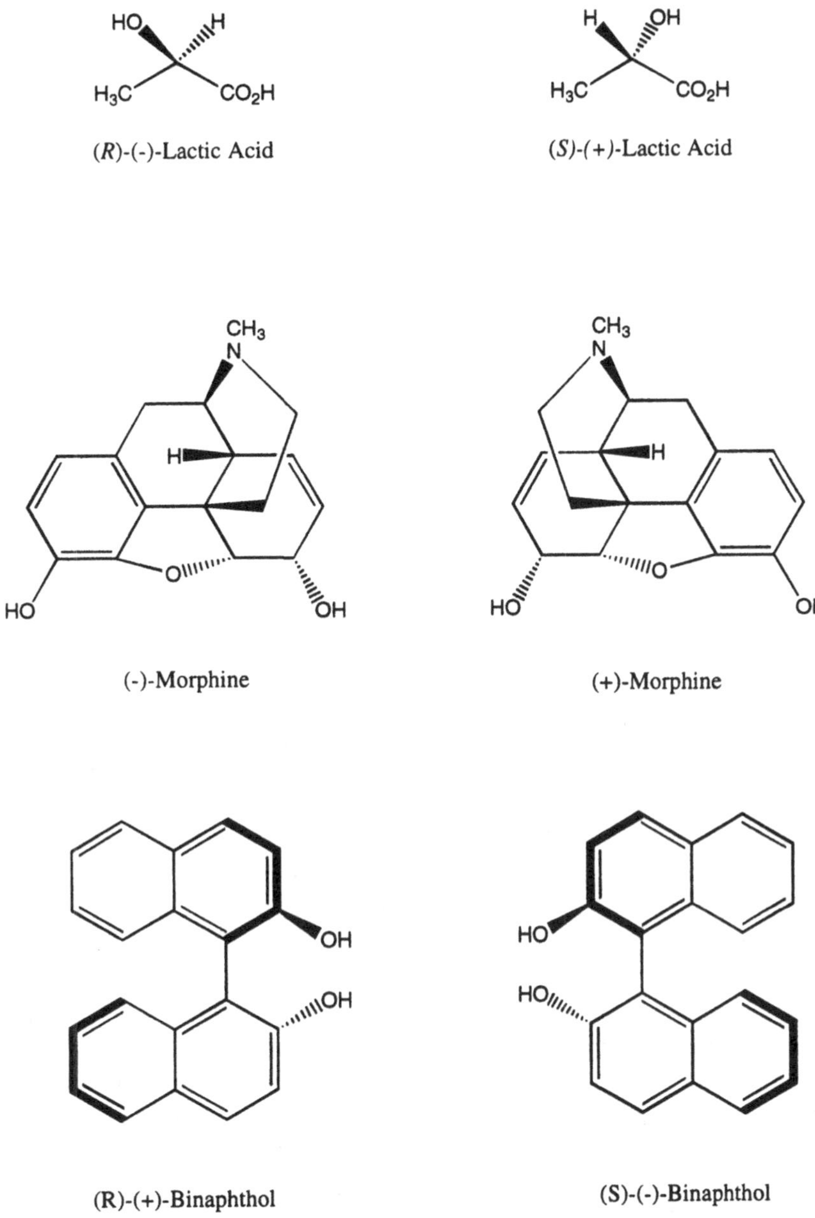

(*R*)-(-)-Lactic Acid

(*S*)-(+)-Lactic Acid

(-)-Morphine

(+)-Morphine

(R)-(+)-Binaphthol

(S)-(-)-Binaphthol

Fig. 6.2. Examples of enantiomers.

centers (such as carbon atoms bearing four different substituents) and those that contain internal axes or planes of asymmetry (Eliel and Wilen 1994, p. 1119). Some examples are shown in Fig. 6.2.

Opposite enantiomers of a given molecule exhibit identical physical and chemical properties in an achiral environment. For example, each member of a pair of enantiomers will have the same melting point, boiling point, density,

all-*trans*-Retinoic Acid 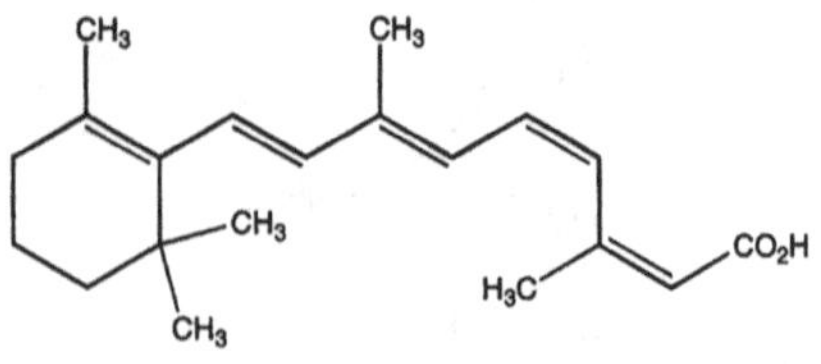   13-*cis*-Retinoic Acid

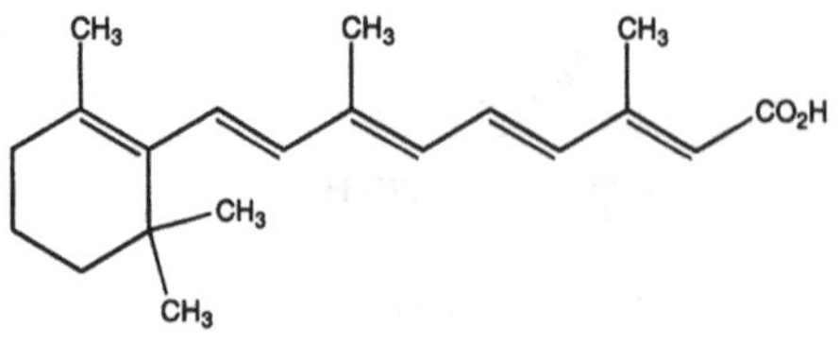

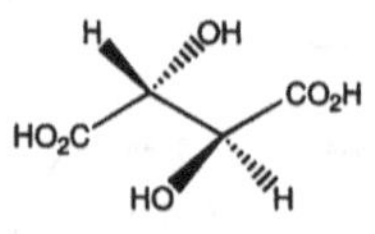

meso-Tartaric Acid 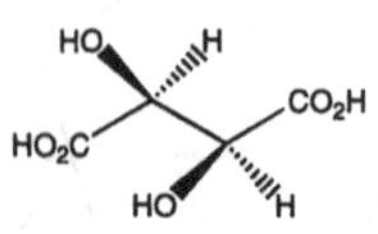   2S,3S-(-)-Tartaric Acid

Fig. 6.3. Examples of diastereomers.

heat of formation, and any other physical property. In addition, each will exhibit the same NMR, IR, UV, and other spectral properties. They will differ, however, on interaction with chiral influences. For example, enantiomers rotate polarized light in opposite directions and can react with chiral reagents at different rates.

Enantiomers can be designated in several ways, the most common being the direction in which they rotate the plane of polarized light (+ or −; *d* or *l*), by analogy to the arbitrarily chosen glyceraldehyde standards (D or L), or by their absolute configuration (R or S). The absolute configuration of a molecule is an unequivocal expression of the spatial arrangement of its atoms based on a set of specific rules (Eliel and Wilen 1994, p. 103).

The enantiomeric, or optical, purity of a substance indicates the amount of one enantiomer present relative to the opposite enantiomer. An equimolar mixture of enantiomers is a racemic mixture. Such a mixture will not exhibit optical activity, as the rotation imparted by one molecule is canceled by the opposite rotation imparted by an enantiomeric molecule. A variety of methods can be used to partially separate (optically enrich) or completely separate (optically purify) racemic mixtures. In general, the process of separating a racemic mixture into the optically pure components (enantiomers) is called resolution. Crystallization-based resolution methods are discussed later.

Diastereomers are stereoisomers that are not enantiomers. The relationships between diastereomers are of various types. Common examples are *cis* and *trans* olefins and isomers possible when more than one chiral center is present in a molecule. Some examples of diastereomers are shown in Fig. 6.3. Diastereomers differ in their physical and chemical properties, as do different molecules.

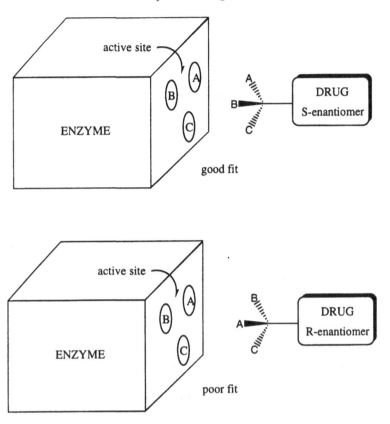

Fig. 6.4. Lock-and-key visualization of drug action. Since enzymes are composed of chiral amino acids, they exhibit chiral sites to which drugs bind.

It is well established that the physiological activity of organic drug compounds can be affected, often dramatically, by their chirality. The lock-and-key visualization of drug action holds that one enantiomer of a drug molecule will fit better into the asymmetric (chiral) enzyme binding site than will the other enantiomer of the same molecule (Fig. 6.4). Even so, most of the best-selling drugs marketed to date have not been marketed in enantiomerically pure form. Although many of these are chiral compounds, they were sold as racemic mixtures.

For example, consider the profen class of nonsteroidal anti-inflammatory drugs (NSAIDs). This is comprised of a series of 2-aryl-substituted propionic acids, each of which contains a chiral center (the carbon atom bearing the carboxylic acid group). In 1986 the 100 most widely prescribed drugs in the USA included seven NSAIDs (Reuben and Wittcoff 1989, p. 98). However, only one of these, naproxen, was sold as the optically pure S-isomer. It was commercialized in the UK in 1973 and in the USA in 1977. Early screening of compounds of this type, carried out by Syntex Corporation in the 1960s, showed that the S-enantiomer is a much more effective anti-inflammatory than is either the R-enantiomer or the racemic mixture (Harrison et al. 1970). S-Naproxen became the fourth largest-selling prescription drug in 1989.

On the other hand ibuprofen, perhaps the best-known of the NSAIDs, has always been marketed as the racemic mixture. During development of this drug, also in the 1960s, it was found that:

1. the R- and S-enantiomers have similar potency;
2. the S-enantiomer is responsible for the desired anti-inflammatory response;
3. the R-enantiomer is converted to the S-enantiomer in vivo.

These facts led to production and sale of ibuprofen as a racemic mixture.

Not always have the physiological differences between enantiomers of a drug been properly recognized. Thalidomide, marketed in Europe in 1961, was sold as a racemic mixture. It proved to be a potent teratogen, resulting in unusually high incidence of birth deformities before being withdrawn from markets worldwide (Kiefer 1997). Subsequent investigations suggest that the S-enantiomer is teratogenic but the R-enantiomer is not (Blaschke, Kraft, and Markgraf 1980).

In 1988 the Food and Drug Administration (FDA) issued a guideline for companies developing chiral compounds as drugs (DeCamp 1989). Since then guideline updates have appeared (see http://www.fda.gov/cder/guidance/stereo.htm). The FDA's Policy Statement for the Development of New Stereoisomeric Drugs, published on 1 May 1996, states the following:

- The stereoisomeric composition of a drug with a chiral center should be known and the quantitative isomeric composition of the material used in pharmacologic, toxicologic, and clinical studies known.
- To evaluate the pharmacokinetics of a single enantiomer or mixture of enantiomers, manufacturers should develop quantitative assays for the individual enantiomers in in vivo samples early in drug development.
- Unless it proves particularly difficult, the main pharmacologic activities of the isomers should be compared in in vitro systems, in animals and/or in humans.
- FDA invites discussion with sponsors concerning whether to pursue development of the racemate or the individual enantiomer.

Basically, the decision to market a pure enantiomer or a racemic mixture must be supported by knowledge of the physiological differences, or lack thereof, among the R-enantiomer, S-enantiomer, and racemic mixture of the compound in question. Rigorous justification is necessary if one hopes to gain approval of a racemate.

As a result of these guidelines, pharmaceutical companies are increasingly interested in the properties of chiral pharmaceuticals. Most have decided to develop single enantiomers of new drug entities (Stinson 1992). In addition, many are actively engaged in "racemic switches" of both prescription and over-the-counter (OTC) drugs. A racemic switch involves development of a pure enantiomer of a drug that is already marketed as a racemate. It was esti-

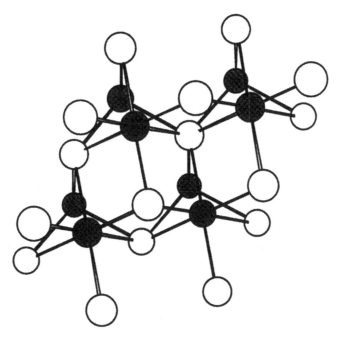

Fig. 6.5. Crystal structure of nickel arsenide. Each nickel atom (shaded circles) is equidistant from six arsenic atoms (open circles) in an octahedral array and each arsenic atom is equidistant from six nickel atoms in a trigonal bipyramidal array.

mated that by the year 2000 the global value of racemic switch products will be about $3.6 billion (DiCicco 1993).

6.2. Structure in the Solid State

Concurrent with the beginnings of stereochemistry in the 1800s was a dawning understanding of the periodic structure of crystalline materials. However, it was not until 1912 that von Laue, the discoverer of x-rays, suggested testing his theory that x-rays had wavelengths on the order of interatomic distances by attempting their diffraction by crystals. In less than a year the structure of sodium chloride was determined by W. L. Bragg using x-ray diffraction. Many inorganic crystals were subsequently examined in this way. This powerful method of determining solid structures was not applied to an organic compound until some ten years later (Dickinson and Raymond 1923).

X-ray crystallographic analyses revealed that solid materials can exist in states of varying degrees of atomic or molecular order. At the extremes are crystals, which exhibit long-range order in three dimensions, and amorphous materials, which resemble liquids in having essentially no long-range order. Ionic and metallic crystals are held together by strong, regular forces that are of consistent strength throughout the lattice. For example, in the NiAs crystal the distances between each nickel atom and adjacent arsenic atoms are identical everywhere in the structure (Fig. 6.5).

Table 6.1. *Comparison of bond energies in organic molecular crystals*

Bond type	Energy range in kcal/mol	Energy range in kJ/mol
Covalent	50–200	200–800
Hydrogen	3–8	12–30
Van der Waals	<0.1 kcal/mol	<4 kJ/mol

Conversely, most organic molecules form molecular crystals. These are composed of regular arrangements of discrete molecules in which intramolecular forces, ionic and covalent bonds, are shorter and stronger than intermolecular ones, which are usually hydrogen bonds and/or van der Waals interactions. This combination of strong and weak forces is responsible for the diverse properties found in molecular crystals (discussed below). Typical energies associated with these bond types are shown in Table 6.1. A typical molecular crystal is illustrated by the structure of hydroxyurea shown in Fig. 6.6.

A significant number of materials, be they elemental, inorganic compounds, or organic compounds, can crystallize in more than one way. Such behavior leads to solids, composed of the same building blocks, with varied properties. Although solids related in this way exhibit different physical properties, if melted, vaporized, or dissolved the resulting liquids, gases, or solutions are identical.

When exhibited by elements, this phenomenon is known as allotropism. One example is carbon which, in addition to an amorphous state, exists as crystalline diamond, graphite, or fullerenes (Fig. 6.7). The physical property differences of these allotropes, at least on comparing diamond with graphite, are obvious and striking. Another example is elemental phosphorus, which occurs in two white forms (α and β), one red form, and one black form. White phosphorus is spontaneously flammable in air and ingestion of approximately 50 mg is likely to be fatal. Red phosphorus, on the other hand, is stable and safe enough to be used in matches.

Inorganic and organic compounds that exist in multiple solid forms are said to be polymorphic. An example of an inorganic compound that is polymorphic is silica, of which there are seven polymorphs known: low and high quartz, low, middle, and high tridymite, and low and high cristobalite. Each of these forms is stable in a particular temperature region. Water has eight or nine polymorphic forms.

Polymorphism occurs in a significant number of organic crystals. The scope and importance of polymorphism as it relates to organic drugs has been reviewed (Byrn 1982; Borka and Haleblian 1990). A recent search of the Cambridge Structural Database found 204 pairs of organic polymorphs for which complete crystal structures have been determined (Gavezzotti and Filippini 1995). Note

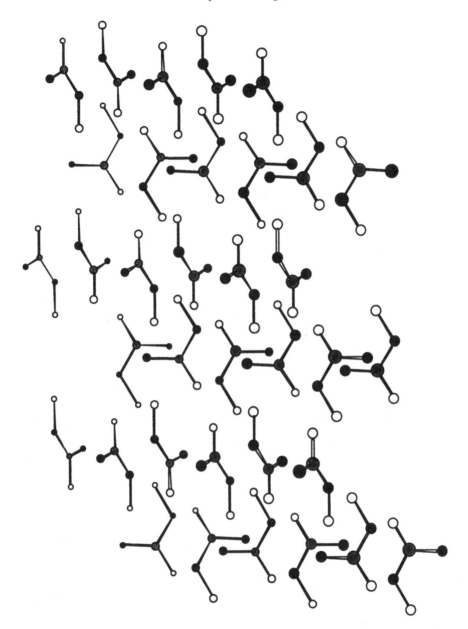

Fig. 6.6. Crystal structure of hydroxyurea. Shaded circles are carbon atoms, black circles are nitrogen atoms, and open circles are oxygen atoms. Covalently bonded molecules (covalent bonds are shown) are held in a regular array by hydrogen and van der Waals bonds (not shown).

also that organics often form crystals with inclusion of water or solvent molecules in ordered fashions in the crystal lattice. These hydrates or solvates exhibit diverse solid properties, as do related polymorphs. Sometimes an organic molecule exists in multiple solid forms. For example, seven forms of dehydroepi-

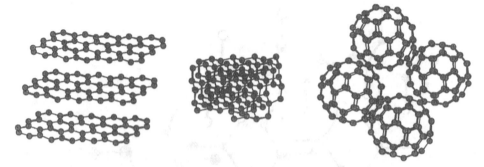

Fig. 6.7. Three allotropes of carbon are graphite (left), diamond (center), and fullerene (right). In graphite, sheets of covalently bound carbon atoms are held together by van der Waals forces, in diamond all of the carbon atoms are covalently bound in tetrahedral arrays, and, in fullerene, spheres of covalently bound carbon atoms are held together by van der Waals forces.

androsterone (DHEA) have been characterized: three polymorphs, three hydrates, and one methanol solvate (Chang, Caira, and Guillory 1995). The crystal structures of four of these, one polymorph, two hydrates, and one methanol solvate, were determined and are shown in Fig. 6.8 (Cox et al. 1990; Caira, Guillory, and Chang 1995).

Different solid forms of the same substance have different energies associated with their crystal lattices. Thus, under any given set of conditions a specific order of stability exists. This order may change as conditions change. Transitions from one polymorph to another may take place in several ways, depending on the structural changes involved. Polymorphic transformations in molecular crystals involve rearrangement of the relatively weak intermolecular bonds. This subject has been reviewed (McCrone 1965).

The two disciplines of stereochemistry and crystal structure analysis allowed the relationships between chiral solid structure and behavior to be rapidly understood. Perhaps the most significant practical consequences of this were improved physical methods for separation of racemic mixtures into their constituent enantiomers (Jacques, Collet, and Wilen 1981a, part 2). More recently, however, the importance of the solid structures of chiral drugs has been recognized.

An aspect of drug development coming under increased scrutiny by the FDA is solid form control. Polymorphism is common in organic compounds, and different polymorphs of a given active drug often differ in properties critical to ease of formulation and rate of in vivo delivery. In addition, patents on new solid forms of an existing drug can be of value. An interesting example of this is a U.S. patent that claims halocarbon solvates of anti-inflammatory steroids (Cook and Hunt 1982). The inventors discovered that, in existing aerosol spray formulations containing steroids and halocarbon propellants, halocarbon solvates of the drugs were formed. Thus their composition-of-matter claim on these solvates provided patent coverage of an established array of products.

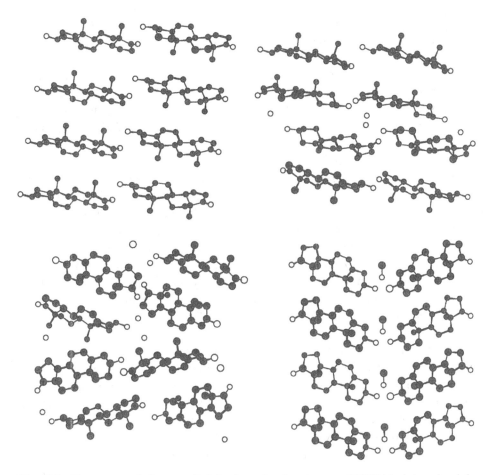

Fig. 6.8. Four crystal forms of dehydroepiandrosterone (DHEA) determined by single-crystal x-ray diffraction. Shaded circles are carbon atoms and open circles are oxygen atoms. An anhydrous polymorph is shown at upper left, a hydrate is shown at upper right, a different hydrate is shown at lower left, and a methanol solvate is shown at lower right. The waters and methanol of crystallization are visible as the atoms that are not covalently bonded to the DHEA molecules.

6.3. Crystallography

Molecules, or any other entity, can be arranged in regular, three-dimensional patterns in many ways. Descriptions of these arrangements are based on the symmetry of the molecules and the patterns which they attain in the crystal lattice. Symmetrical objects are those that are unchanged by operations such as rotation about an axis or reflection through a plane (point symmetry operations). The more symmetrical an object, the more point symmetry operations are evident in its structure. Point groups are sets of point symmetry operations that describe individual molecules. Additional symmetry operations may be present in a collection of objects. Translational symmetry operations are those that generate one object in the collection from another. All the possible three-dimensional arrangements attainable by any object can be described by a finite number of combinations of point and translational symmetry operations, called space groups. There are 230 possible space groups. It is interesting that these were described in the 1890s, long before they were utilized in the solution of crystal structures from x-ray diffraction data (Bunn 1961, p. 267). Many texts dealing with crystallography contain tabulations of the space groups (Bunn 1961, p. 468). While crystals have been found that represent most of the 230 space groups, about 60 percent of the organic crystals studied exist in one of only six (Glusker and Trueblood 1985, p. 97).

It is important to understand the symmetry operations that are critical to the description of chiral molecules. A chiral molecule is one that is not superimposable on its mirror image. This means that a chiral molecule cannot be described by any point symmetry operation that converts a left-handed structure into a right-handed one. Examples of operations of this type, called inverse symmetry operations, include reflection through a mirror plane or inversion through a center. In the same way, space groups may be classified as those that contain inverse symmetry operations (racemic, or centrosymmetric) and those that do not (chiral, or noncentrosymmetric). Of the 230 space groups, 66 are noncentrosymmetric.

Pure enantiomers can only crystallize in noncentrosymmetric space groups. On the other hand, achiral molecules can crystallize in either centrosymmetric or noncentrosymmetric space groups. Stated another way, asymmetric crystals can arise from chiral or achiral molecules. Quartz crystals, such as those studied by Haüy (see above), are examples of the latter. An analogy useful in understanding this situation, presented by Jacques et al. (1981a, p. 7), is that either a left- or right-hand spiral staircase may be built from the same achiral bricks. Another concept to be realized from these relationships is that molecules that crystallize in centrosymmetric space groups can be either chiral or achiral. In order for chiral molecules to crystallize in a centrosymmetric space group, they must be present in equal numbers of R- and S-isomers (a racemic mixture). Therefore, if a crystal is found to be achiral (centrosymmetric), it is composed of either achiral molecules or a racemic mixture of chiral molecules.

If a suitable single crystal of a material is available, an x-ray study can provide a wealth of information. The diffraction pattern yields information about the symmetry of the crystal; unit-cell parameters (the unit cell is the building block from which the entire crystal lattice can be generated) and the space group. Measurement of the intensity of diffracted x-rays at specific spots in the diffraction pattern can be interpreted to give the relative coordinates of the atoms in the crystal. In this way the entire crystal structure is obtained, including both the constitution of the molecules and the way in which they pack into a crystal lattice.

Another x-ray technique that is useful is x-ray powder diffraction (XRPD). In this, x-ray irradiation of a sample of solid powder provides a pattern of diffracted x-ray intensities at distinct angles relative to the incident beam. A specific crystalline lattice exhibits a unique XRPD pattern, which can be considered a fingerprint of that lattice. XRPD analysis is a powerful method of solid-form identification.

6.4. Solid Forms of Chiral Compounds

In discussing the solid forms attained by chiral organic molecules, both the structures of pure enantiomers and enantiomer mixtures must be considered. For compounds that have two enantiomeric isomers, the racemic mixtures can exist in one of three forms: a racemic compound, a conglomerate, or a solid solution. This terminology can be confusing. Remember that "racemic mixture" means an equimolar mixture of R- and S-isomers but "racemic compound" denotes a specific structural arrangement of a racemic mixture.

A racemic compound is a crystalline arrangement wherein equal numbers of R- and S-isomers are present in the crystal lattice. Each occupies a specific, regular spot in the crystal. Consider now the relationships between a solid racemic compound and the solids formed by either of the pure enantiomers that make it up. The racemic compound behaves as a different substance than either of the solid enantiomers. The solid enantiomers, on the other hand, are identical to each other (in the absence of chiral influences). Therefore, the racemic compound when compared to either enantiomer can exhibit a different melting point, heat of fusion, solubility, vapor pressure, density, solid-state infrared spectrum, solid-state NMR spectrum, and XRPD pattern. The racemic compound will crystallize in either a centrosymmetric or noncentrosymmetric space group; either enantiomer will crystallize only in a noncentrosymmetric space group. A well known pharmaceutical that forms a racemic compound under ambient conditions is ibuprofen. The structures of crystalline R,S-ibuprofen (the racemic compound) and crystalline S-ibuprofen (one pure enantiomer) are shown in Fig. 6.9; some of the properties of these materials are compared in Table 6.2.

An interesting consequence of these property differences is that the enantiomer content (optical purity) of a nonracemic mixture can sometimes be determined using an achiral technique. Typically, optical purities are determined using chiral techniques such as high pressure liquid chromatography (HPLC) analyses employing columns containing chiral solid phases or by NMR analyses in the presence of chiral shift reagents. However, if a racemic compound is crystallographically distinct from its constituent enantiomers, each will exhibit a different XRPD pattern and XRPD can therefore be used for optical purity determination (Schlenk 1965). This technique has been applied to ibuprofen. The XRPD patterns of R,S- and S-ibuprofen are shown in Fig. 6.10. By measuring the peak heights of the unique $\{100\}$ diffraction peak ($6.1°2\theta$) of the racemic compound in mixtures of R,S- and S-ibuprofen, a calibration curve was generated. The only sample preparation required was grinding to effect particle-size reduction. Based on the curve, it was calculated that the limit of detection of R,S-ibuprofen in S-ibuprofen was one percent. Therefore R-ibuprofen can be detected in S-ibuprofen (or the reverse) at the 0.5 percent level using an achiral method and solid samples (Stahly et al. 1997).

Table 6.2. *Comparison of the properties of crystalline ibuprofen:*

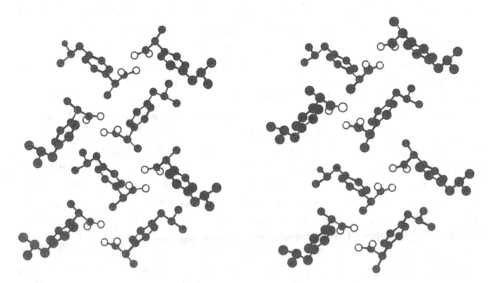

Property	R,S-Ibuprofen (racemic compound)	S-Ibuprofen	References
Melting point (°C)	71	46	Dwivedi et al. (1992)
Heat of fusion (kJ/mole)	26.9 ± 1.0	19.9 ± 0.8	Dwivedi et al. (1992)
Solubility in aqueous HCl–KCl at 25°C (M)	0.943×10^{-3}	1.79×10^{-3}	Dwivedi et al. (1992)
Crystallographic space group	$P2_1/c$	$P2_1$	McConnell (1974) Freer et al. (1993)

Fig. 6.9. Crystal structures of R,S-ibuprofen (the racemic compound) and S-ibuprofen (one pure enantiomer). Shaded circles are carbon atoms and open circles are oxygen atoms. Note the dimers formed by hydrogen-bond pairing of carboxylic acid groups, and the similarity of the molecular arrangements in these structures.

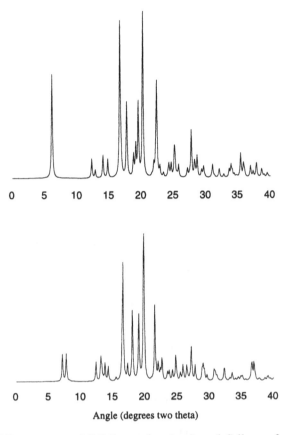

Fig. 6.10. XRPD patterns of RS-ibuprofen (top) and S-ibuprofen (bottom) calculated from single-crystal data.

A conglomerate is a solid racemic mixture that contains equal numbers of R- and S-crystals. It is therefore simply a physical mixture of the two solid enantiomers, which can in theory be separated by triage (manual sorting as carried out by Pasteur). A crystal chosen for x-ray study will be either an R- or an S-crystal; both will yield essentially the same data, including space group, on structure determination. Note, however, that there are differences in the response of enantiomeric crystals to x-rays that provide methods, the most common being that of Bijvoet, to determine the absolute configuration of a molecule in a crystal (Glusker and Trueblood 1985, p. 136). Physical properties of a conglomerate differ from those of the constituent enantiomers; these include melting point, heat of fusion, solubility, and vapor pressure. In a bulk sense a conglomerate behaves as a mixture of two materials. On the other hand, spectral properties, such as revealed by solid-state IR, solid-state NMR, and x-ray powder diffraction, are the same. These reflect internal crystal arrangements, which are the same in both R- and S-crystals. A comparison of properties of enantiomeric and racemic norfenfluramine dichloroacetate, which exists as a conglomerate, is shown in Table 6.3 (Coquerel and Perez 1988).

Table 6.3. *Comparison of the properties of crystalline norfenfluramine dichloroacetate:*

Property	R,S-Norfenfluramine dichloroacetate (conglomerate)	R-Norfenfluramine dichloroacetate
Melting point (°C)	117	139
Heat of fusion (kJ/mole)	39.1 ± 0.4	41.3 ± 0.4
Solubility in 95% ethanol at 22°C (%)	24.9	14.5

In rare cases racemic mixtures will exist as solid solutions, also called pseu-doracemates. These crystalline structures can accept either enantiomer indiscri-minately into identical positions. Therefore both enantiomers coexist in a single crystal without regard to stoichiometry. Most solid solutions exhibit identical physical properties regardless of composition. That is, the racemic mixture, either enantiomer, and any mixture in between behave identically. Sometimes there are physical property variations dependent on enantiomer composition in solid solutions (see the discussion of binary melting point phase diagrams below). As with conglomerates, spectral properties are independent of composition. One pharmaceutical compound that exists as a solid solution under ambient condi-tions is atropine (Fig. 6.11) (Kuhnert-Brandstätter 1993). Note that atropine is the name given to the racemic mixture of this compound; the pure S-isomer is called hyoscyamine.

How can the solid form of a given racemic mixture be determined? If a single crystal can be obtained, an x-ray structure determination will provide the answer. A method that does not depend on single-crystal availability is construction of a binary melting-point phase diagram. This is a plot of the melting points of samples of varying enantiomeric compositions. Each possible racemic form is associated with a unique diagram. Idealized diagrams are shown in Fig. 6.12. The horizontal axes represent compositions; pure R-enantiomer on one side, pure S on the other, and the racemic mixture in the center. The vertical axes represent melting points. Each diagram consists of a plot of the temperatures of melting terminations. The low-melting points are called eutectics. Notice that

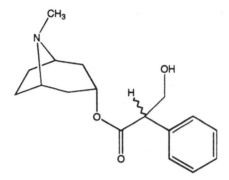

Fig. 6.11. Atropine.

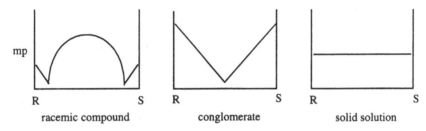

| racemic compound | conglomerate | solid solution |

Fig. 6.12. Idealized binary melting-point phase diagrams.

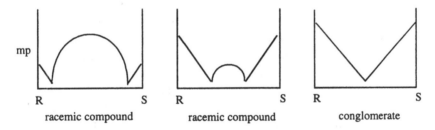

| racemic compound | racemic compound | conglomerate |

Fig. 6.13. More idealized binary melting-point phase diagrams.

these diagrams are symmetric, a consequence of the fact that the enantiomers are thermodynamically identical.

For a racemic compound the eutectic composition is not the racemic mixture. Rather, it is somewhere in between the racemic mixture and a pure enantiomer. The melting point of the racemic compound may be lower or higher than that of a pure enantiomer. Thus a continuum of melting point behaviors is possible for racemic compounds, as illustrated in Fig. 6.13. The eutectic composition exhibited by a conglomerate is always the racemic composition, which may be thought of as representing one extreme of the continuum illustrated in Fig. 6.13. No sharp change in melting behavior, such as the eutectics discussed above, occurs with composition changes for solid solutions.

An easy way to visualize these melting-point behaviors is by analogy to melt-ing-point depression of impure compounds. Organic chemists have long used melting-point depression as a qualitative gauge of compound purity. Since a racemic compound is physically different from its constituent enantiomers, the melting point of a mixture of the racemic compound and either enantiomer is depressed. In a conglomerate the racemic mixture is not a discrete compound, so mixing of the physically enantiomeric R- and S-isomers results in melting-point depression. The eutectic in this case is exactly at the racemic mixture.

Binary melting-point phase diagrams can be constructed by measuring the enantiomeric composition and melting range of a series of mixtures. Alternatively, in many cases the diagram can be calculated given only the melting point and heat of fusion of one pure enantiomer and the racemic mixture (Jacques et al. 1981a, pp. 46 and 90). These data are available simply from differential scanning calorimetric analysis of each compound. That ibuprofen forms a racemic compound is clear from both the calculated and experimentally determined diagrams; furthermore, the calculated and observed eutectic compo-sitions match reasonably well (Manimaran and Stahly 1993).

Most chiral organic molecules form racemic compounds. Of 1308 compounds inventoried in 1981, only 126 were identified as possible conglomerates (Jacques, Leclercq, and Brienne 1981b). It is often stated that 5–10 percent of racemic mixtures exist as conglomerates, and that solid solutions are rare. Interestingly, organic salts appear to form conglomerates with two to three times the frequency of organic covalent compounds (Jacques et al. 1981b).

Much of our knowledge of the solid forms attained by racemic mixtures of pharmaceutical compounds, and all organic compounds for that matter, arose out of studies aimed at separating such mixtures into their component enantio-mers. This process is called resolution or, sometimes, optical purification. The most widely used resolution methods involve crystallizations. To understand how these processes work, one must first realize that there is a close relationship between melting point and solubility behavior. Both melting and dissolution phenomena involve energy changes necessary for disruption of the crystal lattice. Therefore, a plot of solubilities against enantiomeric composition (binary solu-bility phase diagram) will resemble a plot of melting points against enantiomeric composition (binary melting-point phase diagram) for a given set of enantio-mers. The binary phase diagrams shown in Figs 6.12 and 6.13 would be the same if the vertical axes represented, rather than melting point, the amount of a solvent necessary to dissolve a given weight of material. The eutectic composi-tions are the most soluble. A comparison of melting point and solubility beha-vior is shown in Table 6.4 for ibuprofen (Dwivedi et al. 1992).

The effect of these solubility relationships on crystallization behavior is central to the theme of crystallization-based resolutions. If a mixture of enantiomers is crystallized from solution, the composition of the crystal crop will be uphill of the starting composition on the phase diagram. For example, consider a racemic compound exhibiting the binary solubility phase diagram shown in Fig. 6.14 (only half of the symmetrical diagram is depicted). If a racemic mixture were

Table 6.4. *Melting point and solubility behavior of ibuprofen*

Enantiomeric composition	Melting point (°C)	Solubility in aqueous HCl–KCl at 25°C (M)
50% S/50% R (racemate)	71	0.943×10^{-3}
99.7% S	46	1.79×10^{-3}
82% S/18% R (eutectic)	37	1.96×10^{-3}

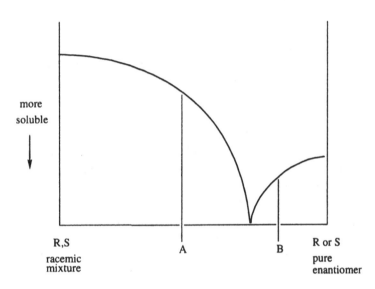

Fig. 6.14. One-half of the melting-point phase diagram of a racemic compound. Crystallization of the racemic mixture will yield a crystal crop containing racemic mixture. Crystallization of a mixture of composition A will yield a crystal crop containing racemic mixture. Crystallization of a mixture of composition B will yield a crystal crop containing pure enantiomer.

recrystallized from the solvent (unspecified in this generalized diagram), the crystal crop would always be racemic because this is the highest point on the diagram. If a solid of the composition A were recrystallized, the crystal crop would be less optically pure (closer to the racemic composition) and the mother liquor would contain a more optically pure composition (closer to the eutectic composition). On the other hand, if a solid of the composition B were recrystallized, the crystal crop would be more optically pure (closer to the pure enantiomer composition) and the mother liquor would contain a less optically pure composition (closer to the eutectic composition). The thermodynamic drive is to leave the eutectic composition in solution. Also, the eutectic composition may

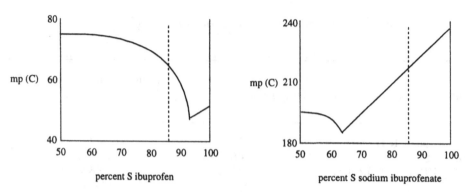

Fig. 6.15. Reproductions of calculated binary melting-point phase diagrams of ibu-profen and sodium ibuprofenate.

not be passed under equilibrium conditions. That is, a crystallization of the sample of composition B that reaches equilibrium cannot yield crystals contain-ing more pure enantiomer than the amount in the eutectic composition.

The practical consequence of this behavior is that a racemic mixture that exists as a racemic compound cannot be resolved by crystallization. However, consider a nonracemic mixture (containing unequal numbers of R- and S-enantiomers) of enantiomers that form a racemic compound. Such a mixture can be considered a mixture of the racemic compound and the enantiomer in excess and may, depending on the position of the eutectic relative to the mixture composition, be optically purified by crystallization. The purified fraction can reside in either the crystal crop or the mother liquor. For example, Fig. 6.15 shows reproduc-tions of binary melting-point phase diagrams calculated for ibuprofen and its sodium salt (Manimaran and Stahly 1993). Recrystallization of a mixture of 85 percent S and 15 percent R ibuprofen from hexane yielded a racemic crystal crop and a mother liquor enriched in the S-isomer, but recrystallization of a similar mixture of sodium ibuprofenate from acetone yielded a crystal crop of > 99 percent S-isomer (Manimaran and Stahly 1993; Stahly 1993). The starting com-positions for these experiments are shown in Fig. 6.15 by the dotted lines.

On the other hand, racemic mixtures that exist as conglomerates can be resolved by direct crystallization. Based on the phase diagram, such a mixture has nowhere to go but uphill. However, as with racemic compounds, under thermodynamic conditions the eutectic composition will be left in solution. Since for a conglomerate this is the racemic composition, no separation will occur at equilibrium. A kinetic crystallization, on the other hand, will lead to optical purification. This is most frequently accomplished by single-enantiomer seeding of a supersaturated solution of the racemate and rapid harvesting of the resulting crystals. This affords a crystal crop enriched in the seed enantiomer and a mother liquor enriched in the opposite enantiomer. By seeding a supersatu-rated solution of the mother liquor with the opposite enantiomer, a crystal crop enriched in this enantiomer is obtained. Subsequent crystallizations of the initial crystal crops will lead to increased optical purities or, if low per-pass recoveries

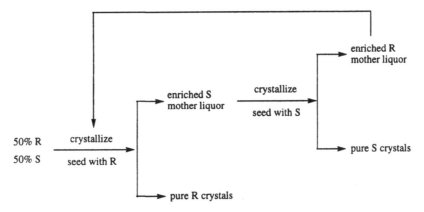

Fig. 6.16. Resolution by direct crystallization of a conglomerate.

are tolerable, essentially pure enantiomer can be obtained in each crystallization. The process is depicted in Fig. 6.16. Occasionally a system is encountered in which direct crystallization can be coupled with a rapid, in-situ racemization. The solution equilibrium of racemic composition is upset as one enantiomer crystallizes out, with the result that a one-pot yield of 100 percent pure enantiomer is possible starting with the racemate. This is called a second-order asymmetric transformation.

Since only 5–10 percent of chiral organic compounds form conglomerates, most resolutions cannot be accomplished by direct crystallization. Typically, racemic mixtures are resolved by formation and separation of diastereomers. This involves reaction of the mixture with an optically pure resolving agent. The result is a 1:1 mixture of diastereomers, which are non-enantiomeric stereo-isomers having different physical properties. A binary melting-point (or solubility) phase diagram of such a mixture is almost always asymmetric because the diastereomers are thermodynamically different (Fig. 6.17).

Crystallization of a mixture of diastereomers will provide a crystal crop enriched in one of the isomers. The level of enrichment depends on the magnitude of both the difference between the diastereomer solubilities and the difference between the eutectic composition and the starting composition. There are many examples of resolutions accomplished by crystallization of diastereomers. The resolution of naproxen in this way is shown in Fig. 6.18 (Felder et al. 1981).

For resolutions that are to be carried out at commercial scale, low-cost methods are desirable. In this regard, direct crystallization can be attractive because it avoids the purchase and processing of a chiral resolving agent. However, as noted earlier, most racemic mixtures exist as racemic compounds and therefore cannot be resolved by direct crystallization. For some compounds of commercial interest, resolution method development has included preparation and solid-state characterization of multiple derivatives in search of a conglomerate. For example, resolution of naproxen by diastereomeric salt crystallization has already been discussed (Fig. 6.18). In a conglomerate search, methyl and ethyl esters

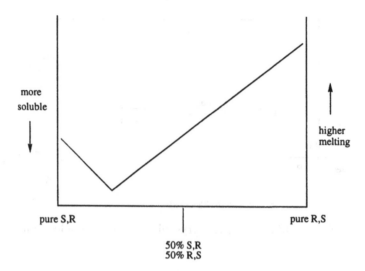

more
soluble

higher
melting

pure S,R

pure R,S

50% S,R
50% R,S

Fig. 6.17. Idealized binary phase diagram of a diastereomer mixture.

of naproxen were found to be conglomerates and, therefore, resolvable by direct crystallization (Kazutaka, Ohara, and Takakuwa 1983).

As stated earlier, for the same substance different crystalline arrangements (polymorphs, hydrates, or solvates) are frequently encountered. This is true for both pure enantiomers and racemic mixtures, often complicating characterization of the solid states exhibited by a chiral material. For example, a racemic compound can exist in polymorphic forms that are themselves racemic compounds. Each will have a different crystalline arrangement, but will consist of equal numbers of R- and S-molecules in their crystal lattice. The permutations of polymorphism related to chiral compounds were reviewed by Jacques et al. and are summarized in Table 6.5 (Jacques et al. 1981a, p. 136).

Sometimes a racemic mixture is capable of existing in different arrangements (Jacques et al. 1981a, pp. 137–44). For example, a racemic compound may transform into a conglomerate (or a solid solution) in response to changes in conditions. In such a case the racemic compound and the conglomerate (solid solution) cannot be considered polymorphic in the strict sense; their crystal lattices contain different (isomeric) molecules. An example is histidine chlorohydrate. Ternary solubility diagrams determined for this compound at various temperatures revealed that the racemic mixture exists as a conglomerate at 45°C but as a racemic compound at 35°C and below (Jacques and Gabard 1972).

The complexity of solid structures of stereoisomers is evident from a study of sobrerol, a compound used in the treatment of respiratory diseases (Bettinetti et al. 1990). Since sobrerol has two chiral centers, it exists as two diastereomeric pairs of enantiomers, as shown in Fig. 6.19.

A solid-state, carbon-13 NMR and solid-state IR spectral survey of these compounds showed that the spectra of the (+)-*trans* (and (−)-*trans*) isomers were indistinguishable from those of the (±)-*trans* racemic mixture, but the

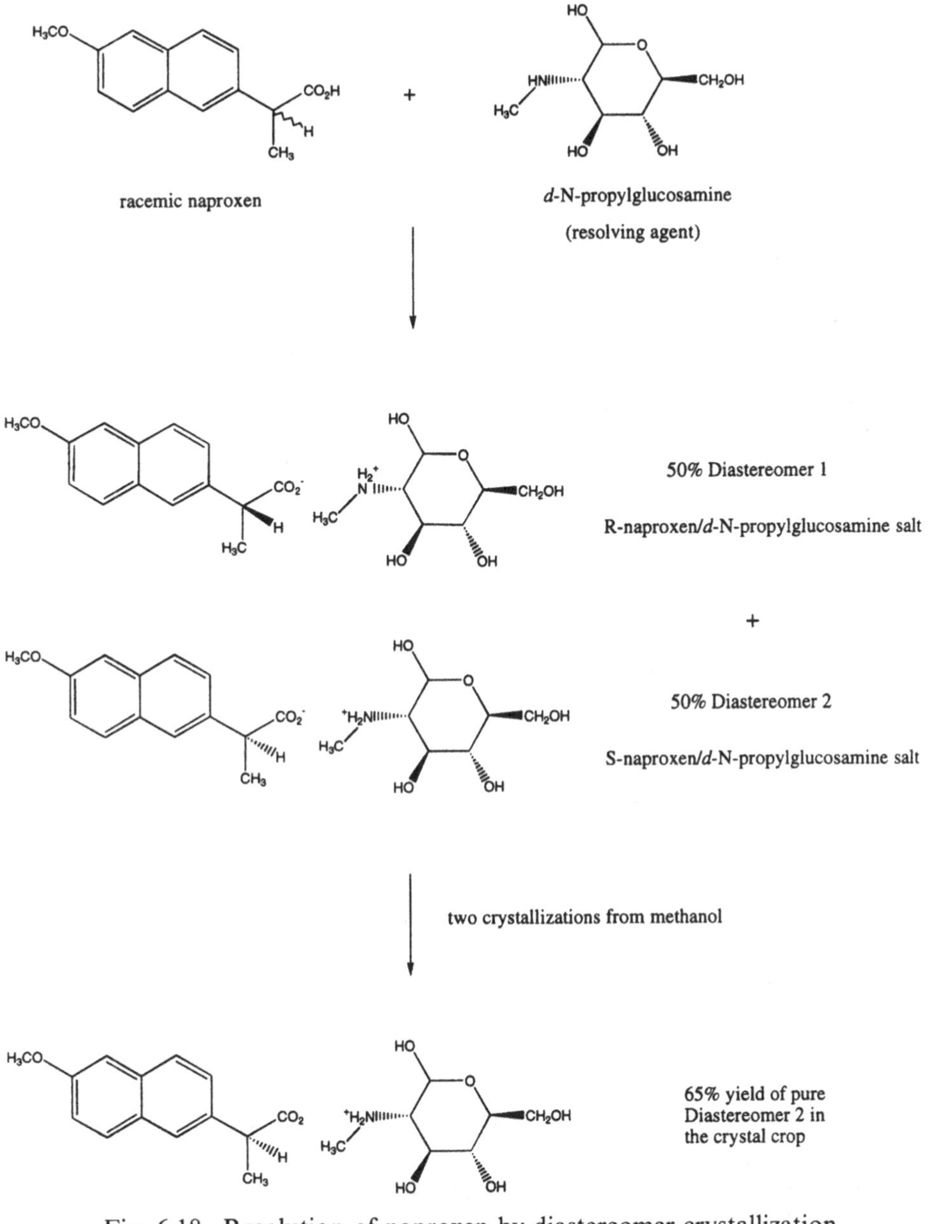

Fig. 6.18. Resolution of naproxen by diastereomer crystallization.

spectra of the $(+)$-*cis* (and $(-)$-*cis*) isomers were very different from those of the $(\pm)$-*cis* racemic mixture. These results suggested that the $(\pm)$-*trans* racemic mixture was a conglomerate and the $(\pm)$-*cis* racemic mixture was a racemic compound. However, thermal analyses revealed that both racemic mixtures formed racemic compounds, but with different eutectic compositions (Table 6.6). Note that the *trans* system is more like a conglomerate (the eutectic composition is closer to 50:50) than is the *cis* system.

Table 6.5. *Possible polymorphic relationships in chiral systems*

Possible relationships	Examples known as of 1981
Polymorphic enantiomers, polymorphic conglomerates	No
Polymorphic enantiomers, one racemic compound	No
One enantiomer, polymorphic racemic compounds	Many
Polymorphic enantiomers, polymorphic racemic compounds	Yes
Polymorphic enantiomers, polymorphic solid solutions	Yes

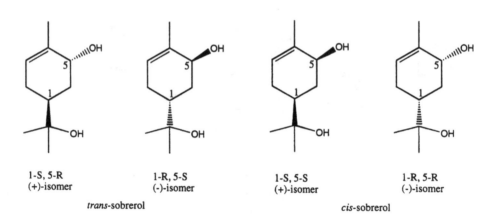

    1-S, 5-R   1-R, 5-S   1-S, 5-S   1-R, 5-R
    (+)-isomer   (-)-isomer   (+)-isomer   (-)-isomer

         trans-sobrerol   *cis*-sobrerol

Fig. 6.19. Sobrerol isomers.

The thermal analyses also suggested the presence of a polymorph of the *cis*-sobrerol racemic compound, which was subsequently isolated. X-ray powder patterns of these five isomers, (+)-*trans*, (±)-*trans*, (+)-*cis*, (±)-*cis* Form A, and (±)-*cis* Form B, were determined and each was different from the others. It was clear from these patterns, however, that there is a high structural similarity between the (+)-*trans* and the (±)-*trans* isomers. This is consistent with the "near-eutectic" nature of the *trans* racemic compound, as well as the similarity of the (+)-*trans* and (±)-*trans* NMR and IR spectra.

6.5. Prediction of Solid Structure

It is currently not possible to predict the solid form a racemic mixture will attain, given only molecular structure. In fact, the more general goal of predicting the

Table 6.6. *Summary of sobrerol data*

Compound	Comparison of (+) and (±) NMR and IR spectra	Eutectic composition	(+) melting point (°C)	(±) melting point (°C)
trans-Sobrerol	Same	36% (+) and 64% (−)[a]	155	131
cis-Sobrerol	Different	18% (+) and 78% (−)[a]	110	97

[a]Or the reverse.

crystal structure expected from an organic compound has been achieved in only a few cases. The problem of predicting the solid structure of chiral compounds is thus a subset of the larger problem of predicting the solid structures of any molecular crystal. The arrangement of molecules in a crystal, the structural differences between polymorphs, and the structural differences between a conglomerate and a racemic compound all depend on small energy differences that result from the numerous and weak intermolecular forces that exist in a particular molecular organization. The appeal of the power to predict solid structure is great; the physical properties of a substance could be estimated as easily as the molecular formula could be drawn.

We are aware of no reports describing attempts to calculate the solid form that will be attained by a chiral molecule. There is, however, literature related to the prediction of crystal structures for achiral organic molecules. The approaches described therein are reviewed below. In addition, some generalizations that may provide approaches to chiral structure prediction are discussed.

Those compounds that are incapable of forming intermolecular hydrogen bonds are held together by van der Waals forces. These are weak attractions that arise primarily from the alignment of electrical dipoles in neutral molecules. The structures of organic crystals held together only by van der Waals forces are characterized by close-packing arrangements. That is, the molecules are as close to each other as allowed by their overall geometries. The "bumps and grooves" on the molecular surfaces are interlocked as much as possible. A measurement of the closeness of packing is given by the packing coefficient, which is the ratio of the molecular volume to the volume of the crystal unit cell. This ratio approaches unity as the density of molecular packing increases. Packing coefficients for organic crystals are usually 0.65–0.77, which is similar to the range of values found for close-packed arrangements of spheres and ellipsoids (Kitaigorodsky 1973, p. 18).

Most organic molecules of interest, particularly those exhibiting pharmaceutical potential, contain a variety of functional groups. Many of these are capable of forming hydrogen bonds and they include alcohol, amine, ether, ketone, and carboxylic acid groups. In addition, organic salts are commonly administered forms of drugs. The intermolecular forces that dominate the crystalline arrangements of such molecules are hydrogen bonds. Since hydrogen bonds are stronger than van der Waals forces (see Table 6.1), when possible, molecules are aligned in a crystal lattice in a way that maximizes the number of hydrogen bonds formed, therefore minimizing the overall free energy. However, to attain the proper alignment of atoms involved in hydrogen bonds, some amount of close packing must often be sacrificed.

Crystal-lattice construction is driven by the simple goals that stability and symmetry be maximized. However, a complex interplay of factors is involved. In general, molecules pack as close to each other as possible in a crystal lattice, minimizing void space. Such arrangements minimize the structural free energy by maximizing intermolecular attractions. Clearly crystal structures represent delicate balances of relatively weak intermolecular forces. It is because of this that polymorphism occurs in many organic solids.

Some generalizations have been noted relative to the molecular structure–solid form relationship in racemic mixtures. Collet (1990) suggested that compounds which pair up about a center of symmetry, such as simple carboxylic acids, form stable racemic compounds, but that compounds which cannot pair up in this way tend to form conglomerates.

In many cases a crystalline racemic compound contains planes or columns of atoms that are identical to columns or planes found in the crystals of its component enantiomers. The difference between the structures arises from different column-to-column or plane-to-plane orientations. This is the case for ibuprofen, which forms molecular columns held together by hydrogen-bond networks. In Fig. 6.9 the columns are oriented vertically. The surfaces of the columns in the racemic compound differ from the surfaces of the columns in the S-enantiomer because of the different orientations of molecules forced by the spatial arrangement about the chiral center. Note in Fig. 6.9 the different front–back orientation of the methyl carbon atoms adjacent to the carboxylic acid groups. Because of the different surface shapes, the racemic compound columns pack against each other more efficiently than do the S-enantiomer columns.

Another example may be found by comparing the structure of the racemic compound *d,l*-valine hydrochloride and the enantiomer *l*-valine hydrochloride. Both of these crystallize in planar arrangements (Fig. 6.20). The planes are oriented horizontally in Fig. 6.20. In the racemic compound the planes are related by a center of symmetry (DiBlasio, Napolitano, and Pedone 1977); in the enantiomer they are related by translation (Parthasarathy 1966; Ando et al. 1967).

Sometimes, however, chirality can have a more significant impact on packing. Consider the structures of the racemic compound and R-enantiomer of the antihistaminic drug chlorpheniramine maleate (Fig. 6.21). Note the alternating

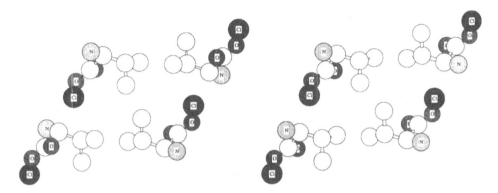

Fig. 6.20. The structures of *d,l*-valine hydrochloride (left) and *l*-valine hydrochloride (right). The horizontal planes in *d,l*-valine hydrochloride are related by a center of symmetry while the similarly packed horizontal planes in *l*-valine hydrochloride are related by translation. These pictures were generated from the data found in Parthasarathy (1966), Ando et al. (1967), and DiBlasio et al. (1977).

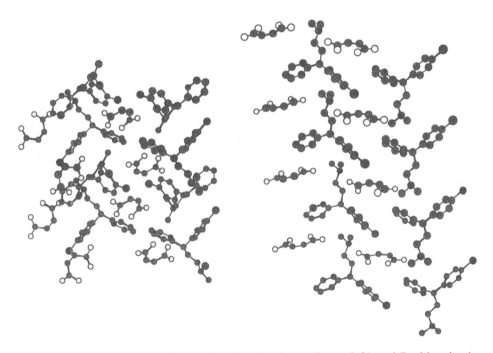

Fig. 6.21. The structures of R,S-chlorpheniramine maleate (left) and R-chlorpheniramine maleate (right).

orientation of molecules in the former, providing a more compact structure than that obtained by the R-enantiomer.

A compilation of the densities of racemic compounds compared to the densities of their pure enantiomers, for the limited number of compounds for which this information is available, showed that compactness alone cannot be responsible for racemic compound formation (Jacques et al. 1981*a*, p. 29; Brock, Schweizer, and Dunitz 1991). Surveys of the frequency of space groups found for chiral organic crystals show that 60–80 percent of racemic compounds crystallize in space groups $P2_1/c$, $C2/c$, or $P\bar{1}$, each of which possesses at least one element of inverse symmetry (Jacques et al. 1981*a*, p. 8). Thus in many racemic compounds opposite enantiomers are paired about a center of symmetry. The advantages of this arrangement have been attributed to maximization of electrostatic interactions (Wittig 1930). As far as thermodynamic factors are concerned, a survey of racemic compounds for which free energies are known showed that about one-third exhibit a $\Delta G_f < 0.4$ kcal/mole and three-fourths exhibit a $\Delta G_f < 0.8$ kcal/mol (Jacques et al. 1981*a*, pp. 94–5).

No simple correlations exist between the parameters discussed above and solid-form preference for chiral molecules, or for organic molecules in general. To be able to predict the solid form that any organic material mixture will attain, the relative stability of various molecular conformations, coupled with the relative stability of crystal lattice energies for various arrangements, must be calculable. This is not a trivial undertaking, involving the small energy differences related to compactness, symmetry, and thermodynamics.

Numerous approaches to the calculation of crystal lattice energies are reported in the literature. In order to understand the relative importance of the factors involved, the goals of initial efforts were to arrive at the known (by single-crystal x-ray analysis) structures of simple aromatic hydrocarbons. These molecules have little conformational mobility and do not form hydrogen bonds. Kitaigorodsky described early calculations of the lattice energies of benzene, naphthalene, and anthracene based on experimentally determined lattice parameters (the atom–atom potential method) (Kitaigorodsky 1973, pp. 170–83). These methods evolved into several computer programs designed to calculate lattice-energy minima, including WMIN (Busing 1981), OPEC (Gavezzotti 1983), and PCK 83 (Williams 1983). In addition, the methods have been applied to more complex molecules (Kitaigorodsky 1987; Gavezzotti 1991; Gavezzotti and Filippini 1994; Schmidt and Englert 1996). Correlations between molecular and crystal properties that allow lattice-energy estimates to be made from molecular parameters of functionalized organic molecules were determined statistically (Gavezzotti 1994). However, even for simple molecules the minimization of lattice energies alone proved insufficient for prediction of solid structure.

The method of "cluster calculations" was used extensively in attempts to predict solid structure from the molecular formula alone. This relies not on lattice parameters, but involves generation of molecular clusters based on basic symmetry operators and calculation of the cluster energies. A disadvantage is that fixed molecular conformations are utilized. Early application of this method to benzene predicted the known herringbone structure but did not provide accurate intermolecular distances and angles (Williams 1980; van de Waal 1981; Oikawa et al. 1985). By combining cluster calculations with statistical analyses of known structures and user judgment, Gavezzotti was able to predict successfully experimental structures in a few cases, but was unsuccessful in the majority of cases attempted (Gavezzotti 1991).

Another approach to crystal structure prediction involves successive packings of molecules onto an axis, axes into two-dimensional layers, and layers into three-dimensional arrays. The best arrangements are selected primarily by minimizing volumes. Again molecular conformation is fixed. The program that was developed based on this process, MOLPAK, closely predicted experimental structures for ten of fourteen organic nitro compounds studied (Holden, Du, and Ammon 1993). More recently, a trial structure of a tetranitro compound predicted by MOLPAK was refined to match experimental single-crystal data acceptably, providing the structure solution (Ammon et al. 1996).

By utilizing logical assumptions based on statistical evaluation of known crystal structures, a program was constructed (ICE9) that was generally successful at predicting organic solid structures with no input of experimental data (Chaka et al. 1996). This program was designed to provide a prediction in a reasonable amount of computer time; none of the examples given required more than two hours of c.p.u. time on a Cray Y-MP. Of sixteen hydrocarbons and aromatic hydrocarbons calculated, ICE9 correctly predicted the experimentally determined structures for ten and provided very close approximations for the other six.

Attempts to predict solid forms, given only molecular formulas, have yielded limited, yet encouraging, successes. Coalescence of randomly oriented molecules into a low-energy arrangement involves calculation of all possible molecular interactions based on a potential energy function coupled with an efficient search methodology to locate the minima. The magnitude of the calculational task is immense, resulting in excessive calculation times even for relatively simple molecules. In spite of this, the known crystal structures of benzene (Gdanitz 1992) and a few other organic molecules (Karfunkel and Gdanitz 1992) were achieved in this way in the early 1990s. Subsequently, extensive efforts to improve the basic method have been under way.

In 1993 F. J. J. Leusen published his doctoral thesis entitled *Rationalization of Racemate Resolution, a Molecular Modeling Study* (Leusen 1993). In this, the efficiency of resolving agents for resolution by diastereomeric salt crystallizations was studied. It was thought that an understanding of the crystal structures of various pairs of diastereomeric salts would provide a method for predicting the efficiency of resolving agents a priori. In several systems for which the resolution efficiency was known from experiment, the structures of the involved salts were determined and analyzed qualitatively (intramolecular bonding patterns and packing modes) and quantitatively (lattice-energy calculations). While qualitative factors could be found that correlated with efficiency for specific salt pairs, the more general computational approach was unsuccessful because the standard molecular mechanics programs then available were not sophisticated enough to reliably reproduce the known crystal structures. Again, the small energy differences that drive solid-structure formation could not be correctly modeled.

Some of the possible sources of error in the calculational work described above were identified (Leusen 1993, p. 172). Since that time efforts have been under way to eliminate such errors and create a program that can be used to calculate lattice energies. This would allow the prediction of solid structures, including whether an organic molecule will exist in polymorphic forms, whether a chiral organic molecule will exist as a racemic compound or a conglomerate, and whether a particular resolving agent will be efficient or not for resolution of a particular racemic mixture. Perhaps the most advanced program currently available is offered by Molecular Simulations, Ltd. as a part of its Cerius2 molecular modeling series. This software has been used to predict the structures of such complex molecules as quinacridone and estrone (Leusen 1996), as well as polymorphic modifications of single molecules. At this writing, however, it is still far from being generally applicable.

No doubt improvements in methods such as those described above will continue to be made. The guarded view of the future expressed by Gavezzotti (1994) in his review may be justified based on the current state of affairs, but the rapid progress made in just the last few years argues that the prediction of solid structure will ultimately be possible. The same procedures developed to analyze various polymorphic arrangements should be applicable to chiral compounds as well, providing a means to predict the structure of racemic mixtures.

References

Ammon, H. L., Du, Z., Gilardi, R. D., Dave, P. R., Forohar, F., and Axenrod, T. (1996). Structure of 1,3,5,7-tetranitro-3,7-diazabicyclo[3.3.0]octane. Structure solution from molecular packing analysis. *Acta Crystallographica*, **B52**, 352–6.

Ando, O., Ashida, T., Sasada, Y., and Kakudo, M. (1967). The crystal structure of L-valine hydrochloride. *Acta Crystallographica*, **23**, 172–3.

Bettinetti, G., Giordano, F., Fronza, G., Italia, A., Pellegata, R., Villa, M., and Ventura, P. (1990). Sobrerol enantiomers and racemates: solid-state spectroscopy, thermal behavior, and phase diagrams. *Journal of Pharmaceutical Sciences*, **79**, 470–5.

Blaschke, G., Kraft, H. P., and Markgraf, H. (1980). Racemattrennung des Thalidomids und anderer Glutaramid-Derivate. *Chemische Berichte*, **113**, 2318–22.

Borka, L. and Haleblian, J. K. (1990). *Acta Pharmacutica Jugoslavica*, **40**, 71.

Brock, C. P., Schweizer, W. B., and Dunitz, J. D. (1991). On the validity of Wallach's rule: On the density and stability of racemic crystals compared with their chiral counterparts. *Journal of the American Chemical Society*, **113**, 9811–20.

Bunn, C. W. (1961). *Chemical Crystallography*. London: Oxford University Press.

Busing, W. R. (1981). Oak Ridge National Laboratory, Oak Ridge, TN.

Byrn, S. R. (1982). *Solid-State Chemistry of Drugs*. New York: Academic Press.

Caira, M. R., Guillory, J. K., and Chang, L-C. (1995). Crystal and molecular structures of three modifications of the androgen dehydroepiandrosterone (DHEA). *Journal of Chemical Crystallography*, **25**, 393–400.

Chaka, A. M., Zaniewski, R., Youngs, W., Tessier, C., and Klopman, G. (1996). Predicting the crystal structure of organic molecular materials. *Acta Crystallographica*, **B52**, 165–83.

Chang, L-C., Caira, M. R., and Guillory, J. K. (1995). Solid state characterization of dehydroepiandrosterone. *Journal of Pharmaceutical Sciences*, **84**, 1169–79.

Collet, A. (1990). In *Problems and Wonders of Chiral Molecules*, ed. M. Simonyi, p. 91. Budapest: Akadémiai Kiadó.

Cook, P. B. and Hunt, J. H. (1982). Chemical Compounds. U.S. Patent 4 364 923.

Coquerel, G. and Perez, G. (1988). Croissance d'un énantiomère en milieu quasi-racémique cas du dichloroacétate norfenfluramine. *Journal of Crystal Growth*, **88**, 511–21.

Cox, P. J., MacManus, S. M., Gibb, B. C., Nowell, I. W., and Howie, R. A. (1990). Structure of 3β-hydroxy-5-androsten-17-one (DHEA) monohydrate. *Acta Crystallographica*, **C46**, 334–6.

DeCamp, W. H. (1989). The FDA perspective on the development of stereoisomers. *Chirality*, **1**, 2–6.

DiBlasio, B., Napolitano, G., and Pedone, C. (1977). DL-Valine hydrochloride. *Acta Crystallographica*, **B33**, 542–5.

DiCicco, R. L. (1993). The future of S-(+)-ibuprofen and other racemic switches. In *Chiral '93 USA*. Stockport, UK: Spring Innovations Ltd.

Dickinson, R. G. and Raymond, A. L. (1923). The crystal structure of hexamethylene-tetramine. *Journal of the American Chemical Society*, **45**, 22–9.

Dwivedi, S. K., Sattari, S., Jamali, F., and Mitchell, A. G. (1992). Ibuprofen racemate and enantiomers: Phase diagram, solubility and thermodynamic studies. *International Journal of Pharmaceutics*, **87**, 95–104.

Eliel, E. L. and Wilen, S. H. (1994). *Stereochemistry of Organic Compounds*. New York: Wiley-Interscience.

Felder, E., San Vitale, R., Pitre, D., and Zutter, H. (1981). Process for the resolution of (+)- and (−)-6-methoxy-α-methyl-2-naphthaleneacetic acid. U.S. Patent 4 246 164.

Freer, A. A., Bunyan, J. M., Shankland, N., and Sheen, D. B. (1993). Structure of (S)-(+)-ibuprofen. *Acta Crystallographica*, **C49**, 1378–80.

Gavezzotti, A. (1983). The calculation of molecular volumes and the use of volume analysis in the investigation of structured media and of solid-state organic reactivity. *Journal of the American Chemical Society*, **105**, 5220–5.

Gavezzotti, A. (1991). Generation of possible crystal structures from the molecular structure for low-polarity organic compounds. *Journal of the American Chemical Society*, **113**, 4622–9.

Gavezzotti, A. (1994). Are crystal structures predictable? *Accounts of Chemical Research*, **27**, 309–14.

Gavezzotti, A. and Filippini, G. (1994). Geometry of the intermolecular X—H$\cdots Y$ (X, Y = N,O) hydrogen bond and the calibration of empirical hydrogen-bond potentials. *Journal of Physical Chemistry*, **98**, 4831–7.

Gavezzotti, A. and Filippini, G. (1995). Polymorphic forms of organic crystals at room conditions: Thermodynamic and structural implications. *Journal of the American Chemical Society*, **117**, 12299–305.

Gdanitz, R. J. (1992). Prediction of molecular crystal structures by Monte Carlo simulated annealing without reference to diffraction data. *Chemical Physics Letters*, **190**, 391–6.

Glusker, J. P. and Trueblood, K. N. (1985). *Crystal Structure Analysis, A Primer*. New York: Oxford University Press.

Harrison, I. T., Lewis, B., Nelson, P., Books, W., Roszkowski, A., Tomalonis, A., and Fried, J. H. (1970). Nonsteroidal antiinflammatory agents. I. 6-Substituted 2-naphthylacetic acids. *Journal of Medical Chemistry*, **13**, 203–5.

Haüy, R. J. (1801). *Traité de Minéralogie*. Paris: Chez Louis.

Herschel, J. F. W. (1822). *Transactions of the Cambridge Philosophical Society*, **1**, 43.

Holden, J. R., Du, Z., and Ammon, H. L. (1993). Prediction of possible crystal structures for C-, H-, N-, O-, and F-containing organic compounds. *Journal of Computational Chemistry*, **14**, 422–37.

Jacques, J. and Gabard, J. (1972). Étude des mélanges d'antipodes optiques. III. Diagrammes de solubilité pour les divers types de racémiques. *Bulletin de la Société de Chimie Française*, pp. 342–50.

Jacques, J., Collet, A., and Wilen, S. H. (1981*a*). *Enantiomers, Racemates, and Resolutions.* New York: John Wiley & Sons.

Jacques, J., Leclercq, M., and Brienne, M.-J. (1981*b*). La formation de sels augmente-t-elle la fréquence des dédoublements spontanés? *Tetrahedron*, **37**, 1727–33.

Karfunkel, H. R. and Gdanitz, R. J. (1992). Ab initio prediction of possible crystal structures for general organic molecules. *Journal of Computational Chemistry*, **13**, 1171–83.

Kazutaka, A., Ohara, Y., and Takakuwa, Y. (1983). Process for Preparing an Optical Active Ester of Naphthylpropionic Acid. U.S. Patent 4 417 070.

Kiefer, D. M. (1997). How an iron-willed FDA Officer averted a birth defect disaster. *Today's Chemist at Work*, **6**, 92–6.

Kitaigorodsky, A. I. (1973). *Molecular Crystals and Molecules.* New York: Academic Press.

Kitaigorodsky, A. I. (1987). *The Atom-Atom Potential Method.* Berlin: Springer-Verlag.

Kuhnert-Brandstätter, M. (1993). Schmelzdiagramme von optisch aktiven Substanzen – Enantiomere und ihre Racemate. *Pharmazie*, **48**, 795–800.

Leusen, F. J. J. (1993). Rationalization of Racemate Resolution, a Molecular Modeling Study. PhD Thesis, Katholieke University, Nijmegen.

Leusen, F. J. J. (1996). Ab initio prediction of polymorphs. *Journal of Crystal Growth*, **166**, 900–3.

Manimaran, T. and Stahly, G. P. (1993). Optical purification of profen drugs. *Tetrahedron: Asymmetry*, **4**, 1949–54.

McConnell. J. F. (1974). *Crystal Structure Communications*, **3**, 73.

McCrone, W. L. (1965). Polymorphism. In *Physics and Chemistry of the Organic Solid State*, Vol. 2, ed. D. Fox, M. M. Labes, and A. Weissberger, pp. 728–31. New York: Wiley-Interscience.

Oikawa, S., Tsuda, M., Kato, H., and Urabe, T. (1985). Growth mechanism of benzene clusters and crystalline benzene. *Acta Crystallographica*, **B41**, 437–45.

Parthasarathy, R. (1966). The structure of L-valine hydrochloride. *Acta Crystallographica*, **21**, 422–6.

Reuben, B. G. and Wittcoff, H. A. (1989). *Pharmaceutical Chemicals in Perspective*, p. 98. New York: John Wiley & Sons.

Salzberg, H. W. (1991). *From Caveman to Chemist*, p. 234. Washington, DC: American Chemical Society.

Schlenk, Jr., W. (1965). Recent results of configurational research. *Angewandte Chemie International Edition English*, **4**, 139–45.

Schmidt, M. U. and Englert, U. (1996). Prediction of crystal structures. *Journal of the Chemical Society Dalton Transactions*, pp. 2077–82.

Stahly, G. P. (1993). Ibuprofen resolution. U.S. Patent 5 189 208.

Stahly, G. P., McKenzie, A. T., Andres, M. A., Russell, C. A., Byrn, S. R., and Johnson, P. (1997). Determination of the optical purity of ibuprofen using x-ray powder diffraction. *Journal of Pharmaceutical Sciences*, **86**, 970–1.

Stinson, S. C. (1992). Chiral drugs. *Chem. and Eng. News*, 28 September, pp. 46–79.

Waal, van de B. W. (1981). Significance of calculated cluster conformations of benzene: Comment on a publication by D. E. Williams. *Acta Crystallographica*, **A37**, 762–4.

Williams, D. E. (1980). Calculated energy and conformation of clusters of benzene molecules and their relationships to crystalline benzene. *Acta Crystallographica*, **A36**, 715–23.

Williams, D. E. (1983). Quantum Chemistry Program Exchange, Indiana University, IN, No. 481.

Wittig, G. (1930). *Stereochemie*, p. 309. Leipzig: Akademische Verlagsgesellschaft.

Index

ab initio molecular orbital calculations, 137
ab initio prediction of crystal packing, 143–51
ab initio quantum-chemistry calculations, 119
ABO_4 type structures, 257
N-(2-acetamido-4-nitrophenyl) pyrrolidine (PAN), 215–16
acetic acid, 217
o-acetoamidobenzamide, 151–2
acicular crystal, 101
additives, enantiomeric distribution of, 188–90
adiabatic processes, 14–15
adjacency matrix, 229, 251
ADP, 257, 298
    adjacency matrix, 244
    aqueous solution grown crystals, 306
    basic crystal graph, 246
    combinatorial properties, 243
    comparison of theoretical and experimental growth forms, 306–7
    conventional unit cell, 244
    crystal graph, 299
    crystal structure and bonds of first coordination sphere, 299
    derivation of the connected nets based on *f*, *g* and *h* bonds, 299
    energy computations of slices, 274–7
    interface behavior and surface–fluid interactions, 298–307
    results of sphere experiments and step-height measurements on pyramidal faces, 305–6
    structural morphology, 270–9
    theoretical growth form, 277–9
aggregation/agglomeration, 107
alanine, 169
(R)-alanine, 220
(R,S)-alanine, 210, 211, 212, 217, 221, 222
alcohols, 120, 220
aliphatic hydrocarbons, 120
alkanes, 218
allotropism, 320
AM1 semi-empirical method, 141, 142
amides, 123–5
ammonia, 217
ammonium chloride (NH_4Cl)
    F faces, 282

fractal features, 283–4
fractal growth pattern, 284
gel-induced fractal habit of, 279–87
growth units, 281
octahedral habit modification, 287
ammonium dihydrogen phosphate *see* ADP
anthracene, 119, 127, 128
anthranilic acid, 102
anti-inflammatory steroids, 322
area shape factor, 100
aromatic hydrocarbons, 120
aromatic size, 110
asparagine, 176
(S)-asparagine, 194–5, 207
asparagine–aspartic acid monohydrate, 203
atom–atom interactions, 118
atom–atom method, 117–18
atom–atom potential-energy functions, 168, 170
atomic force microscopy (AFM), 168, 284
atomic moments, 169
atomic slice configurations, 275
atropine, 328
attachment energy, 126–7, 170, 207, 234
azahydrocarbons, 120
azobenzene, 111

barite type structures, 257
barium sulphate type structures, 251–6
basic crystal graph, 242
    expressed as network, 243–6
BCF model, 97
BCF theory, 96, 97, 98
benzamide, 113, 130, 131, 172, 173
benzene, 341
benzoic acid, 130, 131, 172, 173
benzophenone, 111, 131, 132, 133
benzoquinone, 116
BFDH law, 102, 125, 235–6
BFDH method, 104
BFDH model, 126, 127
bicyclohepta(de,ij)naphthalene, 144–5
bicyclo(4.4.1)undecapentane, 139
binary melting-point phase diagram, 328, 330, 332

binding energy, 170
birth and spread models, 96
blocker additives, 131
blocker molecule, 129
boiling points, 133
Boltzmann constant, 10
Boltzmann distribution, 39, 41
Boltzmann factor, 25, 38
Boltzmann weighting, 25, 27
bond energy, 289, 290, 295, 296
    in organic molecular crystals, 320
bond lengths, 141
bonding, 71–3
    "1–4", 266
Boyle temperature, 36
Bravais–Friedel law, 102
Bravais–Friedel–Donnay–Harker *see* BFDH law
Bravais lattices, 58
p-bromophenyl acetic acid, 159
Buckingham (B) potential, 32–3
bulk diffusion, 97

C (end-centered) lattice, 59
Cambridge Crystallographic Database, 108, 122
    graphical input/output interface, 123
    input query, 123
Cambridge Structural Database, 320
canonical distribution, 5
canonical ensemble, 5, 7, 48
carbon, 320
carboxylic acid, 120, 123–5, 338
    solid solutions of, 193
carboxylic acid dimers, 111
central-difference prediction, 52
centrosymmetric crystals, 184–5, 216
centrosymmetric space groups, 324
centrosymmetric structure, 61
CERIUS2, 104
charge distributions, 24
chemical bonds, 71–3
chemical potential, 10, 20
chiral additives, 213–14
chiral compounds, solid forms of, 325–36
chiral molecules, 324, 340
    solid form of, 337
chiral organic pharmaceuticals, solid-state
    structure of, 313–45
chiral selectivity, 176
chiral shift reagents, 325
chiral systems, polymorphic relationships in, 336
chirality, 213–14, 313
chlorpheniramine maleate, 338–9
E-cinnamamide-2-E-thienylacrylamide, 196
E-cinnamic acid, 194–6
close-packing arrangements, 337
closed system, 9, 13
cluster calculations, 340
clusters, 92
collision rate, 50–1
combination of crystallographic forms, 98
computational chemistry
    crystal structure prediction/determination, 136
    molecular materials, 106–65
computer-based modeling methods, 1
computer simulation, 1

concept of microscopic reversibility, 41
configuration integral, 34, 38
configuration space, 38
conformational isomorphism, 82
conformational polymorphism, 79
conformational synmorphism, 82
conglomerates, 334, 335
    resolution by direct crystallization, 333
connected nets
    nets derived with f and g bonds taken into
        account, 301–4
    nets derived with f, g and h bonds taken into
        account, 302–4
    pictorial presentation, 301–6
connectedness property, 242
continuity equation, 3
convergence, Madelung versus Ewald
    formulations, 264–5
Coulomb energy, 276, 282
Coulomb interactions, 167, 168, 229, 260, 265–6,
    277–9
Coulomb potential, 260
critical surface cluster size, 95
critical transition supersaturation, 97
crystal binding or cohesive energy, 118
crystal chemistry of molecular materials, 108–17
crystal dissolution, 174
crystal energy, 234
crystal engineering, 107, 111, 129, 170–4
crystal faces, 98, 99, 101
    attachment energy, 103
    types, 103
crystal form, calculation, 169–70
crystal graph, 299
    bipartite nature, 251
crystal growth, 55, 89, 94–8, 166, 167
    effect of solvent, 203–13
    imperfect crystals, 96–8
    inhibitors, 167
    perfect crystals, 95–6
    relay mechanism, 209–13
crystal habit, 98, 99, 101–4
    modification, 127–32
    theory, 233
crystal lattice, 56–9, 322, 338, 340
    defects, 84
    geometry, 125
crystal morphology, 98, 127, 167, 168, 197, 237
    effect of additives, 169–70
    effect of solvent, 203–13
    prediction, 293
    quantitative determination, 169
    simulations, 125
crystal packing
    ab initio prediction of, 143–51
    calculations, 135
        Rietveld methods, 153–7
    patterns, 146
    problem, 149
crystal planes (or faces), 98
crystal shape, 98–102, 156
    calculation, 125–7
    and conditions of growth, 101
    prediction, 125–31
crystal structure, 102–4

crystal structure (*cont.*)
   prediction/determination, computational chemistry, 136
crystal surface, "tailor-made" additives on, 199–202
crystal symmetry, 60–3
   lowering through additive occlusion, 199
   reduced, 188–98
crystal systems, 58, 63
crystal texture, "tailor-made" additives, 202–3
CRYSTALDRAW, 279
CRYSTALFORM, 279
crystalline materials, 55
crystalline phase formation, 213–22
crystalline solid solutions, 82
crystalline solvates, 205–7
crystalline state, 55
crystallization, 85–104, 330–2
   basics, 55–105
   bonds, 246
   direct, 332, 333
crystallographic discrepancy factor, 122
crystallographic forms, combination of, 98
crystallographic point groups, 63–6
crystallography, 323–5
   basic concepts, 55–84
crystals, 55
cubic lattice, 56
cyanuric chloride, 117

databases, 122–5
Debye–Scherrer scattering geometry, 154
deformation density maps, 168
degeneracy factor, 5
dehydroepiandrosterone (DHEA), 321–3
diamond, 320
diastereomers, 314, 316–19, 333
dibromobenzene, 316
6,13-dichlorotriphendioxazine, 154–7
dielectric constant, 118
4,8-dimethoxy-3,7-diazatricyclo(4.2.2.2-2,5-) dodeca-3,7,9,11-tetraene, 149–50
dimethylsulfoxide (DMSO), 88
dipeptide glycyl-glycine, 197
dipleiadiene, 144
dipole energy, 234
dipole moments, 141
Dirac delta function, 5
direct chains, 247–8
   computation in terms of paths, 248–9
   computation of F slices in terms of, 249–50
direct crystallization, 332, 333
dislocations, 84, 96
Disordered Flat (DOF) phases, 298
dispersion contribution, 23
dispersion energy, 27–8
displacive transformation of secondary coordination, 75
disrupter molecule, 129
dissolution rate, 100
Donnay–Harker law, 236
donor–acceptor azobenzene, 108
doublet, 239

drug action, lock-and-key visualization of, 317
drug development, 322
dyestuffs, 120

edge dislocations, 84, 85
edge-to-face interactions, 127
electric charge distribution, 24
electric field, 268
electrodynamic potential, 26
electromagnetic forces, 22
electrophotography, 106
electrostatic contributions, 22–3
electrostatic energy, 23–5, 168
electrostatic interactions, 118, 222, 340
electrostatic point charge model in PBC theory, 260–5
electrostatic potentials, 24–5
enantiomeric distribution of additives, 188–90
enantiomers, 314–16, 324, 325, 328, 330, 332, 333, 338, 340
   resolution, 214–15
enantiomorphism, 79–82
enantiomorphous crystals, 220
enantioselective adsorption, 222
energy, long-range intermolecular forces, 23–8
energy calculations, 168
energy change by direct transfer, 16–17
energy computations
   of ADP slices, 274–7
   in ionic structures, 260–70
energy fluctuations, 10–11
ensembles, 4
enthalpy, 17–18
entropy, 12
epitaxial growth, 222
equations of state, 20–1, 35
equilibrium thermodynamics, 12–21
ergodic theorem, 4
Euler's constant, 262
eutectics, 328–32, 335
Ewald formulation
   description of data, 268–9
   site potential for lamina shapes, 267–8
Ewald method, 264–5
exactly defined model, 1
exponent of repulsion, 30
external forces, work done by, 16
external quantities, 9

F faces, 238
   ammonium chloride, 282
   analysis, 270–4
F lattice, 59
F slice cell, 248
F slices, 231–3, 238, 240–1, 258, 269–74, 283
   computation in terms of direct chains, 249–50
   determination, 237
   graph-theoretic approach, 242–3, 247–50, 257
   of gypsum adjacent to vacuum and water solution, 252
   ionic, 251, 256

F slices (*cont.*)
  profusion of, 256–7
  solution-activated, with irreducible borders in
    gypsum, 250–1
F subslices, 247–8
face indices, indeterminacy of overall sign, 241–2
feed-forward neural network, 133–6
FFACE program, 259–60
finite difference methods, 51
first law of thermodynamics, 17
first moment of charge distribution, 24
first-nearest-neighbour interactions, 295–6
first-nearest-neighbour models, 294, 297–8
  versus long-range models, 288
flat faces, 103
flat surfaces, 279–82
flatness condition, 238
Food and Drug Administration (FDA), 318
FORTRAN, 42, 44
fractal features of ammonium chloride, 283–4
fractal morphology, 285
fugacity, 8
fullerenes, 320
functional groups, 338

garnet structure, 298
gas particles, motion of, 33–4
gel-induced fractal habit of ammonium chloride
    (NH$_4$Cl), 279–87
Gibbs distribution, 5
Gibbs free energy, 18, 19, 20, 89, 90, 92, 95
Gibbs–Thomson equation, 92
glide plane, 69
glutamic acid, 176
glyceraldehyde, 315
glycine, 119, 176
α-glycine, 175, 201, 204, 205
γ-glycine, 216, 218
glycyl-leucine additive, 197
grand canonical distribution, 5
grand canonical ensembles, 5, 8–10, 45, 47–8
graphite, 320
gravitational force, 22
grazing-incidence x-ray diffraction (GID), 168,
    199–201, 205, 218
growth layers, 230, 238
  search procedures, 240–1
GSTAT, 123
gypsum
  adjacent to vacuum and water solution, 252
  morphology, 254
  solution-activated *F* slices with irreducible
    borders in, 250–1

habit-controlling energy parameters, 233–5
habit-controlling energy quantities, 266
halocarbon propellants, 322
halocarbon solvates, 322
Hamiltonian, 1, 2
Hankel function, 262
hard-sphere (HS) molecules, 50–1

hard-sphere (HS) potential, 29–30
Hartman–Perdok theory, 103, 228–312
HBD parameter, 136
heat function, 17
heat gained or lost, 16–18
heat of formation, 141
heat of sublimation, 133
Heisenberg uncertainty principle, 5
Helmholtz free energy, 10, 18, 19, 21, 34, 35
hemihydrism, 313
herringbone contacts, 198
heterogeneous nucleation, 93
hexafluoro-isopropyl, 218
hexafluorovaline, 217
hexagonal crystals, 98
hexagonal systems, 66
high-performance liquid chromatography (HPLC),
    189
high-pressure liquid chromatography (HPLC), 325
high-resolution powder diffraction, 107, 154
high-resolution synchrotron x-ray diffraction, 203
homogeneous nucleation, 93
hydrogen bond, 111–17, 120, 123, 136, 211, 229,
    320, 337, 338
  distributions, 269, 277
hydrogen-bonding potential, 118
hydroxyurea, 320, 321
hyoscyamine, 328

I (body-centered) lattice, 59
i–j interaction, 265–6
i–j pair, 265
ibuprofen, 318, 325–7, 331, 332
ICE9, 341
ideal canonical ensemble, 21
importance sampling method
  implementation of, 41–2
  theoretical background, 38–41
  variations, 42
impurity effects, 130
induction contributions, 23
induction energy, 26–7
inorganic materials, 166
integral of interest, 37
integral of motion, 4
interatomic distance, 118
intercell bonds, 245
interfaces, actual bond energies, 289
intermolecular bonding, 127
intermolecular energy, 143
intermolecular forces, 22–36, 193–8, 338
  sources of information, 33–6
intermolecular interactions, 111
intermolecular microscopic forces, 107
intermolecular orientation, 25
intermolecular potential, 35
  modeling, 28–33
internal energy, 16–17, 19
  change in, 17
interplanar distance, 238
interstitial impurities, 84
intracell bonds, 245
inverse *n*-fold axis of symmetry, 61

inverse symmetry operations, 324
ionic bond energies, 296–7
ionic bonds, 72
ionic crystals, 228–312
ionic graph, 251
ionic structures, 229–31
    energy computations in, 260–70
    specialization to, 251–6
ionization potential, 141
irreversible processes, 14
Ising model, 1
Ising temperature, 289
isomorphism, 82

j-subsystem entropy, 13

K faces, 232, 239
K slice, 239, 240
KDP, 257, 270–9, 298
    aqueous solution grown crystals, 306
    comparison of theoretical and experimental
        growth forms, 306–7
    interface behavior and surface–fluid interactions,
        298–307
    results of sphere experiments and step-height
        measurements on pyramidal faces, 305–6
kinetic crystallization, 332
kinetic roughening, 291–2
kinked faces, 103

lactic acid, 315
Laplace's equation, 262
lattice energy, 133
    calculations, 108, 117–22
    versus crystal radii, 119
    versus sublimation enthalpy, 121
lattice parameters, 56
lattice planes, 66–8
Laue method, 74
law of entropy increase, 14
Law of Successive Reactions, 78
layer energy, 170
Lennard-Jones atomic parameters, 275
Lennard-Jones expression, 265
Lennard-Jones molecules, 52
Lennard-Jones potential, 31–2, 36, 51
    depth, 269
    function, 118
(S)-leucine, 201
Lioville equation, 3
lock-and-key visualization of drug action, 317
long-chain hydrocarbons, 218
long-range interactions, 293–6
long-range intermolecular energy, 27
long-range intermolecular forces, energy, 23–8
long-range models versus first-nearest-neighbour
    models, 288

Madelung's equation, 263
Madelung's formulation, 261–3
Madelung's ion lines, 263
Madelung's ion meshes, 263
Maltese cross, 167
Markov chain, 39
    ergodic, 40
Maxwell–Boltzmann distribution, 8, 48, 53
Maxwell–Boltzmann system, 20–1
mean collision time, 50
mean displacement, 42
mean field theory, 296–7
mean internal energy, 9
mechanical invariants, 4
mechanical quantities, 9
melting point, 329–30
metastable solutions, 87
methanol, 206, 207
(S)-methionine, 199–201
2-methoxypyridine-1,4-dimer, 149
Metropolis criterion, 41–2
microcanonical distribution, 5
microcanonical ensemble, 5, 7, 9, 13
Miller indices, 66–8
mirror plane, 61
modeling intermolecular potentials, 28–33
molecular chirality, 184–5
molecular crystals, 72, 167
molecular dynamics
    simulation, 48–53
    studies, 293
molecular information, 136
molecular interactions, quantitative estimates,
    168–9
molecular materials
    computational chemistry, 106–65
    crystal chemistry of, 108–17
molecular mechanics, 137–9
    simulation, 44–8
        within grand canonical ensemble, 47–8
        within *NPT* ensemble, 46
molecular modeling, 1–54, 129
molecular momenta, probability distribution for, 8
molecular nuclei, 143
molecular orbital methods, 140–3
molecular slice configurations, 275
molecular structure, 113, 154, 156
    calculation, 137
molecular-structure/crystal-packing calculation
    algorithm, 137
MOLPAK program, 146, 341
monoclinic structures, 257
Monte Carlo cooling approach, 147
Monte Carlo method, 37–44, 158–9
    origin and fundamentals, 37–8
    simulation of molecular mechanics, 45–8
Monte Carlo simulated-annealing process, 149
MOPAC, 155
morphine, 315
motion of gas particles, 33–4
multilinear regression analysis (MLRA) model,
    133–6
multiplet, 239
multistep transition probabilities of order *t*, 40

naproxen, 334, 335
neural networks, 132
neutron diffraction, 196, 203
next-nearest-neighbour interactions, 294–6
$NH_4H_2PO_4$ *see* ADP
nickel arsenide, 319
nitrobenzene, 154
noncentrosymmetric space groups, 324
noncentrosymmetric structure, 61
nonlinear optics, 190–1
nonsteroidal anti-inflammatory drugs (NSAIDs),
      317
norfenfluramine dichloroacetate, 327
normalization condition, 3
NPT ensemble, 46, 48
nucleation, 89–93, 284
   early stages, 220–2
   theory, 91–3
nucleation barrier, 213
nucleation induction time, 91
nucleation order, 93
nucleation rate, 92–3
nuclei formation, 92
NVT ensemble, 45

N-octylgluconamide, 208
one-component graphs, 242
one-dimensional Fourier series, 262
open systems, 9, 19–20
optical activity, 79
optical birefringence, 192–3
optical purification, 330
optical purity, 325
organic crystals, "tailor-made" etchants for, 174–6
organic materials, 166
organic molecules, 72, 108, 110, 133, 320, 338, 340
orthorhombic crystals, 98, 99
orthorhombic system, 68
Ostwald's step rule, 78
over-the-counter (OTC) drugs, 318
overlap forces, 22
oxohydrocarbons, 120

packing coefficient, 109–11, 337
packing efficiency, 111
packing potential energy (PPE), 133
paclobutrazole, 127, 129
partial atomic charges, 275
partial atomic point-charge sets, 277–9
particle engineering, 107
particle morphology, 107
partition functions, 5, 8, 9
Pauli Exclusion Principle, 22
PBC theory, 102, 228, 230–7, 308
   electrostatic point charge model in, 260–5
   future work, 285
   graph-theoretic formulation, 242–60
   historical setting, 235–6
   morphological predictions, 283
   operational concepts, 237–42
   technical issues, 236–7

PBCs, 102–3, 126, 242, 282
   additional, 237
   analysis, 127
   centrosymmetric structures, 236
   determination, 237
   direction, 236
   evidence for existence of, 231–3
   intersecting, 231, 236
   nonparallel, 238
PCK83, 340
periodic bond chain theory *see* PBC theory
periodic bond chains *see* PBCs
periodic boundary conditions, 49
perylene pigment, 110–11
perylenedicarboximide, 147–8
pharmaceuticals, 313–45
phase space, 2
phase trajectory, 2
phenylalanine, 205
phosphorus, 320
photodimerization, 196–7
phthalocyanines, 106, 116
planes of symmetry, 61, 62
PLUTO, 123
point-charge model, 260–5, 282
point defects, 84
point groups, 63–6
point lattice, 57
polar crystals, 177–84, 207–9, 216
polar habits, 241–2
polyethylene, crystal structure, 70
polymer–polymer systems, 90
polymers, 120
polymorphic forms, 76
polymorphic relationships in chiral systems,
      336
polymorphism, 74–9, 107, 215–20, 320, 322
polymorphs, 320–2
porphyrins, 116
potassium chloride, solubility in aqueous solution,
      87
potassium dihydrogen phosphate *see* KDP
potential energy of interaction, 73
powder diffraction, 74, 154–7
   patterns, 152–9
precipitation, 88
pressure, 15–16
primary nucleation, 91
primitive (*P*) lattice, 58–9
primitive lattice translations, 236, 240
probability distribution, 40
   for molecular momenta, 8
property estimation methods, 132–6
proportionality relation, 289
propylglucosamine, 335
pseudo-Boltzmann factor, 46, 47
pseudo-*F* face, 238
pseudo-*F* slices, 238
pseudo-flat surfaces, 279–82
pseudorandom numbers, 43–4
pseudotwofold symmetry interlinking hydrogen-
      bonded bilayers, 201
pyridazino-(4,5)-pyridazine (PP), 143–4

quartet, 239
quartz crystals, 313

racemic alanine crystals, 220–2
racemic compounds, 330–4, 338, 340
racemic crystals, 81
racemic histidine, 215
racemic mixtures, 80, 317, 322, 324, 325, 328–30, 332, 334
racemic switches, 318
raffinose, 167
random numbers, generation, 42–5
random sampling procedure, 37
random walk, 42, 43
recrystallization, 332
relative growth rates, 230–1, 233–5
resolution, 330
resorcinol, 208
retinoic acid, 316
reversible processes, 14, 15
α-rhamnose monohydrate crystal, 206
rhombohedron (R) lattice, 59
Rietveld refinement method, 152–9
Rietveld refinement program DBWS, 156
rotational disorder transformation, 75
Rough–Flat–Rough (RFR) kinetic phase transition, 292
roughening, 289–93
roughening models, 287–8
roughening transition, 293–4

S faces, 232, 239
S slice, 239, 240
Schrödinger equation, 27
screw axis, 69
screw dislocations, 84, 85, 96
  development of growth spiral, 97
second harmonic generation (SHG), 190
second law of thermodynamics, 14
second moment of charge distribution, 24
second virial coefficient, 35
secondary nucleation, 91, 93
selective solvent binding, 209–13
semi-closed system, 9, 15
semi-empirical quantum mechanics, 146
shape factors, 100
SHELXTL/PC, 157
short-range interactions, 294–6
silica, transformations of, 77
single crystals, 107
  structure, 122
  structure analysis, 136
  study, 157
  x-ray diffraction, 122, 136, 155
singlet, 239
SIREN, 285
slice energy, 126–7, 234, 266
slices, 231–3, 238–41, 269
  classification, 238–9
sobrerol, 334, 336, 337
sodium chlorate, L and D crystals, 81
sodium chloride, 61
  atomic arrangement, 61
  structure, 60

sodium cyanide, 129
sodium ibuprofenate, 332
soft-sphere (SS) potential, 29–30
solid form
  of chiral compounds, 325–36
  of chiral molecules, 337
  control, 322
solid–liquid interface, 174
Solid On Solid (SOS) model, 289–90
solid solutions of carboxylic acids, 193
solid-state chemistry, 107
solid-state structure, 319–22
  of chiral organic pharmaceuticals, 313–45
  prediction, 336–42
solubility, 86–7, 133
Solution Induced Reconstructive Epitaxial Nucleation (SIREN), 279
solutions, 86
solvent–solute interactions, 101
solvent–surface interactions, 203–5
solvents, 86, 203–13
  role in crystal growth and crystal shape, 101
space groups, 71, 324, 340
spinodal curve, 91
spinodal decomposition, 90
statistical averaging, 3
statistical distribution function, 3, 4
statistical distributions, 2–3
statistical equilibrium, 12
statistical-mechanical theories, 293–4
statistical mechanics, 2–11
  and thermodynamics, 9–10
statistical methods, 132
statistical weight, 12
statistically exact information, 1
steady-state condition, 40
stepped faces, 103
stereoisomerism, 313–19, 334
stick and van der Waals (spacefill) representations, 108
stochastic matrix, 40
stoichiometry, 328
strong nuclear forces, 22
structural morphology, 231–3
  ADP-type structures, 270–9
sublayer roughening, 292
sublimation enthalpy, 132–5
subslice, 238
substitutional impurities, 84
β-succinic acid, 119, 127, 128
sucrose, 167
supersaturated solutions, 91
supersaturation, 86–9
supramolecular assemblies, 116
surface diffusion, 96
surface energy, 234
SURFENERGY, 269–70, 275
SURFGRID, 269–70
SURFPOT, 269–70
Sutherland (S) potential, 30–1, 36
symmetry information, 136
symmetry lowering, 186–202
symmetry operations, 323–4
symmetry roughening, 291
synchrotron radiation, 218

Synchrotron Radiation Source (SRS), 154
synchrotron x-ray scattering, 199
system temperature, 15

t-distribution, 37
"tailor-made" additives, 168, 170, 171, 186–202, 216
  on crystal surface, 199–202
  crystal texture, 202–3
"tailor-made" auxiliaries, 167
"tailor-made" etchants for organic crystals, 174–6
tartaric acid, 314, 316
temperature, 15
terephthalic acid (TPA), 88
ternary solubility diagrams, 334
tetragonal system, 67
1,4,5,8-tetramethoxy-pyridazino-(4,5)-pyridazine (TMPP), 143–4
theory of stochastic processes, 39
thermal quantities, 9
thermal roughening, 289–91
thermally isolated system, 15
thermochemical data, 132–3
thermodynamic equilibrium, 15
thermodynamic limit, 21
thermodynamics
  and statistical mechanics, 9–10
  three laws of, 9
thienylacrylamide, 196
three-dimensional symmetry, 187
allo-threonine, 198
(S)-threonine, 220
S-*allo*-threonine, 198
o-toluamide, 172, 173
p-toluamide, 172–4
toluene, 131–3
4-toluene sulfonyl hydrazine, 158–9
transformations, 75
  of silica, 77
transition matrix, 40
transition probability, 40
translational symmetry elements, 68–73
transport coefficients, 48
triclinic system, 67
triethylammonium-6,7-benzo-3-*o*-hydroxyphenyl-1,4-diphenyl-2,8,9-trioxa-4-phospha-1-boratricyclo(3.3.1) nonane, 140–2

trigonal systems, 66
trimesic acid, 114
2,4,6-trinitro-*N*-methyldiphenylamine (CSD), 146–7
trivalent cations, influence on growth mechanism and morphology, 304–5
twinning, 82–3, 220–2, 284
two-dimensional crystal, 103
two-dimensional Fourier series, 262
two-dimensional surface nucleation, 95
two-dimensional symmetry, 187

ultraviolet irradiation, 214
uniform sample mean method, 38
unit cell, 56, 57, 125, 136
  dimensions, 144, 146, 154, 155
urea, 119, 121, 213

vacancies, 84
vacuum sublimation, 157
d,l-valine hydrochloride, 338
l-valine hydrochloride, 338
van der Waals energy, 276
van der Waals equation of state, 36
van der Waals forces, 337, 338
van der Waals interactions, 114, 118, 119, 172, 229, 260, 265–6, 277–9, 288, 320
van der Waals parameters, 168
van der Waals radii, 114
VISTA, 123
volume shape factor, 100

weak nuclear forces, 22
WMIN, 340
work done by external forces, 16
Wulff form, 98–9

x-ray crystallographic analyses, 319
x-ray diffraction, 73–4, 196, 197, 199, 222, 319
x-ray powder diffraction (XRPD), 324–5, 327

zeroth moment of charge distribution, 24